Statistics: A First Course

Walter Gleason
Bridgewater State College

Wadsworth Publishing Company
Belmont, California
A Division of Wadsworth, Inc.

To my father
John B. Gleason

Mathematics Editor: Richard Jones
Signing Representative: John Moroney
Production Editor: Maureen P. Conway
Designer: Judy Gardoni
Copy Editor: Evelyn Tucker
Technical Illustrator: Vantage

© 1981 by Wadsworth, Inc. All rights reserved. No part of this book may be reproduced, stored in a retrieval system, or transcribed, in any form or by any means, electronic, mechanical, photocopying, recording, or otherwise, without the prior written permission of the publisher, Wadsworth Publishing Company, Belmont, California 94002, a division of Wadsworth, Inc.

Printed in the United States of America

1 2 3 4 5 6 7 8 9 10—85 84 83 82 81

Library of Congress Cataloging in Publication Data

Gleason, Walter.
  Statistics, a first course.
  Includes index.
  1. Statistics. I. Title.
QA276.12.G54   519.5   80-19575
ISBN 0-534-00909-3

CONTENTS

Note to the Instructor   ix
Note to the Student   xi

1   INTRODUCING STATISTICS

1-1   A historical note   1
1-2   A sample versus a population   2
1-3   A parameter versus a statistic   4
1-4   Descriptive versus inferential statistics   5
1-5   Summation notation   6
1-6   Measures of central tendency   11
1-7   Discrete versus continuous data   19
1-8   The frequency distribution table   21
1-9   Percentiles   32
        Summary   35
        Can you explain the following?   36
        Miscellaneous exercises   36

2   THE VARIANCE AND THE STANDARD DEVIATION   39

2-1   Introduction   39
2-2   Standardizing data   47
        OPTIONAL
2-3   A further look at the standardizing formula   49
2-4   Chebyshev's theorem   50
2-5   Symmetry   52
        Summary   54
        Can you explain the following?   55
        Miscellaneous exercises   55

3 AN INTRODUCTION TO PROBABILITY 57

- 3-1 Introduction 57
- 3-2 A look at probability 61
- 3-3 Probability curves 70
- Summary 75
- Can you explain the following? 76

4 PROBABILITY 77

- 4-1 Counting 77
- 4-2 Permutations 80
- 4-3 Combinations 85
- 4-4 More about probability 88
- 4-5 Conditional probability 92
- 4-6 Mathematical expectation 95
- OPTIONAL
- 4-7 The birthday problem and the World Series problem 99
- Summary 101
- Can you explain the following? 101
- Miscellaneous exercises 102

5 SAMPLES AND SAMPLING 104

- 5-1 Why sample? 104
- 5-2 Sampling with replacement versus sampling without replacement 104
- 5-3 The random sample 108
- 5-4 Stratified, cluster, and systematic sampling 109
- 5-5 Bias 111
- OPTIONAL
- 5-6 More about bias 113
- Summary 114
- Can you explain the following? 115
- Miscellaneous exercises 115

6 THE NORMAL DISTRIBUTION 117

- 6-1 The normal distribution 117
- OPTIONAL
- 6-2 A deeper look at the standard normal distribution 130
- Summary 130

Can you explain the following?   131
Miscellaneous exercises   131

7 THE SAMPLING DISTRIBUTION OF MEANS FOR LARGE SAMPLES   133

7-1 Introduction   133
7-2 Testing statistical hypotheses about μ 139
7-3 The two types of parameter estimates   156
7-4 A partial answer to the question of sample size   162
Summary   165
Can you explain the following?   166
Miscellaneous exercises   166

8 THE SAMPLING DISTRIBUTION OF MEANS FOR SMALL SAMPLES   168

8-1 Degrees of freedom   168
8-2 The t-distribution   170
8-3 Testing hypotheses about μ 175
8-4 The two types of estimates of μ 180
Summary   182
Can you explain the following?   182
Miscellaneous exercises   182

9 NONPARAMETRIC STATISTICS   184

9-1 Introduction   184
9-2 The sampling distribution of runs   185
Summary   190
Can you explain the following?   191

10 THE SAMPLING DISTRIBUTION OF SUCCESS NUMBERS   192

10-1 The binomial experiment   192
10-2 Approximation by a normal distribution   198
10-3 Hypothesis testing   204
10-4 The sign test   208
OPTIONAL
10-5 Exact binomial probabilities   216
Summary   219

Can you explain the following? 219
Miscellaneous exercises   219

11   THE SAMPLING DISTRIBUTION OF SUCCESS RATIOS   222

11-1   Introduction   222
11-2   Estimating p   225
11-3   Another partial solution to the question of sample size   227
       OPTIONAL
11-4   Some theory   230
       Summary   232
       Can you explain the following?   232
       Miscellaneous exercises   233

12   THE SAMPLING DISTRIBUTION OF THE DIFFERENCE BETWEEN MEANS   234

12-1   Introducing the distribution   234
12-2   (Case I) At least one sample size is small   236
12-3   (Case II) Both sample sizes are large   244
12-4   The Mann-Whitney U-test   249
12-5   A note on experimental design   253
       Summary   254
       Can you explain the following?   255
       Miscellaneous exercises   255

13   THE LINE OF REGRESSION   257

13-1   Introduction   257
13-2   The line of regression   258
       Summary   274
       Can you explain the following?   275
       Miscellaneous exercises   275

14   LINEAR CORRELATION   277

14-1   Introduction   277
14-2   A look at the correlation coefficient   277
14-3   The nonparametric version of the correlation coefficient   289

OPTIONAL
14-4 The sampling distribution of r's 292
Summary 294
Can you explain the following? 295
Miscellaneous exercises 295

15 THE CHI-SQUARE DISTRIBUTION 298

15-1 Introduction 298
15-2 Goodness-of-fit 302
15-3 Independence of classification 309
15-4 Yates' correction 317
15-5 Confidence intervals for σ^2 and σ 323
15-6 More about Chi-square 325
OPTIONAL
15-7 The Chi-square test for population normality 327
Summary 331
Can you explain the following? 331
Miscellaneous exercises 332

16 THE SAMPLING DISTRIBUTION OF F RATIOS 334

16-1 Introduction 334
16-2 The F ratio 335
16-3 Testing the equality of population variances 336
16-4 A historical note 340
16-5 One-way analysis of variance 341
16-6 The Kruskal-Wallis Test 356
OPTIONAL
16-7 Some theory 362
Summary 364
Can you explain the following? 364
Miscellaneous exercises 365

APPENDIX I TABLES 367

I: The standard normal 368
II: The Student-t Distribution 369
III: The Chi-square Distribution 370
IV: The F-Distribution ($\alpha = .05$) 371

V: The F-Distribution ($\alpha = .01$) 373
VI: Squares and Square Roots 375
VII: Random Numbers 384

APPENDIX II ANSWERS AND HINTS TO SELECTED EXERCISES 386

Index 402

NOTE TO THE INSTRUCTOR

Statistics: A First Course can be used in either a one-semester or a two-semester course. It is suitable for both two-year and four-year colleges. The only prerequisites for reading this elementary statistics text are a knowledge of high school algebra and a willingness to work.

The content of this book is different from that of most statistics texts. For example, all statistics books define *standard deviation*. This book also explains why we compute the standard deviation and why we divide by N (the finite population size) when calculating the population variance. In addition to discussing what is meant by a *random sample,* this book explains how to test a sample to see whether we can reasonably believe it was randomly selected. Like all statistics books, this one defines *normal population*. Unlike most statistics books, however, it then shows how to find out whether it is reasonable to believe that the population is normally distributed. The book also contains a detailed analysis of the formula for computing z scores. In short, this book tells students *why*.

The book provides optimum exposure for the important topic of probability. If you want minimal coverage of the subject, you can have students read only Chapter 3, An Introduction to Probability. If you wish more extensive coverage, your students can read Chapter 4, Probability, as well. Even users who want the maximum amount of material on probability should be satisfied; however, anyone who wants to supplement Chapters 3 and 4 should feel free to do so.

Many chapters have optional sections that may require some mathematical background beyond high school Algebra I. These sections are included as a challenge for students who would like to learn a little more, and they are not essential to an understanding of the rest of the book.

NOTE TO THE INSTRUCTOR

This book contains more exercises than any book now available. End-of-section exercise sets have been divided into A and B levels. The B-level exercises are either harder or more time consuming than the A-level ones. Exercises for which a hand calculator is suitable are indicated by a calculator symbol. Application problems are marked with a symbol indicating the field (a key to the symbols can be found on the inside front cover). Review exercises are indicated by the head *Cumulative Review*. Answers to odd-numbered problems and hints or complete solutions to some problems are in Appendix I.

I sincerely thank my colleagues, including some that I don't know personally, who have given me many helpful suggestions.

Walter Gleason

NOTE TO THE STUDENT

This book will give you a sound intuitive introduction to statistics, with an emphasis on inferential statistics. Throughout the book, the stress is on applications, so that when you have completed your study, you will be able to apply statistics to any major field. Examples are drawn from such fields as anthropology, biology, business, chemistry, economics, education, health education, law enforcement, nursing, physical education, psychology, and sociology.

Examples and exercise sets begin with problems that can be easily solved with only a pencil. In this way you can gain computational command of a formula before you have to apply it. The examples and exercises that follow gradually become harder and more applied. A symbol identifies the field for each application problem (you can find a key to the symbols on the inside front cover). Problems that you can solve more easily with a hand calculator are also indicated by a symbol. Answers to odd-numbered problems and hints or complete solutions to some problems are in Appendix I.

I hope that you find this book an enjoyable first voyage into applied statistics.

Walter Gleason

1 INTRODUCING STATISTICS

1-1 A HISTORICAL NOTE

Before the 19th century, statistics was little more than the gathering and recording of data. True, calculus was well developed and ready for use as a tool of theoretical statistics, but high-speed computers capable of digesting mountains of data were not on the scene. At about the same time that computers were coming into their own and were being considered for use in analyzing statistical data, professional statistical firms began to appear. These firms organized well-trained staffs on a national basis. Today at least one such firm, Gallup and his associates, is well known to the average person.

Who has a use for statistics? In medicine, a researcher might wish to check the variability of a patient's white cell count. A manufacturer would like to gauge her probability of succeeding when she markets her new product. An educator would like an estimate of the average reading score in his high school in which he can be 95% confident. A physical education major would like to have a test to measure the difference in stamina between smokers and nonsmokers. A sociologist would like to set up a test to see whether people who receive financial aid in college later work for society. In particular, she is interested in whether or not they do any volunteer work. A biologist would like to test the independence of two biological traits in fish. A police department is trying to find out whether there is any connection between crime rates and the lighting in various areas of the city. An economist is trying to find a statistical formula for an inflationary factor that will let him predict the impact of this factor on the economy of the country during the next decade. These are some of the people who use statistics and some of the ways in which they use it.

For anything beyond a minor statistical undertaking, one person, alone and without computer facilities, is at a disadvantage. Planning an unbiased random process for gathering data is tedious, to say nothing of the actual gathering and evaluation of the data. Thus, while we expect you to be able to do the calculations necessary for evaluating statistical formulas from raw data, we must be realistic about the amount of computation that can be accomplished with only a pencil. We are more concerned that you learn to understand the statistical formulas—why they work and when they work.

1-2 A SAMPLE VERSUS A POPULATION

A population usually is analyzed on the basis of one sample. However, a second sample may be secured if the first yields inconclusive results. It is important to distinguish, at the outset, between a *sample* and a *population*. In the study of statistics a population is not necessarily a population of people. For example, a chemist may be interested in studying the reaction of white blood cells to a certain chemical. The *population* in this case would be all of the injected cells, but since they could number in the millions, she would select a *sample* of them to examine for chemical effects. An efficiency engineer may be interested in checking special machine parts for quality. He could choose a *sample* of parts for examination from a *population* of parts produced in a week's time.

Suppose you are a manager at the Ford Motor Company and your assembly lines have been using a new type of paint on full-size models for two years. You would like to know how the paint withstands the elements and how the customers feel about their paint job. Ideally, you should inspect every car, but this is clearly impractical. Just locating them would be a formidable task, especially financially. Instead, you inspect a few (the sample) of all the cars produced (the population). You then analyze opinions and suggestions from the customers whose cars were inspected (the sample) in order to try to make inferences about the population of cars.

In the next paragraphs we distinguish further between populations and samples. And we introduce two companies that you will meet throughout the book as case studies in applied statistics: Finast Sugar and PVC Plastic Pipe.

Finast Sugar Company, sells 1-lb, 2-lb, and 5-lb packages of sugar. In addition it sells sugar in bulk to industry. We wish to consider the population of 5-lb packages of sugar, and one particular *population parameter* (value calculated from the population)—the average package weight. Since consumers don't like to receive less than the 5 lb of sugar promised by a 5-lb package, and the company doesn't want to give sugar away by packing more than the stated amount, the quality control department must assure management that the actual package weight varies as little as possible from the weight stated on the package. How can the department guarantee that this population, which consists of thousands of packages of sugar, does meet strict company standards? The answer is that it cannot. Then what can be done to control key population parameters?

1-2 A SAMPLE VERSUS A POPULATION

Each week the quality control department selects a sample of 5-lb packages of sugar for testing. The important values measured from this sample are called *sample statistics*. Following is one week's sample. It consists of 31 5-lb packages carefully selected and weighed to the nearest thousandth (.001) of a pound.

5.000	5.000	5.000	5.000	[5.010]	5.000
5.000	5.000	5.000	[4.990]	5.000	5.000
5.000	5.000	5.000	5.000	5.000	5.000
[5.005]	5.000	5.000	5.000	5.000	5.000
5.000	5.000	5.000	[4.995]	5.000	5.000
5.000					

Now, the quality control department begins to calculate key sample statistics. First, it checks to see that the average weight lies between 4.990 and 5.010 lb (the desired tolerance). If it does not, there will be immediate cause for alarm: the corresponding population parameter—and thus the population—may be in danger of deviating from the prescribed 5 lb. Here, the average weight is $(1/31)(155 \text{ lb}) = 5.000$ lb. The sample begins to look good. The fact that only 4 of the 31 values differ from the optimum value is also a good sign.

The quality control department then calculates another sample statistic, the *sample range*. The sample range is simply the largest package weight minus the smallest package weight. The company has predetermined that this statistic must not exceed .030 lb in any given week. Here the sample range is $5.010 - 4.990 = .020$ lb. Further inspection shows only one package with a weight of 5.010 lb and only one package with a weight of 4.990 lb. Because the sample looks good, we infer that the population from which the sample was selected is in good shape.

Definition 1-1   A *population* is the collection of data that we wish to investigate.

We will adopt the convention of denoting populations by capital letters from the beginning of the alphabet, such as A, B, and C.

Definition 1-2 A *sample* is a part of a population.

We will denote samples of a population by capital letters from the end of the alphabet, such as *W, X,* and *Y*.

Next, meet the PVC Plastic Pipe Company, which manufactures, as one of its many products, 4-in. (diameter) plastic pipe for the plumbing industry throughout the United States. Consider this population of 4-in. pipe. Clearly, pipe advertised as 4 in. must be 4 in.; otherwise, it will never connect with existing 4-in. pipe. Any plumber who buys this pipe from PVC would be very unhappy to discover any significant change in this 4-in. measurement. Thus, PVC must constantly monitor important population parameters such as the diameter of this pipe, and does so by regularly selecting pieces of 4-in. pipe for measurement. Some of the sample statistics that the company monitors are the average diameter of the pipe, the average thickness of the pipe wall, and the minimal tensile strength of sample pipe when tested under pressure.

Suppose the plumbers who buy from PVC insist that the measurement of the pipe be within one thousandth (.001) of an inch of 4 in. This means that each piece of pipe will fit into one of two categories: (1) right diameter (success) or (2) wrong diameter (failure). What happens if a piece of pipe is manufactured with an incorrect diameter measurement?

An outsize piece of pipe is an out-and-out failure. PVC's business is not like the clothing industry where mistakes can be sold as "seconds" to recoup basic expenses. If a manufacturing mistake is caught at the plant, PVC suffers a financial loss of material, labor, time, heat, light, and so forth. And if the nonstandard manufactured pipe actually reaches a construction site, PVC also suffers a two-way transportation loss. In an effort to minimize such losses, PVC calculates yet another sample statistic—the ratio of the number of correct pieces of pipe in a sample (successes) to the sample size. The company has predetermined that this sample statistic must be at least .95. Once again, based upon a sample, an inference is made about the entire population of pipe.

You can expect to meet these two companies and their applied problems in future chapters.

1-3 A PARAMETER VERSUS A STATISTIC

A collection of data is *finite* (has an end) if the number of pieces of data in the collection is a natural number—that is, (1, 2, 3, . . .). The desks in a school, the people on a train, the residents of a town, and all the telephone numbers in a telephone book are examples of finite sets. Any collection of data that is not finite is *infinite* (has no end). We cannot count all the elements in an infinite collection of data. The set of even numbers, the set of positive powers of 2, the natural numbers, the points on the number line between 0 and 1 are all examples of infinite collections of data.

You will probably never work with infinite populations. A population may be enormous but still finite, such as all the ears of corn in the state of Iowa or all the fish in the Atlantic Ocean. Of course, some populations are so large that, for practical purposes, they might as well be infinite populations. A sample, however, is always finite.

Definition 1-3 Numerical values calculated from a population are called *parameters*.

Definition 1-4 Numerical values calculated from a sample are called *statistics*.

Let the collection of number grades in a freshman psychology course at your college be a population. The average grade is a parameter of this population. The value of the highest grade minus the lowest grade is another parameter. The most frequent grade is also a parameter, as is the lowest grade value.

Select a sample of 100 freshman psychology students and record their grades as follows. Let the first student selected be the first student alphabetically whose last name starts with A. Let the second student be the first student whose last name starts with B. Go through the entire alphabet this way. Let the 27th student be the second student alphabetically whose last name starts with A, and so forth. The average grade for these 100 students is a statistic. The value of the highest grade minus the lowest grade, the most frequent grade, and the lowest grade value are all statistics. Notice that for each parameter there is a corresponding statistic.

1-4 DESCRIPTIVE VERSUS INFERENTIAL STATISTICS

Statistics can be divided into two categories: descriptive statistics and inferential statistics. *Descriptive statistics* deals with the gathering and recording of data. The recording may be done in various ways—bar graphs, line graphs, charts, tables, and so on. Making basic calculations such as the arithmetic average are also part of descriptive statistics. Descriptive statistics is the main subject of this chapter.

In contrast, statistical inference, *inferential statistics,* deals with making generalizations about a population from a sample(s). It consists of making predictions or educated guesses such as a business makes about gross sales for a new store, and such as a pollster makes about which candidate will win an election. Such statements about the population are based on the data taken from a sample.

1-5 SUMMATION NOTATION

Many of the applied formulas in statistics are based on adding numerical values such as $2^2 + 3^2 + 4^2 + \ldots + 50^2$. Because this happens so often in statistics, we want a compact way of expressing the sum of a collection of numerical values which may, at times, involve many additions. Suppose you want to express the summation (process of adding) of even numbers from 2 to 600 inclusive. If you call the sum S, you can write $S = 2 + 4 + 6 + 8 + 10 + \ldots + 600$. The following is the compact notation for expressing this summation:

$$\sum_{n=1}^{n=300} 2n$$

The upper-case Greek letter sigma, Σ, acts as a "command signal" to add successive numerical values obtained from the formula "$2n$" by first letting $n = 1[(2)(1) = 2]$, and then $n = 2[(2)(2) = 4]$, and then $n = 3[(2)(3) = 6]$, and so forth. The first value to be substituted into the formula is written directly below Σ. Each successive value substituted into the formula is always exactly one larger than the preceding value. The process ends when we have substituted the value at the top of Σ into our formula (here $n = 300$).

The numbers at the bottom and top of Σ are, respectively, starting and stopping command numbers. They are called *index numbers*. The actual choice of letter to be used in a summation formula is immaterial. When we chose n in the formula $2n$, we made an arbitrary choice. For brevity, we generally omit the letter from the upper index. Thus,

$$\sum_{n=1}^{300} 2n$$

The following examples will help to illustrate this important concept further.

1-5 SUMMATION NOTATION

Example 1-1 Find the sum of the numbers of the form n^2 (based on the formula n^2) from $n = 1$ (lower index number) to $n = 4$ (upper index number). Write the sum using summation notation. Evaluate (substitute numbers into the formula).

Solution

$$\sum_{n=1}^{4} n^2 = 1^2 + 2^2 + 3^2 + 4^2$$
$$= 1 + 4 + 9 + 16$$
$$= 30$$

Example 1-2 Find the sum of numbers of the form $2t + 1$ from $t = 1$ to $t = 3$. Write the sum using summation notation. Evaluate.

Solution When $t = 1$, $2t + 1 = 2(1) + 1 = 3$.
When $t = 2$, $2t + 1 = 2(2) + 1 = 5$.
When $t = 3$, $2t + 1 = 2(3) + 1 = 7$.

Thus, $\sum_{t=1}^{3} (2t + 1) = 3 + 5 + 7 = 15$.

Example 1-3 Write a summation based on the formula $3^i - 1$.
Evaluate this formula from the beginning value $i = 1$ to the ending value $i = 3$.

Solution When $i = 1$, $3^1 - 1 = 2$.
When $i = 2$, $3^2 - 1 = 8$.
When $i = 3$, $3^3 - 1 = 26$.

Thus, $\sum_{i=1}^{3} (3^i - 1) = 2 + 8 + 26 = 36$.

Example 1-4 Evaluate $\sum_{j=1}^{3} 1/j$.

Solution

$$\sum_{j=1}^{3} \frac{1}{j} = \frac{1}{1} + \frac{1}{2} + \frac{1}{3} = \frac{6}{6} + \frac{3}{6} + \frac{2}{6} = \frac{11}{6}$$

Example 1-5 Evaluate $\sum_{t=2}^{4} (t^2 - 3t)$.

Solution

$$\sum_{t=2}^{4} (t^2 - 3t) = (2 \times 2 - 3 \times 2) + (3 \times 3 - 3 \times 3) + (4 \times 4 - 3 \times 4)$$
$$= (4 - 6) + (9 - 9) + (16 - 12)$$
$$= 2$$

Example 1-6 Evaluate $\sum_{j=0}^{3} (j - 2)$ and $\sum_{j=0}^{3} j - 2$.

Solution $\sum_{j=0}^{3} (j - 2) = (-2) + (-1) + 0 + 1 = -2$

Notice the effect of omitting the parentheses:

$$\sum_{j=0}^{3} j - 2 = [0 + 1 + 2 + 3] - 2 = 4$$

EXERCISES A

1-1. Suppose that you have complete information about the fire insurance claims filed against an insurance company during November 1980. Under what conditions would you consider such data (a) a sample? (b) a population?

1-2. The Fish and Game Department of Arizona has complete records of how many deer licenses were issued to hunters during each of the years 1965 through 1975. Show how these data could be looked upon as a sample; as a population.

1-3. Your population is the set of all heights of students on your college campus. Your sample is the set of heights of your class. Name two statistics you might get from your sample.

1-4. What is the most frequent value for the data (Section 1-2) of the Finast Sugar Company? Is this answer a surprise? Explain.

For Exercises 1-5–1-14 evaluate the summation and state the final answer as a single real number whenever possible.

1-5. $\sum_{i=0}^{3} 2^i$ (Example 1-3)

1-9. $\sum_{x=0}^{4} x^2$

1-6. $\sum_{t=1}^{5} (t^2 - 7)$ (Examples 1-2, 1-5)

1-10. $\sum_{t=1}^{3} [(t/2) - 1]$

1-7. $\sum_{n=1}^{5} (n + 1)(n + 2)$

1-11. $\sum_{i=1}^{4} (5i - 10)$

1-8. $\sum_{i=1}^{4} (4i^2 - i)$

1-12. $\sum_{j=0}^{3} (j^3 - j)$ (Example 1-5)

1-13. $\sum_{i=1}^{3} i^i$ (Hint: This would be read "i to the ith power.")

1-14. $\sum_{x=0}^{2} (x^2 - 5x + 1)$

B
1-15. Express $x + x^2 + x^3 + \cdots + x^{10}$ in summation notation.
1-16. Express $1 + 3 + 5 + 7 + \cdots + 21$ in summation notation.
(Hint: $2n - 1$ where $n = 1, 2, 3, \ldots$ is an odd number.)
1-17. Express $10 + 15 + 20 + \cdots + 50$ in summation notation.
1-18. Express $3 + 5 + 9 + 17 + 33$ in summation notation.
1-19. Verify $\sum_{k=1}^{n} k = [n(n + 1)]/2$; for $n = 3, 5, 10$.

Example 1-7 Given the collection of data $X = \{-1, 8, 0, 1, 4\}$ with the data "tagged," left to right, $x_1 = -1, x_2 = 8, x_3 = 0, x_4 = 1, x_5 = 4$. Find $\sum_{i=1}^{5} x_i$.

Solution

$\sum_{i=1}^{5} x_i = x_1 + x_2 + x_3 + x_4 + x_5$ (When $i = 1$ we get $x_1 = -1$; when $i = 2$ we get $x_2 = 8$, and so forth.)
$= (-1) + 8 + 0 + 1 + 4$
$= 12$

Assume now that we have a sample X with n values that are not necessarily different. $X = \{x_1, x_2, x_3, \ldots x_n\}$. Summation notation offers us a convenient way of expressing the sum of all the values in the sample X.

$$\sum_{i=1}^{n} x_i.$$

Example 1-8 Given the collection of data $Y = \{5, 2, 2, 2, 3, 6\}$ with data "tagged," left to right, $y_1 = 5, y_2 = 2, y_3 = 2, y_4 = 2, y_5 = 3, y_6 = 6$. Find $\sum_{i=1}^{6} y_i^2$.

Solution

$\sum_{i=1}^{6} y_i^2 = y_1^2 + y_2^2 + y_3^2 + y_4^2 + y_5^2 + y_6^2$
$= 5^2 + 2^2 + 2^2 + 2^2 + 3^2 + 6^2$
$= 25 + 4 + 4 + 4 + 9 + 36$
$= 82$

An important and interesting special case is the one where the letter that appears in the index does not appear as part of the summation formula. In this case, the standard agreement is to write the summation formula each time that we try to substitute a numerical value for the index letter. The next examples illustrate the technique.

Example 1-9 $\sum_{i=1}^{4} 5 = 5 + 5 + 5 + 5 = 20.$

Each time we try to substitute for i. $i = 1$, $i = 2$, $i = 3$, $i = 4$, our formula gives us the constant value 5.

Definition 1-5 If c is a constant, $\sum_{i=1}^{n} c = \underbrace{c + c + c + \cdots + c}_{n \text{ terms}} = cn$

Example 1-10 $\sum_{j=2}^{4} a = a + a + a = 3a$

Example 1-11 $\sum_{i=1}^{n} 1 = \underbrace{1 + 1 + 1 + \cdots + 1 + 1}_{n \text{ terms}} = n$

Example 1-12 $\sum_{i=1}^{3} (t + 1) = (t + 1) + (t + 1) + (t + 1)$
$= 3t + 3$

We are substituting for i and the formula is in the letter t. The result is that "$t + 1$" acts like a constant each time we try to substitute for i.

EXERCISES A

1-20. Given $X = \{2, 4, 5\}$ with $x_1 = 2$, $x_2 = 4$, $x_3 = 5$. (Examples 1-7 and 1-8)
  a. Find $\sum_{i=1}^{3} x_i$.   b. Find $\sum_{i=1}^{3} x_i^2$.

1-21. Given $Y = \{1, 0, -1, 3, 4\}$ with $y_1 = 1$, $y_2 = 0$, $y_3 = -1$, $y_4 = 3$, $y_5 = 4$.
  a. Find $\sum_{i=1}^{5} (y_i + 1)$.   b. Find $\sum_{i=1}^{5} (2y_i - 1)$.

1-22. Given $T = \{4, 2, 5, 3, 2, 7\}$.
  a. Assign numerical values to $t_1, t_2, \ldots t_6$.
  b. Find $\sum_{i=1}^{6} (t_i + 3)$.
  c. Find $\sum_{i=1}^{6} (t_i^2 - 5)$.

1-23. Given $M = \{3, 1, 1, 1, 4, 4, 5\}$.
    a. Assign numerical values to $m_1, m_2, \ldots m_7$.

    b. Find $\sum_{i=1}^{7} (4m_i + 4)$.

    c. Find $\sum_{i=1}^{7} (m_i^2 - m_i + 2)$.

B   **1-24.** $\sum_{t=2}^{6} k$ (Example 1-10)

    1-25. $\sum_{i=1}^{5} 1$

    1-26. $\sum_{i=1}^{n} 1$ (Example 1-11)

    1-27. $\sum_{s=1}^{5} (a + b)$ (Example 1-12)

    1-28. $\sum_{j=1}^{3} (a + b - c)$

    1-29. $\sum_{i=1}^{4} 5$

    1-30. $\sum_{t=3}^{5} 7$

1-6 MEASURES OF CENTRAL TENDENCY

Suppose you want to know all about the batting performance of a ball player on your favorite baseball team. You could look at a daily tabulation of hits, runs, times at bat, and so forth. However, this is both boring and time consuming. You would prefer to look at a single number, say the player's batting average. This single value tells you a lot about a collection of hits (data) accumulated over days of batting.

    Still, it is hard to find a single numerical value that can portray all of the salient features of a collection of data. A player's batting average doesn't tell you whether the player hits singles, doubles, or home runs. This part of a ball player's batting performance might be clearer if you knew the number of a player's most common type of hit. Another value, the number of runs batted in, tells you a lot about a player's ability to hit under pressure.

Finding a numerical value or a combination of numerical values that help to describe a collection of data is the subject of this chapter. Following are some important kinds of descriptive values—or statistics—with definitions and examples.

The sample average is our first, and very important, sample statistic. If $X = \{-1, 8, 0, 1, 4\}$, where $x_1 = -1$, $x_2 = 8$, $x_3 = 0$, $x_4 = 1$, and $x_5 = 4$, we find the average for this sample by adding the five numbers and dividing the sum by 5. It is common practice to refer to the sample average as the *sample mean*.

A sample mean is commonly denoted by placing a bar over the small letter associated with the sample name. Thus, the mean of the sample X is $\bar{x}$, the mean of the sample Y is $\bar{y}$, and so forth. Here we obtain

$$\bar{x} = \frac{x_1 + x_2 + x_3 + x_4 + x_5}{5} \quad \text{or} \quad \frac{\sum_{i=1}^{5} x_i}{5}$$

$$= \frac{-1 + 8 + 0 + 1 + 4}{5}$$

$$= \frac{12}{5} \quad \text{or} \quad 2.4$$

Definition 1-6 For a sample X, of size n, the *sample mean* ($\bar{x}$) is

$$\bar{x} = \frac{1}{n} \sum_{i=1}^{n} x_i \quad \text{or} \quad \frac{\sum_{i=1}^{n} x_i}{n}$$

Example 1-13 Evaluate the mean of the sample $X = \{4, 8, 5, 4, 7, 5, 5, 2\}$.

Solution $\bar{x} = (1/8)(4 + 8 + 5 + 4 + 7 + 5 + 5 + 2)$
$= (1/8)(40)$
$\bar{x} = 5$

Example 1-14 The Clark family received these monthly dental bills for the first half of the year: $74, $50, $32, $87, $40, $63. Find the average (mean) dental bill.

Solution The mean $= (1/6)(\$74 + \$50 + \$32 + \$87 + \$40 + \$63)$
$= (1/6)(\$346)$
The mean $= \$57.67$

1-6 MEASURES OF CENTRAL TENDENCY

Definition 1-7   The *mode* of a collection of data is the value that occurs most often.

Example 1-15   For the sample $T = \{1, 2, 1, 2, 2, 5, 1, 1\}$ find the mode.

Solution   The mode is 1.

A collection of data may have two values that occur equally often and yet more often than other values of the collection. In this case, the collection of data has two modes and is said to be *bimodal*. If a collection of data has three or more potential modes we say it has no mode.

Definition 1-8   When a collection of data has an odd number of values and the data are arranged in increasing (decreasing) order, the middle value is the *median*.

Example 1-16   Find the median for the sample $X = \{2.5, 7, 3, 9.1, .028\}$.

Solution   Place the members of X in increasing order.

$$X = \{.028, 2.5, 3, 7, 9.1\}$$

The median is 3.

Definition 1-9   When a collection of data has an even number of values and the data are arranged in increasing (decreasing) order, the average of the middle two values is the *median*.

Example 1-17   Find the median for the sample $W = \{2, 8, 6, 13\}$.

Solution   Place the values of W in order.

$$W = \{2, 6, 8, 13\}$$

The median is the average of 6 and 8. The answer is 7.

Note that the median does not have to be a value in the data collection; see Example 1-17. In contrast to the mean, median, and mode, we have another numerical value that computes the distance from the smallest value to the largest data value. It is called the range.

Definition 1-10 The *range* of a collection of data is obtained by taking the largest value in the collection and subtracting from it the smallest value.

Example 1-18 For the sample $Y = \{1, 7, 5, 3, 9\}$, find the mode, the median, and the range.

Solution We first place the members of Y in order.

$$Y = \{1, 3, 5, 7, 9\}$$

The median is 5. The range is 8. There is no mode.

Example 1-19 The prices of one style of prefabricated home in a low-income development are as follows:

$29,990 $28,850 $27,900 $30,280 $28,300
30,220   29,340   30,790   28,500   29,200
30,000   30,400   28,650   30,270

Find the mode, the median, and the mean price.

Solution There is no mode. The median value is $29,665. The mean value is $29,477.86. It is hard here to find one value more representative of the data than any other. If in doubt, state both median and mean.

The mean is the single numerical value we use most often in statistics to describe a collection of data and, in general, it does a creditable job. However, in some unusual cases, the mean can be misleading. See the next example.

Example 1-20 From a government file a sociologist had the following incomplete sample information on the annual wage of workers in a South American town:

1-6 MEASURES OF CENTRAL TENDENCY

    a. The sample size is 100.

    b. The sample mean is $10,000.

    c. The workers interviewed were existing on a near-starvation diet.

How can this be? Facts (b) and (c) seem to be contradictory.

Solution Our sociologist investigated. He found that 95 of the 100 workers were paid annual wages on a poverty level. The other five "workers" were actually managers who were also part-owners of the enterprise. They received the equivalent of $50,000, $75,000, $225,000, $98,000, and $150,000. The sample mean was misleading.

If the mode had been quoted, it would have shown the large number of very low annual wages and failed to show the five exceptionally large annual wages.

If the median had been quoted, it, being the middle value, would have been one of the lower 95 annual wage values. It also would have failed to show the five exceptionally large annual wages.

Here is a situation where the range value would be a big help toward describing the nature of the data. Given the combination of three values—range, mode, and median—we would then get a real feel for this set of data; but the mean alone was misleading.

The mode, median, and mean are all called *measures of central tendency*. Because these values are usually different, this practice can be confusing. As a rule of thumb, unless otherwise specified, assume that a person who speaks of a measure of central tendency is referring to the mean.

Example 1-21 Our data are the 31 5-lb package weights of the Finast Sugar Company of Section 1-2. Compute the mean of these weights.

Solution   Mean = $(1/31)(\underbrace{5.000 + 4.990 + \cdots + 5.000}_{31 \text{ terms}})$

              = $(1/31)(155)$

    Mean = 5.000 lb

Although this solution is perfectly correct, it is often faster to work by combining like (identical) data values. Here, for example, we can make the computation easier by considering all 27 of the 5.000's as one group. Thus,

16 INTRODUCING STATISTICS

$$\text{Mean} = (1/31)(5.010 + 5.005 + (5.000)(27) + 4.995 + 4.990)$$
$$= (1/31)(155)$$
$$\text{Mean} = 5.000 \text{ lb}$$

Obviously you save time by evaluating all the like data values at one time. This is called *grouping* the data. Here let's discuss calculating the mean of a sample X by grouping. That is, X is a sample of size n where several data values are the same. We break the sample down into subgroups of identical data values, such as all the 2's, all the 3's, all the 4.15's, and so on. If there are k such groups each with a frequency f_j, then $\bar{x}$ is as follows:

$$\bar{x} = (1/n) \sum_{j=1}^{k} x_j f_j; \; k \leq n$$

Example 1-22 Use grouping to evaluate the mean of the sample $X = \{2, 3, 4, 2, 3, 2, 4, 5, 4, 3\}$.

Solution We make four groups (piles). Each group is tagged with an integer from 1 to 4 inclusive. We tag the group of 2's as group 1. Its frequency is $f_1 = 3$. Also, $f_2 = 3$, $f_3 = 3$, and $f_4 = 1$.

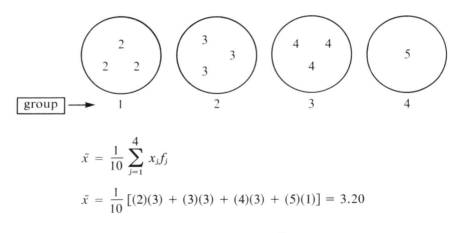

$$\bar{x} = \frac{1}{10} \sum_{j=1}^{4} x_j f_j$$

$$\bar{x} = \frac{1}{10}[(2)(3) + (3)(3) + (4)(3) + (5)(1)] = 3.20$$

For any sample X, of size n, it is true that $\sum_{i=1}^{n} (x_i - \bar{x}) = 0$. The next two examples illustrate this fact.

Example 1-23 For the sample $X = \{1, 3, 5, 7, 9\}$, verify that $\sum_{i=1}^{5} (x_i - \bar{x}) = 0$.

Solution $\bar{x} = 5$. Let $x_1 = 1, x_2 = 3, x_3 = 5, x_4 = 7, x_5 = 9$.

$$\bar{x} = \frac{1 + 3 + 5 + 7 + 9}{5} = 5$$

$$\sum_{i=1}^{5} (x_i - \bar{x}) = \sum_{i=1}^{5} (x_i - 5) = -4 - 2 + 0 + 2 + 4 = 0$$

Example 1-24 For the sample $Y = \{1, 2, 14, 7\}$, verify that

$$\sum_{i=1}^{4} (y_i - \bar{y}) = 0.$$

Solution $\bar{y} = 6$. Let $y_1 = 1, y_2 = 2, y_3 = 14, y_4 = 7$.

$$\sum_{i=1}^{4} (y_i - 6) = -5 - 4 + 8 + 1 = 0$$

English letters are used to denote statistics and Greek letters are reserved to denote population parameters. Our population mean, for example, is denoted by the Greek letter mu (μ). Mu is evaluated in exactly the same way as the sample mean.

Definition 1-11 For any finite population A, of size N, the *population mean* (μ) is computed as follows:

$$\mu = (1/N) \sum_{i=1}^{N} a_i$$

where a_i is the ith value in population A.

Example 1-25 Evaluate μ for the population $A = \{1, 2, 3, \ldots, 10\}$.

Solution $\mu = 1/10 \sum_{i=1}^{10} a_i = (1/10)[55] = 5.5$

Suppose we took the sample $X = \{1, 5, 9\}$ from population A of the preceding example, $\bar{x} = 5$. Observe that our statistic $\bar{x}$ is close in value to the parameter μ. In such cases, we say that $\bar{x}$ is *a good estimator of* μ.

Next, we take four more samples from the population A and compute their means as follows:

$$Y = \{2, 7, 9\} \qquad \bar{y} = 6$$
$$Z = \{2, 5, 8\} \qquad \bar{z} = 5$$
$$U = \{1, 10, 1\} \qquad \bar{u} = 4$$
$$V = \{3, 8, 10\} \qquad \bar{v} = 7$$

Although the preceding samples are all drawn from the same population, we see that the values of the sample mean are quite different.

If a die is rolled forever and the number from 1 to 6 on the top face recorded at each roll, we can generate a population of data whose mean can be shown to be exactly 3.5 ($\mu = 3.5$). The following are six such samples, each consisting of five rolls of a die with the sample mean calculated.

$$X = \{4, 6, 5, 1, 4\} \qquad \bar{x} = 4$$
$$Y = \{2, 4, 2, 5, 3\} \qquad \bar{y} = 3.2$$
$$Z = \{6, 6, 2, 1, 6\} \qquad \bar{z} = 4.2$$
$$W = \{5, 6, 4, 2, 6\} \qquad \bar{w} = 4.6$$
$$U = \{3, 5, 3, 5, 4\} \qquad \bar{u} = 4$$
$$V = \{4, 1, 2, 4, 4\} \qquad \bar{v} = 3$$

Observe again how the value of the sample mean varies.

Use samples X, Y, Z, W, U, and V to construct a table illustrating the variability of the sample statistics: mode, median, and range.

Sample	Mode	Median	Range
X	4	4	5
Y	2	3	3
Z	6	6	5
W	6	5	4
U	3 and 5	4	2
V	4	4	3

TABLE 1-1

1-7 DISCRETE VERSUS CONTINUOUS DATA   19

Notice that *all statistics are variables*. In contrast, because we deal with only one population at a time, the *parameters of a population are constants*.

1-7 DISCRETE VERSUS CONTINUOUS DATA

Data can be divided into two categories: (1) *discrete data* and (2) *continuous data*. Usually you can determine whether your data are discrete or continuous simply by whether you counted to get them (discrete) or measured to get them (continuous). The number of ants in an ant trap, the number of people in nursing homes in Illinois, the number of people in a socioeconomic group, and the white cells in a blood sample are all counts and are examples of discrete data. The strength of pieces of wire, people's weights, daily temperatures, and the amount of pollutant in a sample of air are all measures, and thus are examples of continuous data.

Most discrete collections of data are finite. The one exception to this is a collection of data that can be matched, one to one, with the whole numbers, $W = \{0, 1, 2, 3, \ldots\}$. Such a discrete collection of data is said to be countably infinite. See Example 1-26.

Example 1-26 Gary and Bobby both toss coins. If the two coins land alike (HH, TT), they stop and write the letter S for success. If the coins land not alike, the boys write the number 1 and repeat the process; if not alike a second time, the number 2, and so forth. They continue this procedure until the coins fall alike. Now this procedure could go on forever with the following discrete collection of data arising.

Possible data: 1, 2, 3, 4, . . .

The presence of fractions or decimals in data does not necessarily imply that the data are continuous. An example in which decimals appear but the data are actually discrete is in competitive diving. Here contestants receive scores by halves. A score of 5.5, but not of 5.8, is possible.

Example 1-27 Discuss whether the following collections of data are discrete or continuous.
a. The number of eggs produced by every chicken in the United States on a given day.
b. The amount of milk given by every cow in the United States on a given day.

Solution
a. Here we have a discrete collection of data. To be specific, a chicken can't lay half an egg.
b. In theory, cows can yield an amount of milk corresponding to any point in an interval of the number line beginning at zero. This is a continuous collection of data.

EXERCISES A

1-31. Classify the following data as discrete or continuous. (Example 1-27)
   a. The number of automobile accidents per year in New York.
   b. The length of time needed to play nine holes of golf.
   c. The number of building permits issued per month in San Francisco.
   d. The weight of grain produced per acre in Nebraska.

For the sample sets in Exercises 1-32, 1-33, and 1-34 evaluate the mean, mode, median, and range.

1-32. $X = \{5, 7, 9, 11\}$ (Examples 1-13 and 1-18)
1-33. $Y = \{2, 3, 4, 2, 5, 2, 6, 7, 3, 7, 2, 5\}$
1-34. $Z = \{4, 6, 5, 3, 2, 4, 9, 5, 1, 5, 6, 7, 10, 2, 3, 3, 5, 10, 6, 4\}$
1-35. Do the computation to find the mean of the data in Example 1-19.
1-36. You have a sample of size 5 with a range of 0. Describe this sample.
1-37. The mean of a set of 12 numbers is 8; what is the sum of the numbers?

1-38. The number of cavities found at dental checkups in a group of individuals is $\{2, 10, 4, 7, 8, 15, 12, 11, 5, 3, 0, 2, 1, 5\}$. Find the mean number of cavities. (Example 1-14)

1-39. Ten cars were timed passing a police checkpoint. The speeds were $\{53, 48, 50, 47, 43, 58, 42, 46, 51, 38\}$. For this set of data find the mean, median, mode, and range speed.

1-40. The following data represent the price per pound of ground chuck bought at one market during 1979: 69, 69, 78, 79, 79, 88, 88, 89, 89, 89, 97, 98, 99, 99, 99, 99, 119, 119, 119, 129. All prices are given in cents. Find the mean, median, mode, and range of this set of data to the nearest cent.

1-41. Twelve people attended a weight-watching club. After six weeks, they reported their weight losses as negative integers and their gains as positive integers. The 12 integers are $\{6, 12, -3, 5, 8, -7, 1, -1, 3, 7, 9, 5\}$. Compute the mean weight loss (gain).

1-42. Three sets of numbers have respective means of 12, 20, and 30 on the basis of 10, 30, and 50 observations, respectively. What is the mean if the three sets of numbers are combined into a single set of numbers? [Hint: $(10)(12) = 120$; $(30)(20) = 600$; $(50)(30) = 1,500$.]

1-43. The mean yearly income of workers in the Ajax Show Company is $11,800. If each of the workers is given a raise of $800, what is the new average wage? If each worker is given a 6% raise, what is the new mean?

1-44. There are six children in the Jones family. The twins are 2 years old and the ages of the other children are 8, 11, 12, and 13. Find the mean, median, and modal ages. What will the mean, median, and modal ages of these same children be in 13 years?

1-45. During the past year the O'Toole family had the following electric bills: $43.02, $38.60, $36.50, $44.10, $43.98, $39.45, $41.15, $40.86, $37.50, $38.40, $40.61, $41.30. Find the value of the mean electric bill for this family.

1-8 THE FREQUENCY DISTRIBUTION TABLE

B   In Exercises 1-46, 1-47, and 1-48, first group the data (separate the data into piles of the "same value"), and then determine the value of the mean, median, mode, and range. (Examples 1-21, and 1-22)

 1-46. The following data values are all milk prices in cents per quart: $M = \{.47, .48, .50, .52, .52, .53, .54, .54, .54, .55, .57, .57, .59, .59, .59, .60, .60, .62, .63, .70\}$.

 1-47. The collection F represents 20 flour prices recorded in cents: $F = \{77, 89, 89, 95, 95, 97, 98, 98, 99, 99, 103, 105, 108, 109, 109, 109, 111, 115, 119, 129\}$.

 1-48. The Maury High School seniors helped with the local Red Cross drive. The numbers of memberships secured by 50 members of the senior class are listed as follows:

5	4	12	6	3	6	13	4	6	15
6	5	6	9	5	6	9	11	13	11
10	6	4	11	8	5	7	16	2	6
18	8	6	5	7	4	6	8	1	4
11	6	13	5	2	15	5	9	7	10

Determine the mean, mode, median, and range values.

 1-49. A small gift shop reports the following sales to tourists during two weeks' time.

$38.63 $72.61 $45.44 $50.10 $61.38

$60.71 $48.40 $38.22 $15.89 $19.15

$48.61 $61.60 $51.20 $18.08

Find the mean, median, and range values.

1-50. In Exercise 1-32 show $\sum_{i=1}^{4} (x_i - \bar{x}) = 0$. (Examples 1-23, and 1-24)

1-51. In Exercise 1-33 show $\sum_{i=1}^{12} (y_i - \bar{y}) = 0$.

1-8 THE FREQUENCY DISTRIBUTION TABLE

Once statisticians gather data they are faced with the problem of presenting their data in a form that is easy to read. It is at best difficult to analyze rows of *raw data,* the actual numbers. In this section we discuss several ways of presenting data in a managable and useful form.

The following data represent the speeds, in miles per hour (mph), of cars traveling northbound on U.S. Route 95 in Massachusetts.

```
47  52  50  54  59  63  67  57  55  48
54  36  62  58  59  53  47  61  54  53
70  49  46  57  45  52  53  38  42  55
60  56  61  49  51  53  55  55  56  37
40  53  51  52  48  60  44  49  68  57
62  48  50  53  51  53  57  45  59  47
58  48  50  52  54  44  52  50  43  58
48  55  56  58  52  51  59  51  60  61
56  55  49  60  39  56  55  57  58  59
57  48  53  57  59  58  57  60  41  58
```

Chances are you will find little meaning in these rows of numbers. A much better overall picture of such data results from grouping the speeds in a table.

Speed (mph)	f, Observed Frequency	Number of Cars
66–70	111	3
61–65	⊤⊦⊦⊦ 1	6
56–60	⊤⊦⊦⊦ ⊤⊦⊦⊦ ⊤⊦⊦⊦ ⊤⊦⊦⊦ ⊤⊦⊦⊦ ⊤⊦⊦⊦ 1	31
51–55	⊤⊦⊦⊦ ⊤⊦⊦⊦ ⊤⊦⊦⊦ ⊤⊦⊦⊦ ⊤⊦⊦⊦ ⊤⊦⊦⊦	30
46–50	⊤⊦⊦⊦ ⊤⊦⊦⊦ ⊤⊦⊦⊦ 111	18
41–45	⊤⊦⊦⊦ 11	7
36–40	⊤⊦⊦⊦	5
		100 = sample size

TABLE 1-2

With the sample data thus tabulated, you—or the state police—are in a better position to analyze the driving habits of Massachusetts drivers using this highway. The posted speed limit is 55 mph. The table shows that 60 of the 100 drivers stay below this limit. Indeed, the slowest five drivers don't exceed 40 mph, and may be a hazard to other motorists. Of the 40 drivers who do exceed the speed limit, 31 do so by no more than 5 mph. Only three of the drivers sampled exceeded the speed limit by more than 10 mph. The table offers us considerable insight into the driving habits along Route 95.

This kind of table is called a *frequency distribution table*. The observed frequency (f) in each group is the number of pieces of data in that group. For example, the number of drivers (frequency) who were observed traveling from 51 to 55 mph inclusive is 30. Data are recorded by tally in groups called *classes*. For example, each class of drivers in the auto-speed study just described represented a speed range of 5 mph. In general, our data will be broken down into from 5 to 15 classes of more or less equal range.

Class ranges do not have to be equal, but equal ranges are more convenient, particularly when you want to present a frequency distribution in a graph. Actually, however, you determine the number of classes and the class range to be used by a common-sense inspection of the data.

The smallest and largest values within any class are called the *lower class limit* and the *upper class limit*, respectively. For the class 46–50, 46 is the lower class limit and 50 is the upper class limit. In Table 1-2, the lower class limits are 36, 41, 46, 51, 56, 61, and 66. The upper class limits are 40, 45, 50, 55, 60, 65, and 70.

A *class boundary* is a point midway between the upper class limit of one class and the lower class limit of the succeeding class. The number 40.5 is the lower boundary of the 41–45 class; 40.5 is also the upper boundary of the 36–40 class. Consider 45.5. It is the upper boundary value for the 41–45 class and also the lower boundary value of the 46–50 class.

In Table 1-2, the lower boundaries are 35.5, 40.5, 45.5, 50.5, 55.5, 60.5, and 65.5. The upper boundaries are 40.5, 45.5, 50.5, 55.5, 60.5, 65.5, and 70.5.

Example 1-28 A dentist has recorded the number of patients coming to her office daily for the past 40 days.

10	8	5	7	8	12	5	7	9	10
16	13	7	8	5	4	14	8	15	12
11	11	8	7	10	9	15	18	16	5
9	9	10	12	8	11	14	4	6	8

a. Make a frequency distribution table from this set of data, starting with lightest patient days and working upward. The first two classes then might be 4–6 patients and 7–9 patients.

b. Discuss the limits and boundaries of the 10–12 class.
c. Compute the sample mean directly from the raw data.

Solution **a.**

Classes	Frequency, f		Class Mid-point
16–18	111	3	17
13–15	++++	5	14
10–12	++++ ++++	10	11
7–9	++++ ++++ ++++	15	8
4–6	++++ 11	7	5
Total		40	

Observe that the "lowest class" (4–6) does not necessarily contain the smallest frequency, nor does the "highest class" (16–18) contain the highest frequency.

b. For the 10–12 class: the lower class limit is 10; the upper class limit is 12; the lower class boundary is 9.5; and the upper class boundary is 12.5

c. Using the raw data we obtain:

Mean = (1/40)[10 + 8 + 5 + 7 + 8 + 12 + 5 + 7 + 9
+ 10 + 16 + 13 + 7 + 8 + 5 + 4 + 14
+ 8 + 15 + 12 + 11 + 11 + 8 + 7 + 10
+ 9 + 15 + 18 + 16 + 5 + 9 + 9 + 10
+ 12 + 8 + 11 + 14 + 4 + 6 + 8]
= (1/40)(384)
= 9.60

Once data are completely tallied, statisticians often treat the data within each class as lying entirely at the midpoint of that class, called the *class mark*. This rationale allows us to evaluate the mean directly from a frequency distribution table. High and low values within a class tend to cancel each other. For a large collection of data, calculating the mean directly from a frequency distribution table will generally save considerable time. This will be illustrated in succeeding examples.

Example 1-29 Tally the sample data of $X = \{2, 8, 15, 30, 17, 23, 14, 16, 25, 20, 5\}$ in a frequency distribution table.

Record the observed frequency, the class mark, and the upper boundary of each class.

Solution Because the sample size is small ($n = 11$), we will set up only five classes. Let's set our lowest class to include the values 0 to 6 inclusive. Let the next lowest class include the values 7 to 13 inclusive, and so on. These class ranges are reasonable, but they certainly aren't the only possible choices.

Class	Observed Frequency	Class Mark	Upper Class Boundary
28–34	1	31	34.5
21–27	2	24	27.5
14–20	5	17	20.5
7–13	1	10	13.5
0–6	2	3	6.5

TABLE 1-3

Let's make some comments about Example 1-29. Could we have selected the lowest class range as 1 to 5 inclusive and the second lowest class as 6 to 10 inclusive? Yes. What we always desire is that the smallest data value be contained in the lowest class and the largest data value be contained in the highest class. Specifying two successive class ranges always determines all the other class ranges. Is it clear that using 1–5, 6–10, and so forth, will demand that we use six classes to record the data of Example 1-29? Investigate this claim with your pencil.

The *relative frequency* of any class is defined as a ratio of the observed class frequency f divided by the sample size n. For the ith class, the relative frequency is denoted f_i/n: $0 \le f_i/n \le 1$. In practice a relative frequency value of 0 is unusual and a relative frequency value of 1 is not desirable. Do you see why? Further, for the case of k classes, it is always true that $\sum_{i=1}^{k} f_i/n = 1$. In Table 1-2 the relative frequency of the 46–50 class is $18/100 = .18$, which means that the 46–50 class contains .18 or 18% of the measurements.

Example 1-30 Redo Example 1-29, letting the lowest class range be 1–5 inclusive and the second lowest class range be 5–9 inclusive. In addition, complete a column entitled relative frequency.

Class	Observed Frequency	Class Mark	Upper Class Boundary	Relative Frequency
29–33	1	31	33	1/11
25–29	1	27	29	1/11
21–25	1	23	25	1/11
17–21	2	19	21	2/11
13–17	3	15	17	3/11
9–13	0	11	13	0
5–9	2	7	9	2/11
1–5	1	3	5	1/11

TABLE 1-4

Note the relative frequency value of 0.

Solution Observe that successive upper and lower class limits are equal; that is, the upper boundary of one class equals the lower boundary of the class above it. Further boundary values will now equal class limits. Should a data value equal a boundary value, we must decide which class will receive the datum value. For example, as you can see from the table, 17 creates a problem. In such a case let's agree to place the datum value in the larger of the two possible classes. There is no special reason for this choice. It is simply that a choice must be made.

Example 1-31 a. Compute the sample mean directly from the raw data of Example 1-29.
b. Compute the mean of sample X by using the data as recorded in the frequency distribution table of Example 1-29.
Compare.

Solution
a. When we use the raw data:
$\bar{x} = (1/11)[2 + 8 + 15 + 30 + 17 + 23 + 14 + 16 + 25 + 20 + 5]$
$= (1/11)[175]$
$= 15.91$
b. To compute $\bar{x}$ from the frequency distribution table, assume that all of the data within a class lie at the class mark. For example, we assume all five values in the class 14–20 are 17's. Although this is not technically accurate, the gains and losses tend to offset each other.

$$\bar{x} = (1/11)[(1)(31) + (2)(24) + (5)(17) + 1(10) + 2(3)]$$
$$= (1/11)[180]$$
$$= 16.36$$

Observe that the two solutions are only approximately equal.

For a deeper intuition into our work here, let's look at the geometric representation of frequency distribution. We will discuss two types of graphs that can be formulated directly from frequency distribution tables using as the data driving speeds on U.S. 95 in Massachusetts, Table 1-2.

The first type of graph is called a *histogram*. It is simply a bar graph. Mark off the class boundaries along a horizontal axis. Mark off the observed frequencies (f) on a vertical axis. Try to do this with as little distortion as possible. Again, however, statisticians choose the vertical scale more or less for graphing convenience. Above each class draw a bar with height equal to the observed frequency of that class. If a class has an observed frequency of zero, imagine a bar of height zero representing the data from this class. The final result is a graph consisting of a series of adjacent bars.

A histogram of Table 1-2 is given in Fig. 1-1.

Figure 1-1

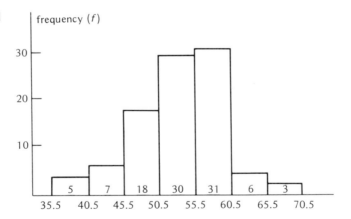

The second kind of graph used to portray a frequency distribution is called a *frequency polygon*. This is simply a line graph. It is drawn by finding the midpoint of the top of each histogram bar plus two imaginary bars of height zero at either end to keep the graph from floating, and then joining successive points by line segments.

Figure 1-2 is a histogram of Table 1-2. Figure 1-3 shows both the histogram and the frequency polygon of the frequency-distribution table in Example 1-29.

Figure 1-2

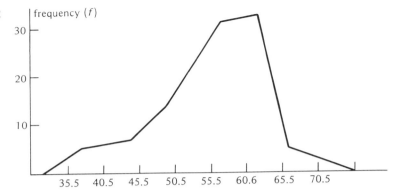

Figure 1-3

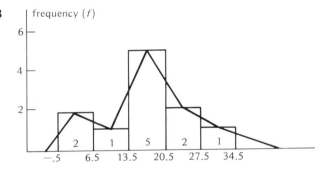

Example 1-32 The following data represent a business person's workday snack expenses over an eight-week period. What is the largest expense? The smallest? What is the range? Use the range as an aid in setting up a frequency distribution table for this set of data. In the table include columns for observed frequency, class mark, upper class boundary, and relative frequency.

1.20	1.15	.50	2.50	.75	1.20	1.15	.50	2.50	.75
.85	.60	.50	2.90	1.90	.85	.60	.70	2.10	.95
.85	1.65	1.70	1.15	1.40	.90	1.60	1.65	.90	1.00
.80	.75	.80	3.20	1.45	2.80	1.45	1.90	1.55	1.80

Solution   The largest value is 3.20. The smallest value is .50. The range is 2.70. In general, we want about 10 classes. By inspection, we see that 9 will divide evenly into our range value, giving nine classes with a convenient class range of 30. Our bottom class must contain .50. We will set up our table so that the upper boundary of one class equals the lower of the class above it. For the sake of convenience, let's agree that when a data value equals a boundary value we will always record it in the larger of the two possible classes.

Classes	Observed Frequency (f)	Class Mark	Upper Class Boundary	Relative Frequency
3.00–3.30	1	3.15	3.30	1/40
2.70–3.00	2	2.85	3.00	2/40
2.40–2.70	2	2.55	2.70	2/40
2.10–2.40	1	2.25	2.40	1/40
1.80–2.10	3	1.95	2.10	3/40
1.50–1.80	5	1.65	1.80	5/40
1.20–1.50	5	1.35	1.50	5/50
.90–1.20	7	1.05	1.20	7/40
.60– .90	11	.75	.90	11/40
.30– .60	3	.45	.60	3/40
Totals	40			1

TABLE 1-5

The data we are recording in a frequency distribution table are discrete data. A histogram serves as a bridge that lets us represent discrete data in a continuous graphical manner, by the use of a series of adjacent rectangles. Specifically, the histogram in Fig. 1-1 portrays the highway speed data in a continuous manner along the number line from 35.5 to 70.5.

EXERCISES A In Exercises 1-52 to 1-55 use the frequency distributions given to construct: (a) the resulting histogram, (b) the resulting frequency polygon.

1-52.

Classes	Frequency (f)
31–35	2
26–30	4
21–25	8
16–20	12
11–15	10
6–10	5
1–5	2

1-53.

Wages	Frequency (f)
$ 50.00–$ 59.99	8
60.00– 69.99	15
70.00– 79.99	24
80.00– 89.99	30
90.00– 99.99	40
100.00– 109.99	34
110.00– 119.99	28
120.00– 129.99	21

1-54. Times between equipment failures:

Failure Interval (hours)	Frequency (f)
At least 0 but less than 50	4
At least 50 but less than 100	7
At least 100 but less than 150	15
At least 150 but less than 200	19
At least 200 but less than 250	23
At least 250 but less than 300	21
At least 300 but less than 350	18
At least 350 but less than 400	10
400 or more	1

1-8 THE FREQUENCY DISTRIBUTION TABLE   31

 1-55.

Weight of Males (lb)	Frequency (f)
250 or greater	3
225–249	4
200–224	6
175–199	13
150–174	12
125–149	8
100–124	4
less than 100	1

1-56. Explain why, in general, a relative frequency of 1 would not make good sense.

B 1-57. Construct a frequency distribution table for the following data. Let the lowest class be 5–9 inclusive and the second lowest class be 10–14 inclusive. Include columns for observed frequency (f), class mark, upper class boundary, and relative frequency. (Example 1-30) The following data represent the percentage of iron in water in various wells in a given area.

```
22   8  15  19  13  23  23   9   5  13
20  17  11  11  13  17  11  10  21   7
19  26  17  23  14  24  21  17  12  16
15  14  21  20  10  26  13  11  15  14
```

1-58. Using the data of Exercise 1-57, construct a histogram and a frequency polygon.

A 1-59. Construct a frequency distribution table for the following data. Let the lowest class be 3–5 inclusive and the second lowest class be 6–8 inclusive. Include columns for observed frequency (f), class mark, upper class boundary, and relative frequency. The following data represent the number of accidents per year at 48 intersections in Chicago.

```
33  22  32  15   5  10  13  31  11  25  12  20
14  11  13  21  15  15   4  34  22  16  17   8
17  18  23  19  21  24  26   7  23  18  16  15
28  28  16  26  14  20  20  19  19  17  25  11
```

1-60. Using the data of Exercise 1-59, construct a histogram and a frequency polygon.

1-61. Calculate the mean in Exercise 1-59 directly from the frequency distribution table. (Do not use the raw data.) (Example 1-31)

1-62. The guidance department at Edison Regional High School kept a daily list of students who appeared for counseling during a period of 10 consecutive weeks. Let the bottom two classes be 1–3 and 4–6.

30	28	16	23	22	10	8	15	16	23
29	30	22	15	24	9	5	4	15	24
21	24	16	14	21	13	6	2	23	22
24	21	27	13	20	8	4	3	24	20
22	18	23	23	17	7	6	1	16	17

From the preceding data construct a frequency distribution table.

1-63. a. Compute the mean of the raw data of Exercise 1-62.
b. Compute the mean of the data of Exercise 1-62 directly from the frequency distribution table. Compare the two answers.

1-64. The following are lengths (in centimeters) of 50 sea trout. Group this collection of data in a frequency distribution table having a class range of 1 cm. Let the bottom two classes be 16–17 and 17–18. Find all class marks and relative frequencies.

19.5	19.8	18.9	20.4	20.2	21.5	19.9	21.7	19.5	20.9
18.1	20.5	18.3	19.5	18.3	19.0	18.2	23.9	17.0	19.7
20.7	23.1	16.6	19.4	18.2	18.5	20.1	22.4	21.7	18.6
19.4	21.8	19.5	20.9	16.8	19.1	22.6	19.8	20.4	21.7
19.5	22.0	20.8	19.3	22.2	19.6	21.5	20.2	22.7	23.1

1-9 PERCENTILES

When trying to interpret data you may find it illuminating to record the data in increasing order of size and divide them into percentiles (percentages).

Definition 1-12 When a large collection of data is arranged in increasing order of size the *p*th *percentile* is a value such that $p\%$ of our data lie below this value and $(100 - p)\%$ lie above it.

The Scholastic Aptitude Test (SAT) represents an everyday use of percentiles. If Mary's SAT score is in the 78th percentile, 78% of all the students who took this exam scored lower than Mary. Of course, 22% of the students who took this exam scored higher than Mary.

Suppose that you had data recorded in a frequency distribution table such as Table 1-2, and wanted to calculate percentiles, say the 78th percentile. You would proceed as follows:

1. Calculate 78% of the sample size.
   $(.78)(100) = 78$
2. By inspection, determine which class contains the 78th largest data value.
   Here it is in the 56–60 class. Sixty pieces of data lie below the lower class boundary of the 56–60 class, 55.5.
3. For the class we are now working in, find the distance from the lower class boundary to the upper class boundary.
   $60.5 - 55.5 = 5.00$ or just 5
4. The total number of pieces of data accumulated from the classes 36–40, 41–45, 46–50, 51–55 is $5 + 7 + 18 + 30 = 60$. Because we are interested in the 78th of 100 pieces of data, we are interested in 18 pieces of data ($78 - 60 = 18$) in the class 56–60. It is assumed that the 31 pieces of data in the 56–60 class are spread uniformly throughout this class. Thus, 18/31 of 5 = $(18/31)(5) = (.58)(5) = 2.90$.
5. Add the value computed in part 4 to the lower class boundary value, 55.5, and you have your answer.
   Answer The 78th percentile for speed among drivers on Highway 95 in Massachusetts is $55.50 + 2.90 = 58.40$ mph.

Example 1-33   Compute the 95th percentile for the data in Table 1-2.

Solution $(.95)(100) = 95$
91 pieces of data lie below 60.5. Therefore, our answer lies somewhere in the 61–65 class.
Answer $60.50 + (4/6)(5.00) = 60.50 + 3.33$
$= 63.83$

The lower (Q_1), median (Q_2), and upper (Q_3) *quartiles* are the 25th, 50th, and 75th percentiles, respectively.

The 10th percentile is called the first *decile*, the 20th percentile is called the second decile, and so forth.

Example 1-34 a. Find the value of the median quartile for the data of Table 1-5.
b. Compare the answer to part (a) with the median value of these data.

Solution
a. $(.50)(40) = 20$
The 20th piece of data is in the .90–1.20 class.
.90 + (6/7)(.30)
.90 + .26
1.16
b. When all 40 pieces of data are arranged in increasing order, the median value is 1.15. The two values differ by only .01, which is what we intuitively expect should happen.

As Definition 1-12 states, percentiles are best employed when dealing with large collections of data. In working with percentiles we encounter these two drawbacks:

1. In order to use percentiles to visualize a collection of data, it is often necessary to calculate many values.

2. In general, percentiles are unequal units of measure and as such they cannot be added, subtracted, multiplied, or divided.

EXERCISES A   Exercises 1-65 to 1-71 are based upon the businessperson's data in Table 1-5.

1-65. Find the 85th percentile. (Example 1-33)

1-66. Find the 58th percentile.

1-67. Determine the value of Q_1 (the lower quartile).

1-68. Determine the value of Q_3.

1-69. Find the value of the 64th percentile.

1-70. Find the value of the seventh decile.

1-71. For any collection of data we define the *semi-interquartile* range to be: $Q = (1/2)(Q_3 - Q_1)$. Is it clear that as the difference between Q_3 and Q_1 increases the value of Q will also increase? Compute the value of Q for the data of Table 1-5.

1-72. Compute the semi-interquartile range value, Q, for the data of Table 1-2.

Exercises 1-73–1-78 deal with the frequency distribution table of wages (Table 1-6).

1-73. Find the 78th percentile.

1-74. Find the 64th percentile.

1-75. Find the eighth decile.

1-76. Determine the value of Q_1.

1-77. Determine the value of Q_3.

1-78. Determine the value of Q, the semi-interquartile range.

Daily Wages	Frequency (f)
$ 50.00–$ 59.99	8
60.00– 69.99	15
70.00– 79.99	24
80.00– 89.99	30
90.00– 99.99	40
100.00– 109.99	34
110.00– 119.99	28
120.00– 129.99	21

TABLE 1-6

SUMMARY

The collection of data analyzed is called the *population*. Because of restrictions of time and money, only part of our population called a *sample* was analyzed. By analyzing a sample, we hope to be able to make inferences about our population. Hence, the term *inferential statistics*. In contrast, *descriptive statistics* deals primarily with tabulating data, and the construction of charts and tables for these data. A numerical value calculated from a population is called a *parameter* (denoted by a letter of the Greek alphabet); a numerical value calculated from a sample is called a *statistic*.

Some important statistics are: (1) the *mean* (or average), (2) the *median* (middle value when the data are arranged in order of magnitude), (3) the *mode* (most frequent occurring data value), and (4) the *range* (largest data value, less smallest data value). The mean, median, and mode are all referred to as measures of *central tendency*. For a sample called "X," it is customary to denote the mean x̄, "x bar." For the population from which the sample is drawn, it is customary to call the mean μ, "mu."

Summation notation was introduced in this chapter because many important formulas of statistics are expressed in this convenient notation.

Data are considered to be either *discrete* or *continuous*. Roughly speaking, the former type of data arises from counting and the latter type arises from measurements.

Frequency distribution tables which are a compact way of recording data were also discussed. Based on a frequency distribution a *frequency polygon* (a line graph) of data or a *histogram* (a bar graph) of data may be created.

The chapter concluded with a discussion of *percentiles*. The "pth percentile" for a collection of data is a value such that p-percent of the data lies below this value.

Can You Explain the Following?

1. population
2. sample
3. parameter
4. statistic
5. descriptive statistics
6. inferential statistics
7. summation notation (Σ)
8. population mean (μ)
9. mean for a sample X ($\bar{x}$)
10. measure of central tendency
11. median
12. mode
13. range
14. raw data
15. frequency distribution table
16. relative frequency
17. class interval
18. class limit
19. class mark
20. frequency polygon
21. histogram
22. percentile
23. decile
24. quartile
25. semi inter-quartile range

MISCELLANEOUS EXERCISES

1-79. Distinguish between the words *statistic* and *parameter*.

1-80. Classify the following data as discrete or continuous.
   a. The number of false alarms at station 5.
   b. The amount of time necessary to complete a manufacturing step on the assembly line of the Brockton Gear Works.
   c. The weight of a robin's egg.
   d. The number of daily blood donations at Memorial Hospital.

 1-81. Describe the population(s) and the sample(s).
   a. An economist working for the Bureau of Labor Statistics in Springfield, Mass., knows the number of workers unemployed last month. This month she surveys 1000 Springfield workers to discover whether the unemployment rate has changed.
   b. A political scientist is investigating the difference in the proportion of registered Republicans among upper-income families versus lower-

income families in California. He finds that among 500 upper-income voters, 26.7% vote Republican, whereas of 500 lower-income voters, 22.4% vote Republican.

1-82. $\sum_{p=0}^{2} (3p - 2) = ?$

Given: $X = \{2.0, .4, .5, 1.2\}$; in Exercises 1-83 and 1-84 evaluate:

1-83. $\sum_{i=1}^{4} (x_i + 1).$

1-84. $\sum_{i=1}^{4} (x_i^2 - 1).$

1-85. Find the mean, median, mode, and range of the following data: 14, 5, 8, 7, 6, 12, 10, 5, 8, 9.

1-86. The following five figures represent "profit per unit" on five types of canned goods sold by Federal Food Stores (all values are in cents): .12, .23, .17, .08, .20. Find the mean, median, and modal values.

1-87. Based upon the following frequency distribution, construct (a) a histogram and (b) a frequency polygon.

Hours Worked	Number of Employees
45–46	4
43–44	5
41–42	16
39–40	17
37–38	5
35–36	3
	50

1-88. Return to Exercise 1-87 and estimate the mean value by using class marks.

1-89. The following data were recorded by the chief actuary of Henson Insurance Company and represent the age at death of 100 policy holders. The ages have been rounded to the last birthday.

```
75  37  47  59  67  46  59  44  63  59
41  84  51  63  58  63  38  57  74  79
81  63  74  61  74  58  55  68  86  48
44  67  42  69  36  27  60  78  71  76
70  75  68  72  75  69  40  42  62  70
78  55  68  78  74  66  62  59  49  67
62  61  86  54  61  55  77  86  65  84
68  81  61  74  70  89  45  35  44  47
69  78  61  74  62  80  44  63  55  75
71  73  77  78  79  53  46  71  78  68
```

  a. Construct a frequency distribution table for this set of data with the bottom classes 25–34 and 35–44.
  b. Find the relative frequency of each class.
  c. Make a histogram based on the frequency distribution table of part (a).

 1-90. Return to Exercise 1-87 and compute:
  a. The 38th percentile.
  b. The fourth decile.
  c. Q_1
  d. Q_3
  e. Q

1-91. State three (by trial and error) numbers to illustrate:
  a. How the mean and median of a set of data can be equal.
  b. How the mean can be less than the median.
  c. How the median can be less than the mean.

1-92. Evaluate $\sum_{i=a}^{i=b} 1$. (Hint: Use the index values, a and b.)

2 THE VARIANCE AND THE STANDARD DEVIATION

2-1 INTRODUCTION

We now return to the Finast Sugar Company and its population of 5-lb packages of sugar (Chapter 1). As you saw, the company will tolerate little variation in actual package weight from the true mean weight of the population, and it works to keep the mean weight in agreement with the weight advertised on the package.

Ideally, the company would like zero variability. However, it knows that this is unrealistic, for it implies that every package weighs exactly 5.000 lb. How do members of Finast's quality control department measure the package variability relative to the mean package weight? They compute the mean weight of the 31-package sample, 5.000 lb. They then compute the distance squared between the mean package weight and each package weight in the sample, as in the following table.

31 Package Weights	Distance to the Mean	Squared Distance to the Mean
5.010	.010	.0001
5.005	.005	.000025
5.000	.000	.000
.	.	.
.	.	.
.	.	.
5.000	.000	.000
4.995	.005	.000025
4.990	.010	.0001
		Total .000250

Next, they divide the total of the squared distances by 30. They got 30 by subtracting 1 from the sample size.

$$\frac{.000250}{30} = .0000083$$

This quotient, .0000083, is called the *sample variance,* and is denoted by a small letter s, squared (s^2). It is this quotient that the quality control department uses to determine the variation in package weights. Here the researchers are happy with the result; although s^2 isn't zero, it is small. If they wanted to be able to express the variation in the original unit, pounds (lb), they could simply "reverse" what they did in squaring the distance to the mean, and take the positive square root of the sample variance $\sqrt{.0000083} = .0029$. This value is called the *sample standard deviation.*

Like the Finast Sugar Company, we are interested in the spread of data relative to the mean value. Specifically, we are interested in knowing whether our data tend to confine themselves (as a collection) to a rather small range about the mean value or whether they tend to be spread at a considerable distance from it. Our problem is to determine the distance of a set of data from its mean. It is customary to speak of this spread as the deviation of the data. It is understood that, from here on, the deviation will always be taken from the mean.

Logically, the easiest way to solve this problem is simply to compute the average distance of our data from the mean. However, when this idea is pursued in more advanced texts it is soon abandoned because of extreme mathematical difficulty. A simple alternative (and the one we use) is: *Compute the sum of the squares of the differences between each datum and the mean value, and then divide this sum by the sample size less one* $(n - 1)$. The resulting numerical value is called the *sample variance* and is defined here for emphasis.

Definition 2-1 The *variance* (s^2) of a sample X, of size n, is

$$s^2 = \frac{1}{n-1} \sum_{i=1}^{n} (x_i - \bar{x})^2 \quad \text{or} \quad \frac{\sum_{i=1}^{n}(x_i - \bar{x})^2}{n-1} \quad n \neq 1$$

The reason for dividing by $n - 1$ rather than by n is not easily explained. Intuitively, it can be shown that if we take many samples from a given population, we find the sample variance, s^2, for each sample, and average each of these variances, this average will not equal the population variance, σ^2, unless we use $n - 1$ as the denominator. A formal explanation lies in the concept of degrees of freedom, explained in detail in Chapter 8.

Example 2-1 Compute the sample variance for the sample $T = \{2, 4, 6, 8\}$.

Solution $\bar{t} = 5$, $n - 1 = 4 - 1 = 3$

$$s^2 = \frac{(2 - 5)^2 + (4 - 5)^2 + (6 - 5)^2 + (8 - 5)^2}{3}$$

$$s^2 = \frac{20}{3}$$

Definition 2-2 The *standard deviation* (s) of a sample X (size n) is the positive square root of the sample variance.

$$s = \sqrt{s^2} = \sqrt{\frac{\sum_{i=1}^{n}(x_i - \bar{x})^2}{n - 1}} \qquad n \neq 1$$

Example 2-2
  a. Calculate s^2 for the sample $X = \{1, 2, 3, 4, 5\}$.
  b. Calculate s.

Solution
  a. $\bar{x} = 3$
  $s^2 = (1/4)[(1 - 3)^2 + (2 - 3)^2 + (3 - 3)^2 + (4 - 3)^2 + (5 - 3)^2]$
  $= (1/4)[4 + 1 + 0 + 1 + 4]$
  $= 10/4$ or $5/2$
  b. $s = \sqrt{5/2}$

Example 2-3
  a. Calculate s^2 for the sample $Y = \{7, 7, 7\}$.
  b. Calculate s.

Solution
  a. $\bar{y} = 7$
  $s^2 = (1/2)[0 + 0 + 0]$
  $s^2 = 0$
  b. therefore,
  $s = 0$

42 THE VARIANCE AND THE STANDARD DEVIATION

The preceding example illustrates a general truth—namely, the variance is zero if, and only if, all the pieces of data have the same value. Is it equally clear that the smallest possible variance is zero? Notice that s^2 can never be negative ($s^2 \not< 0$).

Example 2-4 As part of a drive to eliminate waste, R and D Lumber Yard has the following odd lengths of prime lumber. $Y = \{4\text{ ft}, 8\text{ ft}, 5\text{ ft}, 6\text{ ft}, 1\text{ ft}, 4\text{ ft}, 3\text{ ft}, 3\text{ ft}, 2\text{ ft}, 4\text{ ft}\}$.
a. Find the mean.
b. Find the variance.
c. Find the standard deviation.

Solution
a. $\bar{y} = 4$ (or 4 ft)
b. $s^2 = (1/9)(0 + 16 + 1 + 4 + 9 + 0 + 1 + 1 + 4 + 0)$
   $= (1/9)(36)$
   $= 4$
c. $s = 2$

Because in algebra a negative number squared is always positive, just as a positive number squared is, *we can interpret each of these squared differences from the mean as squared distances.* Suppose we consider a data value of 8 from a sample whose mean is 4; the distance squared from 4 to 8 is $(8 - 4)^2 = 16$. For a data value of 1, the distance squared would be $(1 - 4)^2 = 9$.

In case you doubt the value of knowing the variance, consider the courageous general who learned that the enemy army was camped across a nearby river. The general decided to cross the river and strike a decisive blow to the enemy flank. His scouts assured him that the average depth of the river crossing was 3.50 ft. At dawn the next day the entire army marched bravely into the river, and half of them drowned. The general failed to consider variability.

In trying to visualize a collection of data, the overall inference we can draw from the variance and the standard deviation tends to be about the same. Then why bother with the standard deviation? The answer is that it is part of a simple but indispensable formula in applied statistics that you will meet in the next section.

Although using the definition in the examples just given is a perfectly acceptable way of calculating s^2, we do have another formula. It is derived through algebraic manipulation of Definition 2-1. This formula has two advantages over the definition (formula) for s^2:

1. It makes the arithmetic easier for noninteger data such as 4.375 or 2.998
2. It lets us compute s^2 without calculating the mean. As you have seen, when you work from the definition of sample variance, you must calculate the mean first.

A computational formula for s^2:

$$s^2 = \frac{\sum_{i=1}^{n} x_i^2 - \frac{\left(\sum_{i=1}^{n} x_i\right)^2}{n}}{n-1}$$

Example 2-5 For the sample $T = \{2, 4, 6, 8\}$;
a. Calculate s^2 directly from the definition.
b. Calculate s^2 by using the computational formula.

Solution

a. $s^2 = (1/3) \sum_{i=1}^{4} (t_i - \bar{t})^2; \quad \bar{t} = 5$
$= (1/3)[9 + 1 + 1 + 9] = (20/3)$

b. $s^2 = \dfrac{\sum_{i=1}^{4} t_i^2 - \dfrac{\left(\sum_{i=1}^{4} t_i\right)^2}{4}}{3}$

$= \dfrac{[4 + 16 + 36 + 64] - (400/4)}{3} = (20/3)$

Compare this answer with the answer in Example 2-1.

Example 2-6 Suppose that in an air pollution study you, as city health officer, want to estimate the average daily emission of sulfur oxides by a large industrial plant. You also want to calculate the sample variance, s^2. Compute these values from a five-day sample: 6 tons, 3 tons, 8 tons, 5 tons, 2 tons.

Solution The mean is 4.80 tons. Using the definition for s^2:

$$s^2 = \frac{(6 - 4.8)^2 + (3 - 4.8)^2 + (8 - 4.8)^2 + (5 - 4.8)^2 + (2 - 4.8)^2}{4}$$

$$= \frac{22.80}{4} = 5.70$$

You can also solve this example by using the computational formula.

Example 2-7 Five swimmers in a life-saving course were graded according to a 100-point scale on their ability to detect potential danger signs. Evaluate s^2 and s for the following five grades: 58, 70, 92, 88, 95.

Solution (using the computational formula):

$$s^2 = \frac{\sum_{i=1}^{5} x_i^2 - \frac{\left(\sum_{i=1}^{5} x_i\right)^2}{5}}{4}$$

$$s^2 = \frac{33{,}497 - 32{,}481.8}{4}$$

$$s^2 = 253.8 \quad \text{and} \quad s = 15.93$$

Here, we might mention, and answer, two classic student questions.

1. Can s be greater than s^2, and if so when? The answer: When $0 < s^2 < 1$, $s > s^2$.
2. What would be a small standard deviation (variance)?

Granted, we would probably agree that $s = .01$ is small and $s = 100$ is large, but we don't have a specific boundary value. Small and large tend to be relative to the nature of the data.

The *standard deviation is the basic unit of measure* in statistics. For data plotted on a number line, we measure the distance between data values in standard deviations. Think of a new scale of measure imposed upon the old number line with 0 standard deviations located at the mean. Here a value of +3 means a point located 3 standard deviations to the right (on the number line) of zero. Similarly, a value of −1.50 means a point located 1 1/2 standard deviations to the left of zero. The standard deviation unit is an unusual unit of measure because it varies. Clearly, two samples could yield 2 different standard deviations. The standard deviation of a sample is a statistic, and we have already pointed out that a statistic varies in value from sample to sample. This is quite unlike such a familiar linear unit of measure as the foot, which is always the same in every problem. See Example 2-8.

Example 2-8 A major airline is investigating the relationship between people who fly out of 14 northeastern cities and those who stay at hotels in the 14 cities. As part of this investigation, the airline found itself calculating the variance for the passenger numbers (X) flying out of these 14 cities on a given day in May. Calculate the variance for the airline.

$X = \{3,756; 5,100; 4,950; 894; 480; 1,908; 5,388; 240; 468; 1,662; 96; 5,334; 786; 648\}$

Solution

$$s^2 = \frac{\sum_{i=1}^{14} x_i^2 - \frac{\left(\sum_{i=1}^{14} x_i\right)^2}{14}}{13}$$

$$s^2 = \frac{130{,}858{,}020 - \frac{(31{,}710)^2}{14}}{13}$$

$$s^2 = \frac{130{,}858{,}020 - 71{,}823{,}150}{13}$$

$$s^2 = 4{,}541{,}143.80$$

Just as a sample has a variance, our population also has a variance, which is denoted σ^2 and is defined in the following. Note that we divide by the population size, N. This is an important contrast to the sample variance where we divide by $n - 1$ (sample size less one).

Definition 2-3 For any finite population A, of size N, with mean μ the *population variance* is

$$\sigma^2 = \frac{\sum_{i=1}^{N} (a_i - \mu)^2}{N}$$

where a_i corresponds to the ith value in population A.

Definition 2-4 For any finite population A, of size N, the *population standard deviation* is $\sigma = \sqrt{\sigma^2}$.

EXERCISES A   Each of the first four exercises contains two samples. Try to determine, just by looking, which sample has the larger variance. Next, check your answer by actually calculating the two sample variances.

46 THE VARIANCE AND THE STANDARD DEVIATION

2-1. $X = \{3, 5, 7\}$; $Y = \{1, 5, 9\}$. (Example 2-1)
2-2. $X = \{3, 5, 7\}$; $T = \{9, 11, 13\}$.
2-3. $Y = \{1, 5, 9\}$; $W = \{1, 1, 9, 9\}$.

2-4. $M = \{2.51, 3.07, 4.53, 1.89\}$; $N = \{12.18, 5.87, 7.93, 10.02\}$.

[Hint: Use the computational formula.] (Examples 2-5, 2-7)

2-5. Machine A fills 8-oz cups with $\mu = 7.50$ oz and $\sigma = .10$. Machine B fills these cups with $\mu = 7.50$ oz and $\sigma = .20$. Which machine is more likely to overfill cups? Explain.
2-6. Is it possible for a person who is 5 1/2 ft tall to drown in a pool whose average depth is 3 ft? Explain.
2-7. For $W = \{8, 8, 8, 8\}$, compute $\bar{w}$ and s. (Example 2-3)

For Exercises 2-8, 2-9, and 2-10 compute s.

2-8. $X = \{5, 10, 15\}$. (Example 2-2)

2-9. $Y = \{2, 3, 4, 2, 5, 2, 6, 7, 3, 7, 2, 5\}$.
2-10. $U = \{4, 6, 5, 3, 2, 4, 9, 5, 1, 5, 6, 7, 10, 2, 3, 3, 5, 10, 6, 4\}$.
2-11. The following is a sample of IQ scores from 10 elementary school children: 106, 97, 98, 100, 102, 100, 94, 98, 102, 103. Evaluate s^2 and s.
2-12. Calculate the value of s^2 and s for Exercise 1-41 of section 1-7.

In Exercises 2-13, 2-14, and 2-15 compute s.

2-13. Twenty 14-oz bottles of mouthwash were taken from a grocery store shelf, and the amount or volume of liquid in each bottle was measured in cubic centimeters (cc). The results are as follows:

Sample	Volume	Sample	Volume
1	423	11	422
2	426	12	422
3	421	13	426
4	428	14	422
5	420	15	430
6	418	16	425
7	423	17	426
8	426	18	420
9	427	19	418
10	428	20	421

2-14. A chemist was concerned about the length of time it took to complete a new reaction. She decided to take 12 different pieces of raw material and time the reaction for each piece. She recorded the following times:

17 12 11 10 10 9

14 11 10 10 9 9

2-15. A specialist in internal medicine is studying the blood pressure of former athletes. She records the following diastolic pressures (lowest pressure in the arteries as the heart fills with blood) from her first group of eight patients: 98, 86, 88, 97, 88, 72, 90, 88.

B

2-16. If half the measurements in a set of data are 1's and the other half are 3's, what is the value of s^2?

2-17. Can two different samples of data of the same size have the same mean? If your answer is yes, give an example.

2-18. Can two different samples of the same size have the same variance? If your answer is yes, give an example.

2-19. a. State a collection of numbers whose variance is zero.
b. State an example of a collection of numbers whose variance is infinite. [Hint: You must pick an infinite collection.]

2-2 STANDARDIZING DATA

In this section we find immediate use for the standard deviation. As we travel from one research project to another, the sample obtained from each project will have (most likely) a different mean, a different variance, and a different standard deviation. Similarly, the populations themselves will differ in their means, variances, and standard deviations. Thus, it would be a great help to learn a way of standardizing—gaining some degree of uniformity—some of the data obtained from a research project.

In this section we introduce a very important formula (equation) of statistics. This formula can take *any collection of data* (either a population or a sample) and turn it into *a new collection called "Z"*, of the same size, but having two important features: (1) a mean of zero, and (2) a variance of 1. Of course if the variance is 1, then the standard deviation is also 1. The numerical values in this new collection of data, Z, are called *z scores*. Our next definition will formally present the important formula that we use to standardize a collection of data.

Definition 2-5 For a sample X of size n, the ith *z score* corresponding to the ith x score of X is given by the formula:

$$z_i = \frac{x_i - \bar{x}}{s}$$

where $s \neq 0$ and i goes from 1 to n.

The following two examples have such simple values that you can easily see for yourself the properties of the Z set of data.

THE VARIANCE AND THE STANDARD DEVIATION

Example 2-9 For the sample $X = \{1, 3, 5\}$, calculate the corresponding set Z.

Solution $\bar{x} = 3$, $s = 2$

$$\begin{cases} \text{if } x = 1, \text{ then } z = (1 - 3)/2 = -1 \\ \text{if } x = 3, \text{ then } z = (3 - 3)/2 = 0 \\ \text{if } x = 5, \text{ then } z = (5 - 3)/2 = +1 \end{cases}$$

$Z = \{-1, 0, +1\}$

Example 2-10 For the sample $Y = \{1, 1, 2, 3, 3\}$, calculate the corresponding set Z.

Solution $\bar{y} = 2$, $s = 1$

$Z = \{-1, -1, 0, +1, +1\}$

When you find the standardized set Z in an exercise you may be tempted to insert plus signs between successive elements of Z. This is wrong. Z is a special new collection of data with two properties: mean = 0, and variance = 1.

What does a z score tell us? It tells us the distance of the datum relative to the mean in terms of standard deviations. For example, an educator has a population of grades with $\mu = 50$ and $\sigma = 5$. A grade of 60 has a z score of $(60 - 50)/5 = 2$. This tells us that the grade 60 is located 2 standard deviations to the right of 50 on the number line. Is it clear that a grade of 45 has a z score of -1? Thus, 45 is located 1 standard deviation to the left of 50 on the number line.

For a population A, of size N, the standardizing formula is

$$z_i = \frac{a_i - \mu}{\sigma}$$
$$\sigma \neq 0, \ 1 \leq i \leq N$$

In assessing the difference between the largest package weight in a sample and the mean weight, the Finast Sugar Company computes the difference between these two values in terms of standard deviations. Thus, 5.010 is the largest package weight; and $5.010 - 5.000 = .010$ lb represents the difference in terms of the given unit, pounds. How many times is $s = .0029$ (see Section 2-1) contained in .010 lb?

$$.010/.0029 = 3.448$$

Thus, the largest package weight 5.010 is 3.448 standard deviations from the mean. See Fig. 2-1.

2-3 A FURTHER LOOK AT THE STANDARDIZING FORMULA (optional)

Figure 2-1

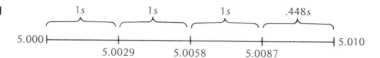

EXERCISES A

2-20. a. Verify that the z scores of Example 2-9 have a mean = 0 and $s = 1$.
b. Verify that the z scores of Example 2-10 have a mean = 0 and $s = 1$.

In Exercises 2-21, 2-22, and 2-23 find the corresponding set of z scores.

2-21. For $G = \{2, 4, 6, 8, 10\}$. (Examples 2-9, 2-10)
2-22. For the data, $X = \{6, 5, 4, 8, 5, 6, 7, 7, 6\}$.
2-23. For $Y = \{4, 8, 5, 6, 1, 4, 3, 3, 2, 4\}$.

B

2-24. We are given the sample data, $X = \{0, 2, a\}$, and told that $s^2 = 4$. What is the value of a?
2-25. Prove that there are two values for which $s^2 = s$.

2-3 A FURTHER LOOK AT THE STANDARDIZING FORMULA (optional)

Analyzing the Numerator

The effect of the numerator in the standardizing formula is to leave the distances between elements unchanged. It merely moves the entire collection of data left or right. In no way is the geometric shape of the distribution altered. Here is a breakdown of cases.

1. If $\bar{x}$ already $= 0$, we have no movement. The location of the collection of data is geometrically unchanged.

2. If $\bar{x} > 0$, the data move in a negative direction, that is, to the viewer's left. The shape of the distribution remains unchanged.

3. If $\bar{x} < 0$, the data move in a positive direction, that is, to the viewer's right. The shape of the distribution remains unchanged.

Analyzing the Denominator

The denominator stretches (when $s < 1$) or shrinks (when $s > 1$) the distance of each piece of data from the mean. In particular, if $s = 1$ the geometrical shape of the distribution is unaltered. Finally, as we previously pointed out, it completes the conversion of the data into standard deviation units. Here is a breakdown by cases:

50 THE VARIANCE AND THE STANDARD DEVIATION

1. If $0 < s < 1$, we have a stretching effect on the entire collection. The effect is outward from the mean with the mean value remaining constant. Clearly, the range is then increased.

2. If $s = 1$, the geometric shape of the distribution remains the same. Relative distances between pieces of data are unaltered.

3. If $s > 1$, we have a shrinking effect on the entire collection. The effect is inward from both extremes with the mean value remaining constant.

Example 2-11 Consider the sample collection $X = \{2, 1, 3, 4, 3, 2, 4, 2, 5, 4\}$. Use little squares to make a simple sketch of the sample $(X - \bar{x})/s$.

Solution $s^2 = 14/9 = 1.56$ $s = 1.25$.

2-4 CHEBYSHEV'S THEOREM

Theorem 2-1 (*Chebyshev's Theorem*) We are given a population A, of size N. There is, outside of an interval of $k\sigma$ ($k \neq 0$) about μ, *at most* $1/k^2$ of the N pieces of data.

Proof Because N is a natural number, we can subscript all data values (if any) outside the interval $[\mu - k\sigma, \mu + k\sigma]$. Let r denote the number of pieces of data outside of this interval. Of course, r could be zero. These r pieces of data will be subscripted i, where $1 \leq i \leq r$. Next subscript all those data values inside or on the boundary of this same interval (if any), $(r + 1) \leq i \leq N$. See the following sketch:

$1 \leq i \leq r$	$(r + 1) \leq i \leq N$	$1 \leq i \leq r$
$|a_i - \mu| > k\sigma$	$|a_i - \mu| \leq k\sigma$	$|a_i - \mu| > k\sigma$
$\mu - k\sigma$	μ	$\mu + k\sigma$

Because of the way we constructed the interval, about μ, for $1 \leq i \leq r$; $(a_i - \mu)^2 > k^2 \sigma^2$. Thus,

$$\sigma^2 = 1/N \sum_{i=1}^{N} (a_i - \mu)^2$$
$$= 1/N[(a_i - \mu)^2 + \cdots + (a_r - \mu)^2]$$
$$+ 1/N[(a_{r+1} - \mu)^2 + \cdots + (a_N - \mu)^2]$$

In the preceding line, drop those terms with subscripts $r + 1 \leq i \leq N$. Replace $(a_i - \mu)^2$, $1 \leq i \leq r$ by $k^2 \sigma^2$ to obtain:

$$\sigma^2 \geq \frac{rk^2\sigma^2}{N} \qquad \sigma \neq 0$$

therefore, $\qquad \frac{r}{N} \leq \frac{1}{k^2}$

The theorem tells us that *at most* $1/k^2$ of our total number of pieces of data lie outside an interval of k standard deviations about the mean. Of course, it also tells us that inside this same interval lies *at least* $1 - 1/k^2$ of our data. For example, outside an interval of 2 standard deviations about the mean lie at most one-fourth of our data. For $k = 3$, at most one-ninth of our data lie outside this interval.

Observe that whereas Chebyshev's Theorem is not wrong for $k \leq 1$, it is of no practical value. Specifically, if $k = 1/2$, the theorem could place, at most, 400% of our data outside this interval.

Example 2-12 If $k = 2.5$ in Chebyshev's Theorem, at least what percent of our data lie inside an interval of 2.5 standard deviations (s) about the mean?

Solution

$$1 - \frac{1}{(2.5)^2} = 1 - .16 = .84$$

Example 2-13 How many standard deviations about the mean (i.e., from the mean in both directions) should we go to capture at least 96% of our sample? (This result will only be meaningful when the n, our sample size, is very large.)

Solution We must have at most 4% outside the interval: $1/k^2 < 4/100$, which implies $1/k < 1/5$, which implies $k > 5$.

EXERCISES A

2-26. What does Chebyshev's Theorem tell us about the percentage of our data outside on interval of 4 standard deviations about the mean? What percentage is inside? (Theorem 2-1)

52 THE VARIANCE AND THE STANDARD DEVIATION

A

2-27. State a natural number, k, of standard deviations such that the interval $[\mu - k\sigma, \mu + k\sigma]$ will contain at least 99% of the population. (Example 2-13)

2-28. What does Chebyshev's Theorem tell us about the percentage of our data inside an interval of 5 standard deviations about the mean? Of 5 1/2 standard deviations? Of 6 standard deviations?

B

2-29. The average cost of an overseas toll call is $20 with a standard deviation of $5. Based upon 100,000 toll calls, use Chebyshev's Theorem to determine the minimum number of toll calls that can be expected to fall between $10 and $30.

2-30. What does Chebyshev's Theorem tell us about the percentage of our data outside an interval of 3.5 standard deviations about the mean? Of 3.75 standard deviations about the mean?

2-31. How many standard deviations about the mean should we go to be sure that we have at least 97% of our sample inside this interval? Give an integer solution. [Hint: see Example 2-13.]

2-5 SYMMETRY

The mean is our basic measure of central tendency. The standard deviation is our basic measure of variation relative to the mean value. Now we want to measure something a little different. We want to answer the question: How symmetrical, if at all, is our collection of data about a line drawn through its mean and perpendicular to the axis upon which the data are plotted? For small samples, you may be able to plot all of your data and get, just by looking, a reasonable estimate of line symmetry. But for large samples with many data values, looking is at best a difficult and time-consuming process.

We start to solve our problem by cubing the difference between pieces of data and the mean value. Data values far to the right of the mean will yield large positive values and data values far to the left will yield large negative values. If the negative cubes dominate, our data are clustering to the right of the mean. If the positive cubes dominate, data are clustering to the left of the mean. At this point, we would be done if the label of our answer doesn't bother us. For example, if the data are in feet, then the average of differences cubed bears the title (ft)(ft)(ft). Awkward! To avoid this we divide by the standard deviation cubed. Our final result has no title (lb, ft, in., etc.). Our notation for it will be a_3. We call the formula for determining this symmetry, given in the next definition, a measure of the *skewness* of a sample.

Definition 2-6 For a sample X, of size n, the skewness of a sample is denoted a_3 and given by the following formula:

2-5 SYMMETRY   53

$$a_3 = \frac{1/n \sum_{i=1}^{n} (x_i - \bar{x})^3}{s^3} \quad s \neq 0$$

Example 2-14 Calculate the skewness of the sample
$X = \{5, 4, 5, 3, 5, 3, 5, 4, 1, 4, 5, 4\}$.

Solution $s = 1.21$ $s^3 = 1.75$ $\bar{x} = 4$

$$a_3 = \frac{(1/12)[(1 - 4)^3 + 2(3 - 4)^3 + 5(5 - 4)^3]}{1.75}$$

$$a_3 = \frac{-2}{1.75} = -1.14$$

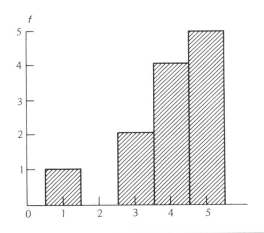

Example 2-15 For the sample $T = \{5 \text{ in.}, 2 \text{ in.}, 5 \text{ in.}, 1 \text{ in.}, 4 \text{ in.}, 1 \text{ in.}, 3 \text{ in.}, 5 \text{ in.}, 4 \text{ in.}, 1 \text{ in.}, 2 \text{ in.}\}$, evaluate a_3.

Solution $\bar{t} = 3$

$$a_3 = \frac{(1/11)[3(1 - 3)^3 + 2(2 - 3)^3 + 2(4 - 3)^3 + 3(5 - 3)^3] \text{ in.in.in.}}{s^3 \text{ in.in.in.}}$$

$a_3 = 0$ because the numerator is zero and $s \neq 0$.

When a_3 is 0, as in Example 2.15, there is perfect symmetry of the data.

Example 2-16 For the sample $V = \{1, 1, 1, 1, 2, 2, 2, 3, 2, 5\}$:
  a. Can you guess the algebraic sign of a_3?

54 THE VARIANCE AND THE STANDARD DEVIATION

    b. Calculate a_3.
    c. Can you pick out one data value that seems to be controlling the algebraic sign?

Solution

a. $s = 1.25$ $s^3 = 1.96$ $\bar{v} = 2$, positive.

b. $a_3 = \dfrac{(1/10)[(-1)(4) + 1 + 27]}{1.96} = 1.22.$

c.   Yes. The value 5.

EXERCISES A

Calculate the value of a_3 in Exercises 2-32, 2-33, and 2-34.

2-32. $X = \{6, 5, 4, 8, 5, 6, 7, 7, 6\}$. (Example 2-15)
2-33. $Y = \{6, 6, 7, 5, 6, 7, 6, -1, 7, 4, 5, 2\}$. (Examples 2-15, 2-16)
2-34. Your data are the diastolic blood pressure readings of Exercise 2-15.
2-35. A secretary typed a sample of six business letters for her supervisor. The number of typing errors was: 6, 0, 2, 0, 1, 3. Compute a_3.
2-36. For the data of Exercise 2-14:
    a. Find the value of a_3.
    b. Find the corresponding set of z scores.
2-37. By inspection, what is the value of a_3 for the data of Exercise 2-16?
2-38. For the case that $s = 0$, what would be a reasonable definition of a_3?
2-39. The number of minutes that a commuter waited for a train during one week were 10, 6, 15, 11, and 3. Calculate:
    a. the mean
    b. s^2
    c. s
    d. a_3

B
2-40. Start with the formula for s^2 in Definition 2-1 and derive the computational formula for s^2.

SUMMARY

This chapter discussed how to measure the deviation (spread) of a collection of data relative to its mean value. This is accomplished by computing either the *variance* or the *standard deviation*. The variance of a sample is denoted s^2; the variance of a population is denoted σ^2. The standard deviation of a sample is denoted s; the standard deviation of a population is denoted σ. The smallest possible variance (standard deviation) is zero. This occurs when all the data of the collection have the same value. The greater the value of the variance the more inclined we are to believe that our data, as a collection, deviates (spreads out) from the mean.

Also discussed was the *standardizing formula* that allows us to turn any collection of data into a new collection of data with two properties: (1) a mean of zero, and (2) a variance of one. Of course if the variance is one, the standard deviation must also be one. Although it is true that the variance and the standard deviation serve to measure the data deviation from the mean, we do need the standard deviation for it is the standard deviation that appears in our important standardizing formula.

Chebyshev's Theorem was also discussed. This theorem is valuable because it applies universally to every collection of data.

The chapter ended with a discussion of the *skewness* (symmetry) of our data about a line drawn through the mean.

Can You Explain the Following?
1. sample variance (s^2)
2. population variance (σ^2)
3. sample standard deviation (s)
4. population standard deviation (σ)
5. computational formula for s^2
6. the standardizing formula
7. z score

MISCELLANEOUS EXERCISES

2-41. Explain what is implied if the sample variance, s^2, equals 0.

2-42. Following are three sample sets of data. (a) Compute the variance of each sample. (b) Show the variance of sample X less than the variance of sample Y. (c) Show the variance of sample Y equal to the variance of sample W.

$X = \{5, 7, 9\}$
$Y = \{4, 7, 10\}$
$W = \{9, 12, 15\}$

2-43. If the wholesale prices of beef, pork, lamb, and veal are, respectively, $2.65, $2.45, $2.75, and $2.15 per pound, what is the average of these four sample prices? What is the variance?

2-44. Standardize the data: $\{2, 8, 2, 8\}$.

2-45. Standardize the data: $\{1, 3, 5, 7, 9\}$.

2-46. An educator has a population of grades in which $\mu = 70$ and $\sigma = 4$.
   a. Use the standard deviation (as a measure) to describe a grade of 82 with respect to μ.
   b. Use the standard deviation to describe a grade of 64 with respect to μ.
   c. Would we intuitively expect many grades below 40 in this population? Explain.

2-47. Find three scores whose mean is 10 and whose variance is 1.

2-48. A study of milk purity counted the number of coliform bacteria (the kind usually found in feces) per milliliter in 11 specimens of milk bought in stores. Because the U.S. Public Health Service recommends no more than 10 coliform bacteria per milliliter, the data look good. Find the sample variance and sample standard deviation of this set of data for the U.S. Health Service.
$\{5, 8, 6, 7, 8, 3, 2, 4, 7, 8, 6\}$

2-49. The average cost of a sirloin steak dinner in a restaurant is $7.00 with a standard deviation of 25 cents. For a sample of 10,000 such dinners, use

Chebyshev's Theorem to determine the minimum number of dinners that can be expected to fall between $6.50 and $7.50.

2-50. Both the sample range and the sample variance measure the dispersion (spread) of a collection of data. State three simple numbers to illustrate that:
   a. The range of a sample may be smaller than the variance of the sample.
   b. The variance of a sample may be smaller than the range of the sample.

2-51. Explain why Chebyshev's Theorem is of no value for the case of an interval of 1σ about μ.

3 AN INTRODUCTION TO PROBABILITY

3-1 INTRODUCTION

Probability may be fascinating, but why study it in a statistics course?

Statistical answers often appear in probability form. We will say that we are 95% or 99% sure of something. It is a probability assurance that the Finast quality control department gives its management. Weather forecasters tell us our *chance* of rain. A doctor knows the probability that a tumor is malignant. Entrepreneurs enter a new venture where they see high probability of success.

We use statistics to estimate population parameters such as μ and σ. The very fact that we estimate them shows that we don't know them. How, then, can we tell whether our estimates are accurate? We cannot, with certainty. But we can calculate the probability of being arbitrarily close to the correct value. This may not be perfect, but it gives us something to work with. Applied statistics, in short, rests on a probability foundation. Because it does, you must master the basic concepts of probability. That is what this chapter is all about.

Events observed under unchanging conditions do not always come out the same. For example, we think of 1/2 as a reasonable likelihood for heads on a single toss of a coin. Fifty, then, would be a reasonable estimate of the number of heads to expect in a sample of 100 tosses. Actually, you can't count on getting exactly 50 heads every time. It is such events, or phenomena, that we call *random* and that we try to quantify in probability.

Sarah and Brian cannot agree on which television show to watch. To settle the dispute they decide that each will hide one hand and, on the word "go," extend one, two, or three fingers. Sarah will yell "go," and Brian will call out "even" or "odd," depending on whether he thinks the sum of extended fingers is even or odd. If Brian is right, he wins; if not, Sarah wins. What the two children are playing is an old Indian game called Tong.

Brian knows no probability theory, and probably won't go over in his mind the set of all possible outcomes:

$$[(1,1)\ (1,2),\ (1,3),\ (2,1),\ (2,2),\ (2,3),\ (3,1),\ (3,2),\ (3,3)]$$

If he did, he would discover that he has the better chance of winning. Do you see why? The answer is that he gets to choose between odd and even, and of the nine possible outcomes five yield even sums whereas only four yield odd. Thus, if Brian says even, he should win five out of every nine times. And for this one game his chances are five in nine or 55.5%.

In probability the game of Tong is an example of an *experiment*. If we record the life of an electric light bulb, we are doing an experiment. When we take a survey to learn the color of newborn babies' eyes, we are doing an experiment. When we record the day's rainfall in inches, when we roll a die, when we roll two dice—these are all examples of experiments. For each experiment that we do, we ask what the probability is of a particular happening.

Definition 3-1   The set of all possible outcomes of an experiment is called the *sample space* of the experiment and denoted by a capital S.

Example 3-1   Supply a sample space, S, for the experiment of rolling a die.

Solution $S = \{1, 2, 3, 4, 5, 6\}$

Example 3-2   Write a sample space, S_1, for the experiment of tossing a penny twice and recording whether it lands heads or tails.

Solution $S_1 = \{HH,\ HT,\ TH,\ TT\}$

The sample space of the last example consists of exactly four members called *sample points*. The four sample points are considered equally probable. How do we know that for sure? We don't. We simply feel intuitively that they are, and we have no reason to believe otherwise.

Most of the time in this book we will work with a sample space of equally likely outcomes. However, this is not the only kind of sample space. Let's assume that we are interested only in the number of heads in the two tosses of our penny. Then we would have the following sample space:

$$S_2 = \{0H,\ 1H,\ 2H\}$$

Observe that this sample space has three sample points. They are not equally probable since if the four points of our first sample space, S_1, for this experiment are equally probable, then for the sample space S_2, the sample point $1H$ (one head) must be twice as likely to appear as $0H$ or $2H$.

Example 3-3 A biologist randomly selects two seeds from a package that contains three seeds for red flowers and two seeds for white flowers. With the aid of subscripts, list S for this experiment.

Solution $S = \{(R_1,W_1), (R_1,W_2), (R_2,W_1), (R_2,W_2), (R_3,W_1),$
$(R_3,W_2), (W_1,W_2), (R_1,R_2), (R_2,R_3), (R_1,R_3)\}$

In short, we do recognize the uses of sample spaces whose sample points aren't equally likely. However, from now on we will restrict ourselves to those with equally likely points.

In Example 3-1, rolling a die, {2, 4, 6} is an *event*. This event corresponds to our die's landing even. The collection of values {1, 2, 3, 4} corresponds to another event, the die landing with a value less than 5 showing. Clearly, many events can be created from this sample space. By convention, we use capital letters for events. Following are some other possible events in the die-rolling experiment.

$$A = \{1, 3, 5\}$$
$$B = \{1, 6\}$$
$$C = \{4\}$$
$$S$$

S (the entire sample space) is an event.

Consider a most unusual event—an event that has no members. Such an event is called an *empty* (or null) *event*. It is commonly denoted by $\emptyset$ (a circle with a slash). This event might arise if you recorded all the fractional values or all the 7's rolled by a single die. It might result if the biologist of Example 3-3 recorded all pairs of seeds with at least one seed for a green flower. In mathematics, $\emptyset$ can be an event in any sample space.

Definition 3-2 Any part (including $\emptyset$) of S will be called an *event*.

Example 3-4 List all the events of $S = \{a, b, c\}$.

Solution

$$
\begin{array}{ll}
S & \{a\} \\
\{a, b\} & \{b\} \\
\{a, c\} & \{c\} \\
\{b, c\} & \varnothing
\end{array}
$$

If the number of members of S is finite or as many as the sequence 0, 1, 2, 3, . . . we say S is a discrete *sample space*. See Example 1-26. If S contains members equivalent to the number of points in an interval of the real line, we say S is a *continuous sample space*.

At this point, we would like to discuss two expressions that sometimes appear in the statement of an event—*at least* and *at most*. Experience has shown that these two expressions cause students some trouble at first. To say that you have at least $5 in your pocket means that you have $5 or more. A person with a million dollars has at least $5. To say you have at most $5 in your pocket, means that you have $5 or less. We consider a person who is flat broke to have at most $5.

We learn that Ellen has at most two cars in her family. This means that Ellen's family owns two cars, one car, or no car. If, on the other hand, Ellen's family has at least two cars, this could mean that they have two cars, three cars, or a whole fleet of cars that they rent out to visitors.

Consider the experiment of tossing three coins and recording heads (H) or tails (T). If you are interested in obtaining at least one head, it means that you are interested in $1H$, $2H$, or $3H$. To say "at most one head" means $0H$ or $1H$.

EXERCISES A

3-1. The experiment is to choose a letter from the first half of the alphabet. Write S. (Example 3-2, 3-3)

3-2. a. The numbers 1 through 10 inclusive are written on slips of paper, which are placed in a bag and thoroughly shuffled. Write S for the experiment of drawing one slip of paper from this bag.
b. State the event that corresponds to the number on the slip of paper being odd, less than 11, greater than 5, and less than 8.

3-3. For the experiment of rolling a die, state the event that corresponds to rolling a number less than 4, greater than 5, a prime number. [Hint: 1 is not a prime number.]

3-4. List all the possible events for the sample space S_2 following Example 3-2. [Hint: There are eight.] (Example 3-4)

3-5. a. Write S for the experiment of recording the sex of the first three children born to a family. [Hint: S has eight members.]
   b. State the event that corresponds to having all girls, at least one girl, exactly one girl, at most one girl.

3-6. In Example 3-2 consider the sample space S_1. State the event that corresponds to at least $1T$, exactly $1T$, at most $1T$.

3-2 A LOOK AT PROBABILITY

HISTORICAL NOTE

CULVER PICTURES
BLAISE PASCAL

BETTMANN ARCHIVES
PIERRE FERMAT

Historically Blaise Pascal (1623–1662) and Pierre Fermat (1601–1665) are regarded as cofounders of probability theory. This important theory grew out of a gambling problem entitled "a problem of points." In this epoch-making problem the first of two players to obtain a specified number of points wins. The problem was first presented to Pascal by the gentleman gambler, Chevalier DeMere. Pascal communicated the problem to Fermat and both men solved the problem independently.

On a particularly dark night you approach your front door with your house key on a ring with five other similar keys. Operating solely on luck, what are your chances of choosing the right key on the first try? The answer is one out of six.

Two people decide to cut a deck of playing cards to settle an issue. The first person cuts a king. At this point, they reshuffle the deck. What chance does the second person have of winning the cut? Because the second person needs an ace to win, that person's chances are 4 out of 52.

Definition 3-3 If S is a nonempty finite sample space with equally likely sample points and $A \subseteq S$ (this means A can be $\emptyset$, part of S, or all of S), then the probability of an event A is a ratio of the number of sample points of A divided by the number of sample points of S. The probability of event A is denoted $P(A)$.

Figure 3-1 gives a breakdown of the teaching staff at State University by rank and by sex. The college staff consists of 130 males and 70 females, for a total of 200. We are interested in randomly selecting a member of the teaching staff. If A corresponds to the event of choosing an associate professor, there are 50 staff members (32 males and 18 females) in this category. $P(A) = 50/200 = 1/4$ or .25. If B corresponds to choosing a female member of the staff, then $P(B) = 70/200 = 7/20$ or .35. If C corresponds to selecting a staff member who is female and also an associate professor, then

$P(C) = 18/200 = 9/100 = .09$. If D corresponds to selecting a staff member who is female or an associate professor ("or" here is to be interpreted as including all of both groups, females and associate professors), then $P(D) = 102/200 = .51$.

Figure 3-1

	Male	Female
Professor	13	12
Associate professor	32	18
Assistant professor	70	30
Instructor	15	10
Totals	130	70

Example 3-5 Calculate the probability of each event for the experiment of rolling a die, where $S = \{1, 2, 3, 4, 5, 6\}$.
   a. The probability of rolling an even number.
   b. The probability of rolling an odd number.
   c. The probability of rolling a prime number.
   d. The probability of rolling a number greater than 4.
   e. The probability of rolling a number less than 7.
   f. The probability of rolling the number 7.

Solution
   a. $3/6 = 1/2$ b. $3/6 = 1/2$ c. $1/2$
   d. $1/3$ e. 1 f. 0

It follows immediately from Definition 3-3 that if $A = \emptyset$, $P(\emptyset) = 0$. The probability of the empty event is 0. See part (f) of Example 3-5. For any event A, $0 \le P(A) \le 1$. Probabilities are real numbers between 0 and 1 inclusive. If $A = S$, then $P(S) = 1$. See part (e) of Example 3-5.

We will restrict ourselves to problems that lead to finite sample spaces. For example, the experiment of drawing a number from the set of natural numbers clearly yields an infinite sample space; thus, we would not try to deal with such a problem. However, this restriction still gives a rather wide range of elementary probability problems.

Example 3-6 The experiment is to choose one student from a class that is 60% female and 40% male.
   a. What is the probability of choosing a female student?
   b. What is the probability of choosing a male student?

Answer .6, .4

Example 3-7 For the game of Tong used to introduce probability, what is the probability of having the sum of the extended fingers even? Odd?

Answer $P(E) = 5/9$, $P(O) = 4/9$

Example 3-8 A history class consists of 36 men and 24 women. Twenty-eight of the men and 20 of the women received grades of C or better. If a student is chosen at random from this class, what is the probability that the student has at least a grade of C or is a woman?

Solution There are 24 women in the class (all grades), and there are 28 men with a grade of C or better. Thus,

$$\frac{24 + 28}{60} = \frac{13}{15}$$

Example 3-9 Consider the experiment of recording the sex of the first three children born to a family.

$$S = \{BBB, BBG, BGB, BGG, GBB, GBG, GGB, GGG\}$$

Find the probabilities of the following events.
   a. P(exactly one boy).
   b. P(exactly two boys).
   c. P(exactly three boys).
   d. P(no boys).
   e. P(at least one boy).
   f. P(at least two boys).
   g. P(exactly four boys).

Solution
   a. 3/8 b. 3/8 c. 1/8 d. 1/8
   e. 7/8 f. 1/2 g. 0

The following *tree diagram* shows how we systematically find the eight sample points of S in Example 3-9.

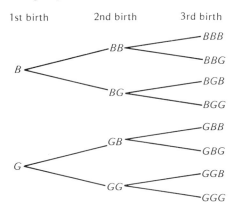

Example 3-10 Three candidates are running for sheriff. Candidates A and B are given about the same chance of winning but candidate C is given twice the chance of either A or B. What is the probability that C wins? What is the probability that A loses?

Solution   Let $P(A) = x$ Let $P(C) = 2x$
Therefore,
$x + x + 2x = 1$ and $x = 1/4$
$P(C \text{ wins}) = 1/2$
$P(A \text{ loses}) = 3/4$

EXERCISES A

3-7. What is the probability that a card drawn from a deck of cards numbered 1 to 50 inclusive is divisible by 5? (Example 3-5)

3-8. In Example 3-9 what is the probability of having at most one boy? At most two boys?

3-9. What is the probability of drawing a single card from a deck of playing cards and getting: (a) a spade? (b) a red card? (c) an ace? (d) a jack or a queen? (Example 3-6)

3-10. Choose an integer between 3 and 12 inclusive at random. What is your probability of getting: (a) an even number? (b) a prime number? (c) an odd number? (d) a number that is both even and odd at the same time?

3-11. (a) Write S for the experiment of recording the sex of the first four boys born to a family. (b) Write S for the experiment of recording H or T when four coins are tossed. (c) Compare parts (a) and (b). (d) What would S be like for the experiment of recording T or F on a four-question quiz?

3-12. A woman has seven pennies, three nickels, four dimes, and six quarters in her purse. Her son randomly selects one coin. Find the probability that he selects a penny, a nickel, a dime, or a quarter.

3-13. Choose a letter at random from the letters of the word *equation*. What is your probability of drawing a *t*? A vowel? A consonant?

B

3-14. If your probability of success in an event is 3/5, what is your probability of failure? (Example 3-7)

3-15. You randomly select one month out of the 12 months of the year. What is the probability that the month selected has exactly 31 days?

3-16. In a single throw of two dice, what is the probability of throwing: (a) a 10? (b) a 6? (c) a 7? (d) at most a sum of 5?

3-17. If three coins are tossed, what is the probability of getting: (a) exactly one head? (b) exactly two heads? (c) exactly three heads? (d) at least one head? (Example 3-9)

3-18. A man has four coins in his pocket and he wishes to tip his newsboy. The coins are a penny, a nickel, a dime, and a quarter. What is the probability that his tip exceeds 10 cents?

3-19. In Example 3-3, what is the probability that the two seeds drawn by the biologist correspond to: (a) two red flowers? (b) a red and a white flower?

3-20. To avoid detection at customs, a traveler has placed two narcotics tablets in a bottle containing four vitamin tablets that look much like the narcotic tablets. If the customs official selects one tablet at random for analysis, what is the probability that she will choose a narcotics tablet?

3-21. Four people, A, B, C, and D, line up in a straight line in front of a theatre box office. What is the probability that A and B are standing beside each other in line?

3-22. What is the probability that Pilar is one of four girls from whom a committee of two is chosen?

3-23. Two boys and two girls are placed at random in a straight line for a picture. What is the probability that girls and boys will alternate in the line?

3-24. A singer knows just six songs. On any given night he sings only two of them. What is his probability of singing two specific songs on a given night?

3-25. Choose a letter from the word *gamble*. (a) What is your probability of getting a vowel? (b) A consonant? (c) A letter before k in the alphabet?

3-26. A bag contains five jelly beans—three red, one green, and one white. If you reach in for two jelly beans without looking into the bag, what is your chance of getting two red ones? One red and one white? One green and one white?

3-27. Last night you tossed two pairs of shoes on the floor of your dark closet. This morning you pick up the first two shoes that you feel. What is your probability of getting a matched pair?

3-28. Consider the following record of the number of wrong answers on a math test given to the freshman class at Easy University.

Wrong Answers	Number of Students
0	20
1	80
2	120
3	250
4	260
5	190
6	80

a. What is the probability of three wrong answers?
b. All of the right answers?
c. At least two wrong answers?
d. At most two wrong answers?

 3-29. A medical survey of the cause of death among a group of 110 adults was categorized as to the cause of death, and the age of the subject at the time of death.

		Age at Death	
Cause of Death	21–40	41–60	61–80
Heart disease	4	9	14
Cancer	2	4	8
Stroke	1	2	4
Flu and pneumonia	0	1	3
Diabetes	1	0	1
Tuberculosis	0	1	0
Other	17	14	24

If one of these subjects is chosen at random, what is the probability that she or he: (a) died of cancer? (b) was over 61 at death? (c) was over 41 and had no listed disease? (d) died of tuberculosis at the age of 61? (e) died of a stroke or of heart disease.

B

In Exercises 3-30, 3-31, 3-32, and 3-33 a die marked x, x, y, y, z, z is rolled twice.

3-30. What is the probability of rolling at least one y?
3-31. P(at most one y shows)?
3-32. P(at least one x and at least one y show)?
3-33. P(no y's turn up)?
3-34. A hat contains five slips of paper numbered 1 to 5. If you draw two slips without putting any back, what is the probability that both numbers are less than 3?
3-35. Three balls numbered 1, 2, and 3 are placed in a bag, mixed, and drawn out one at a time. What is the probability of drawing the balls in the order 1, 2, 3?
3-36. Every day the Massachusetts Lottery prints a randomly generated four-digit number. The digit zero, 0, may be used in all four places of the number. (a) What is the probability of guessing the four-digit number on a given day? (b) If you can also win by guessing the first or last three-digit sequence, what is the probability of guessing the right three-digit number on a given day? (Assume only one guess.)

3-2 A LOOK AT PROBABILITY 67

We can create new events within S by doing specified operations on given events of S. Consider the following definitions.

Definition 3-4 The intersection of two events A and B, denoted by A ∩ B, is the event containing all the sample points common to A and B. (see Fig. 3-2).

Figure 3-2

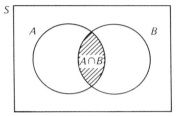

To determine the probability of such an event, we work as before.

Example 3-11 For the experiment of rolling a die, let A = {1, 2, 3, 4} and B = {3, 4, 5, 6}. Find the probability of A ∩ B.

Solution

$$A \cap B = \{3, 4\}; \quad S = \{1, 2, 3, 4, 5, 6\}$$

$$P(A \cap B) = 2/6 = 1/3$$

Example 3-12 The experiment is to toss two coins and record heads (H) or tails (T). Let A = {HH, HT} and B = {HT, TT}.
Find P(A ∩ B).

Solution

$$S = \{HH, HT, TH, TT\}$$

$$A \cap B = \{HT\}$$

$$P(A \cap B) = 1/4$$

Definition 3-5 The union of two events A and B, denoted by the symbol A ∪ B, is the event containing all the sample points that belong to A or to B or to both. A sample point common to A and B is placed in the union event exactly once. See Fig. 3-3.

Figure 3-3

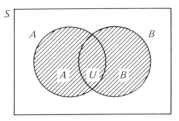

Example 3-13 Let $S = \{l, m, n, o, p, q, r, s, t, u, v\}$, $A = \{p, q, r, s, t\}$, and $B = \{l, m, n, p, t, o\}$. Find $P(A \cup B)$.

Solution

$$A \cup B = \{p, q, r, s, t, l, m, n, o\}$$

$$P(A \cup B) = 9/11$$

Example 3-14 For the experiment of rolling a die, let $A = \{1, 2, 3, 4\}$ and $B = \{3, 4, 5, 6\}$. Find the probability of A union B.

Solution

$$A \cup B = \{1, 2, 3, 4, 5, 6\} = S$$

$$P(A \cup B) = 6/6 = 1$$

Example 3-15 The experiment is to toss two coins and record H or T. Let $A = \{HH, HT\}$; $B = \{HT, TT\}$. Find $P(A \cup B)$.

Solution

$$S = \{HH, HT, TH, TT\}$$

$$A \cup B = \{HH, HT, TT\}$$

$$P(A \cup B) = 3/4$$

Example 3-16 Consider the experiment of drawing a card from an ordinary deck of playing cards (52 cards). Let A be the event of drawing a heart. Let B be the event of drawing an ace. Let C be the event of drawing a red card. Find $P(A \cup B)$, $P(A \cap B)$, $P(A \cup C)$, $P(A \cap C)$, $P(B \cup C)$, $P(B \cap C)$.

Solution One ace is a heart. Thus, $P(A \cup B) = 16/52 = 4/13$
$P(A \cap B) = 1/52$.

All the hearts are red cards. Thus, $P(A \cup C) = 26/52 = 1/2$
$P(A \cap C) = 13/52 = 1/4$.
Two of the aces are red cards. Thus, $P(B \cup C) = 28/52 = 7/13$
$P(B \cap C) = 2/52 = 1/26$.

Definition 3-6 The complement of an event A with respect to S is the set of all members of S that are not in A. We denote the complement of A by A'. (See Fig. 3-4.)

Figure 3-4

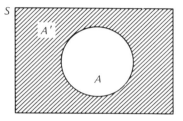

Example 3-17 Consider the experiment of rolling a die. If $A = \{1, 2\}$, then $A' = \{3, 4, 5, 6\}$. This means $P(A') = 4/6 = 2/3$.

Example 3-18 Let $S = \{a, b, c, \ldots, x, y, z\}$. If $A = \{a, b, c, d\}$, then $A' = \{e, f, g, \ldots, x, y, z\}$, thus, $P(A') = 22/26 = 11/13$.
Notice that $P(A) + P(A')$ always $= 1$.

Example 3-19 The experiment is to draw a card from an ordinary deck of playing cards. Let A be the event that we draw a diamond.
   a. Find $P(A)$.   **b.** Find $P(A')$

Solution Thirteen of the 52 cards are diamonds.
   a. $P(A) = 13/52 = 1/4$.
   b. $P(A') = 1 - 1/4 = 3/4$.

Example 3-20 In the experiment of choosing a letter from the word *history*, let A be the event of choosing a vowel and B the event of choosing a letter before s in the English alphabet. Find $P(A)$, $P(A')$, $P(B)$, $P(B')$, $P(A \cup B)$, $P(A \cap B)$, $P[(A \cap B)']$.

Solution We have $S = \{h, i, s, t, o, r, y\}$; $A = \{i, o\}$; $B = \{h, i, o, r\}$; $A \cup B = \{h, i, o, r\}$; $A \cap B = A = \{i, o\}$; and $(A \cap B)' = \{h, s, t, r, y\}$.
Therefore, $P(A) = 2/7$, $P(A') = 5/7$, $P(B) = 4/7$, $P(B') = 3/7$, $P(A \cup B) = 4/7$, $P(A \cap B) = 2/7$, $P[(A \cap B)'] = 5/7$.

EXERCISES A

3-37. The probability of getting an A in statistics class is .125. What is the probability of not getting an A?

3-38. A card is drawn from an ordinary deck of playing cards. What is the probability of drawing: a. a red card or an ace? b. a red card and an ace? (Example 3-16)

3-39. In Example 3-20 if $A = \{i, o, y\}$ and $B = \{h, s, t, y\}$ find: (a) $P(A \cap B)$. (b) $P(A \cup B)$. (c) $P(A')$. (d) $P(B')$. (Examples 3-11, 3-15, 3-17)

3-40. A and B are events in a sample space S. $P(A) = .4$, $P(B) = .3$, and $P(A \cap B) = .2$. Find: (a) $P(A \cup B)$. (b) $P(B')$. (c) $P(A' \cap B)$. [Hint: Draw a Venn diagram.]

3-41. A nurse has four pints of blood stored in identical bottles. He does not realize that one of the bottles is contaminated. What is his probability of randomly selecting two bottles that are not contaminated?

3-42. A three-digit number is formed by randomly choosing three of the digits 1, 2, 3, 4, 5 without replacement. What is the probability that the number formed is even? Is odd? Is a multiple of 5?

3-43. In rolling two dice, what is the probability of rolling a sum less than 8 and greater than 6? Less than 6 or odd? Less than 4 and greater than 10?

3-44. Draw a number from a bag containing the numbers 1 to 50 inclusive. (a) What is the probability of drawing a number that is divisible by 4? (b) Not divisible by 4? (c) Divisible by 4 and also by 5? (d) Divisible by 4 or by 5?

3-45. For the experiment of rolling a die what is the probability of rolling a number that is even and prime? Even or prime?

3-46. The experiment is to toss a coin and then to roll a die. (a) State S for this experiment. (b) $P(H$ and even number$) = ?$ (c) $P(H$ or even number$) = ?$ (d) $P(T$ and less than 5$) = ?$ (e) $P(T$ or less than 5$) = ?$

3-3 PROBABILITY CURVES

In the first two sections of this chapter, we dealt with sample spaces of discrete data (see Section 1-7). In this section, we turn our attention to continuous data.

Suppose our experiment is to record the temperature at 11:00 A.M. on July 18, then we could theoretically record a real number of degrees from, say, 60 to 120 degrees. If our experiment is to record people's heights, we would find heights corresponding to any real number between 66 in. and 78 in. If we are measuring shot-put throws, we can represent all possible shot-put distances by real numbers in a finite interval.

When the values representing the possible data of our experiment can be portrayed accurately by an interval(s) of the real-number line, mathematicians can often draw *probability curves* above the axis upon which a collection of data is plotted. Such an S has an infinite number of points and each interval (segment of the real number line) represents a continuous set of values. In this situation, we compute probabilities by computing areas above part or all of S and below the probability curve. See Fig. 3-5.

Obtaining these probability curves and using the integral calculus to evaluate areas is beyond the present scope of our work. But we can still gain a geometric feeling for the concept of these curves so indispensable in applied statistics.

Figure 3-5

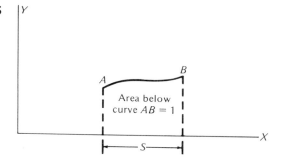

Definition 3-7 (S is a subset of the real line.) A curve is a probability curve if it has the following properties:

1. It lies entirely above or on the X-axis.
2. Lines drawn parallel to the Y-axis intersect the curve at most once.
3. The area below the curve and above S is exactly 1. Thus, mathematicians create the area to match the maximum possible probability.

If x is greater than or equal to 2 ($x \geq 2$) and x is also less than or equal to 5 ($x \leq 5$), we can express this as $2 \leq x \leq 5$. To write $0 \leq x \leq 1.87$ means that x is greater than or equal to 0 and also less than or equal to 1.87 To write $a \leq x \leq b$ means that x is greater than or equal to a and less than or equal to b.

Definition 3-8 If a and b are real values in S such that $a < b$, then the probability associated with the interval $[a, b]$ corresponds to the area directly above this interval and below the probability curve. This probability is denoted by $P(a \leq x \leq b)$.

In order to have a probability value greater than zero, we must have an interval of the real line with finite length. A point has no length. The area above a point and below a probability curve is zero. Thus, we associate with a single real value (a point) of S a probability value of zero when we are dealing with a probability curve. This fact has two immediate consequences: (1) In our statements on probabilities (Definition 3-8), we may weaken the "less than or equal to" statement ($\leq$) to "strictly less than" ($<$). (2) We may now deal with a probability curve (Definition 3-7) that intersects the X-axis without changing our definition of a probability curve.

Measuring probabilities using areas is quite unlike the techniques of the first two sections of this chapter.

Example 3-21 Given the probability curve in Fig. 3-6, consisting of two straight-line segments: (a) State S. (b) Show that the entire area above S and below the curve is 1. (c) Compute $P(0 \leq x \leq 1/3)$. (d) Compute $P(1/8 \leq x \leq 1.)$

Figure 3-6

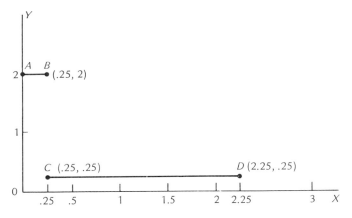

Solution
a. $S = [0, 2.25]$.
b. The area below AB is $(2)(.25) = .5$ and the area below CD is $(2)(.25) = .5$; $.5 + .5 = 1$.
c. Since $1/3 = .33$ and $.33 - .25 = .08$, we have $.5 + (.08)(.25) = .52$.
d. $(.125)(2) + (.75)(.25) = .4375$

Example 3-22 We are given the probability curve ABC (Fig. 3-7).
  a. State S for this probability curve.
  b. Evaluate the following probabilities:
     $P(0 < x < 1)$; $P(0 < x < 1/4)$; $P(1/2 < x < 3/4)$;
     $P(x < 0)$; $P(x < 1)$.

Figure 3-7

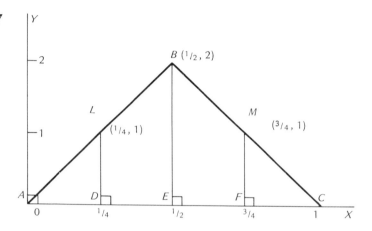

Solution
  a. [0, 1]
  b. $P(0 < x < 1)$ = the area of triangle ABC = $(1/2)(1)(2) = 1$

  $P(0 < x < 1/4)$ = the area of triangle ADL = $(1/2)(1/4)(1) = 1/8$

  $P(1/2 < x < 3/4)$ = the area of trapezoid $BEFM$ = $(1/2)(1/4)(2 + 1) = 3/8$

  $P(x < 0)$ = the area below our curve to the left of $x = 0$. Because there is none, the answer is zero.

  $P(x < 1)$ = the area below our curve to the left of $x = 1$. Because this is our entire curve, the answer is 1.

EXERCISES A

3-47. For the probability curve AB in Fig. 3-8. (a) State S. (b) Find $P(0 < x < 1/2)$. (c) Find $P(1 < x < 2)$. (Example 3-21)

3-48. We are given the probability curve $ABCD$ in Fig. 3-9. (a) State S. (b) Verify that the total area below this curve is 1. (c) Find: $P(1 < x < 2)$, $P(0 < x < 1)$, $P(1/2 < x < 1)$, $P(x < 1)$, $P(x > 1)$, $P(x > 3)$. (Example 3-22)

Exercises 3-49 and 3-51 represent two applied probability curves that will be dealt with extensively in later chapters. For now, we are only interested in obtaining a feeling for the concept of the area below these curves.

Figure 3-8

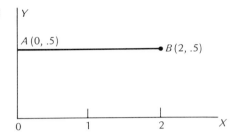

Figure 3-9

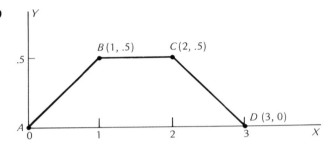

3-49. The *standard normal probability curve* in Fig. 3-10 is plotted on a horizontal axis customarily called the Z-axis. The standard normal probability curve is symmetric about the line $z = 0$. Two area measures below this curve are given in the illustration. Find: $P(0 < z < 2)$, $P(-1 < z < 0)$, $P(z > 2)$, $P(z < -2)$, $P(-1 < z < 1)$, $P(z < -2 \text{ or } z > 2)$.

Figure 3-10

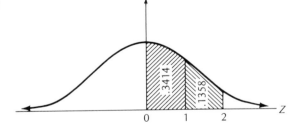

3-50. For the probability curve AB in Fig. 3-11 find:
    a. $P(1 < x < 3)$
    b. $P(3 < x < 4)$
    c. $P(\phantom{3 < } x < 5)$
    d. $P(\phantom{3 < } x > 5)$

Figure 3-11

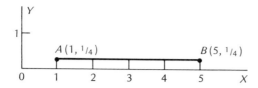

3-51. We are given the probability curve in Fig. 3-12. This is one curve in a family of *chi-square* curves, χ^2, find:
  a. $P(3.940 < \chi^2 < 23.209)$
  b. $P(\ 0\ \ < \chi^2 < 23.209)$
  c. $P(\quad\quad \chi^2 < 18.307)$
  d. $P(\quad\quad \chi^2 > 3.940)$

Figure 3-12

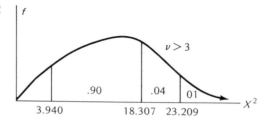

3-52. For the probability curve *AB* in Fig. 3-13 find:
  a. $P(1 < x < 5)$
  b. $P(1 < x < 3)$
  c. $P(3 < x < 5)$
  d. $P(\quad x < 1)$

Figure 3-13

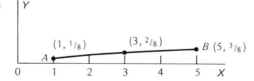

SUMMARY

Because statistical theory is based on probability, a knowledge of probability is necessary. The word *experiment*, in probability, was left undefined. The collection of all possible outcomes for an experiment is called the *sample space* of the experiment and denoted, S. Any subset of S (part of S) is called an *event* and denoted by a capital letter. The smallest possible event is ∅ (the empty event); the largest possible event is S (the sample space). Each member of S is called a *sample point*. The probability of an arbitrary event A is denoted P(A), and is defined to be the number of sample points in A divided by the number of sample points in S. From this definition follows the immediate fact that $0 \leq P(A) \leq 1$. New events were created by discussing the *union of events* (∪) and the *intersection of events* (∩).

This chapter ended with a discussion of *probability curves* (or probability distributions). These curves are important because much of our applied statistics is built upon them. A probability curve is intentionally constructed so that the area below it is exactly one.

Can You Explain the Following?

1. sample space
2. sample point
3. event
4. ∅
5. S
6. "at least"
7. "at most"
8. definition of $p(A)$
9. intersection of events
10. union of events
11. empty event
12. probability curve (or distribution)

4 PROBABILITY

This chapter is really a continuation of the previous chapter. It is designed to provide maximum exposure to the important subject of probability. Here you will meet the topics of counting, permutations, combinations, independent events, conditional probability, and mathematical expectation. You may further enrich your probability knowledge by reading some or all of this chapter. However, none of these topics need be covered in order to understand the rest of the text.

4-1 COUNTING

A baseball team has eight pitchers and three catchers. In how many ways can a coach choose a pitcher and catcher for a ball game? The pitcher can be chosen in eight ways and for each of these eight ways the coach can choose a catcher in three ways. There are then $8 \times 3 = 24$ ways of choosing a pitcher and a catcher for the game. See Fig. 4-1.

A penny and a nickel are tossed. In how many ways can they land? The penny can land in two ways (heads or tails) and the nickel can land in two ways. There are then a total of $2 \times 2 = 4$ ways in which the two coins can land.

Maria plans to travel from town A to town C through town B. She knows that there are three roads from town A to town B and four roads from town B to town C. By how many possible routes could Maria travel from town A to town C? The answer is 12, because, for each of the three roads from town A to town B, there are four roads from town B to town C. There are then $3 \times 4 = 12$ possible routes from which Maria can choose. From how many possible round-trip routes can Maria select? For each of the 12 ways of traveling from town A to town C, there are 12 ways of returning home. Thus, there are 12×12 or 144 ways for Maria to travel round-trip.

78 PROBABILITY

Figure 4-1

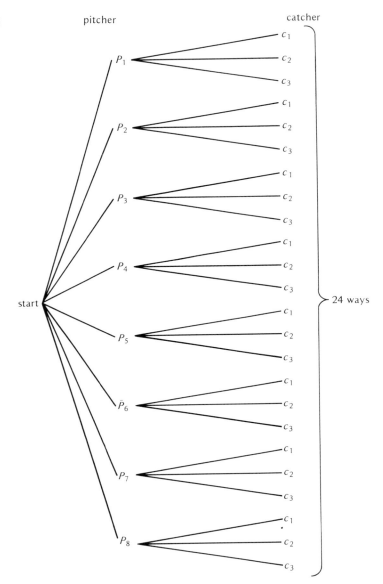

These are problems of counting. You will find the *multiplication principle* helpful in solving counting problems.

| **The Multiplication Principle** | If a selection consists of k steps such that the first step can be done in n_1 ways, and after it is done in any of these ways, the second step can be done in n_2 ways, and after the second step a third step can be done in any of n_3 ways, |

and so forth through k steps, then the entire selection process may be executed in

$$n_1 \times n_2 \times n_3 \times \cdots \times n_k \text{ ways}$$

Example 4-1 If a new-car buyer has a choice of five body colors, four different engines, and 12 interior colors, she can then select her car in $5 \times 4 \times 12 = 240$ different ways.

Example 4-2 In a recent political survey voters were classified by sex, by one of six income categories, and by one of five educational categories. How many possible classifications were there?

Solution 2(for sex) × 6(for income) × 5(education) = 60 classifications

Example 4-3 A student who has not studied must take a multiple-choice exam in which each question has five possible answers. The exam consists of 20 such questions. In how many ways can he mark his exam?

Solution

$$\underbrace{5 \times 5 \times 5 \times \cdots \times 5}_{20 \text{ questions}} = 5^{20}$$

Unfortunately for the student, only one of these corresponds to all 20 right answers.

Our next two sections will deal with two specialized topics in counting: *permutations* and *combinations*.

EXERCISES A

4-1. In how many ways can five students select a seat in a room with 20 desks? (Examples 4-1, 4-2)

4-2. A true–false quiz has 10 questions. In how many different ways can a student mark her answer sheet? (Example 4-3)

4-3. A restaurant offers 12 meals with a choice of four desserts. A customer may order tea, coffee, milk, or tonic with any meal. In how many ways may a customer select a meal, a dessert, and a drink? (Example 4-1)

4-4. A boy has six sweaters, four pairs of slacks, and two pairs of shoes. If he can wear them in any arrangement, how many possible arrangements does he have to choose from?

4-5. A coin is tossed 10 times. How many ways can it land?

4-6. If license numbers consist of one letter in the English alphabet followed by three digits chosen from 0, 1, 2, 3, 4, 5, 6, 7, 8, and 9, how many different license numbers may be made?

4-7. How many different four-digit numerals can be formed from the set of digits 0, 1, 2, 3, 4, 5, 6, 7, 8, 9 if:
a. Digits can repeat?
b. Digits cannot repeat?

4-8. The extension numbers of the office telephones in a large university's history department all have four digits. All the numbers are arrangements of the digits 0, 2, 5, and 7. How many variations (no repetition) are possible if zero is not used as the first digit?

B

4-9. In how many ways can six people be seated in an automobile if only two can drive?

4-10. A woman wishes to give a tip. She has a $1 bill, a $2 bill, a $5 bill, and a $10 bill. In how many ways can she give the tip?

4-11. A sailor has four flags, each of a different color. If a signal consists of at least two flags, how many possible signals are there?

4-12. We have three letters to mail and there are five mailboxes.
a. In how many ways can we mail these letters if each letter must be placed in a different mailbox?
b. In how many ways can we mail these letters if each letter may be placed in any of the five mailboxes?

4-13. a. In how many ways can a student arrange five different textbooks on a shelf?
b. In how many ways can a student arrange five different textbooks on a shelf so that they are not in order?

4-14. In how many ways can four coins land and not be all heads or all tails?

4-15. An optics kit contains five concave lenses, four convex lenses, and two prisms. In how many different ways can one choose a concave lense, a convex lense, and a prism from this kit?

4-16. In a primary election, there are five candidates for mayor, four for council member, and three for treasurer. In how many different ways can you vote for one of the candidates for each office? In how many different ways can you vote if you are allowed to vote for any candidate (repetition allowed) for any office?

4-17. A student is going to take a multiple-choice exam with 10 questions. Each question has five possible answers. This student knows the correct answer to exactly three of the 10 questions. In how many ways can he mark his answer sheet?

4-2 PERMUTATIONS

Suppose three fabulous prizes are to be given for solving a puzzle. Alice (*A*), Barbara (*B*), and Camille (*C*) are trying to win one of these prizes. Though all three prizes are wonderful, the most fabulous prize will go to the first girl to solve the puzzle and the least fabulous prize will go to the last girl to solve the puzzle. We assume there are no ties. The order of finishing the puzzle, then, is important.

Order is important in other situations: in being in line to obtain a ticket to a Houston Oilers game, in test scores when a scholarship is at stake, in a race, and in the sequence of numbers on a lottery ticket. A problem that deals with ordering objects is called a problem in *permutations*.

Let's return to Alice, Barbara, and Camille. In how many ways (permutations) can they finish their puzzle? *ABC* is one order of finish and *BAC* is another order of finish. In fact, there are six possible ways in which the girls could finish:

$$ABC, ACB, BAC, BCA, CAB, CBA$$

In addition to trial-and-error inspection, there are two other intuitive approaches to solving the number of ways in which the girls could finish the puzzle. We can think of having three positions (first, second, and third) such that exactly one letter from *A*, *B*, or *C* must be placed in position number one. This can be done in three ways. For each of these three ways, the second position may be filled in two ways. For the third, and last, position only one girl remains and so there is only one way of filling this position. There are then $3 \times 2 \times 1 = 6$ ways in which the girls can finish the puzzle.

Another way of determining the number of possible ways in which the girls may finish the puzzle is to construct a *tree diagram*. Thus:

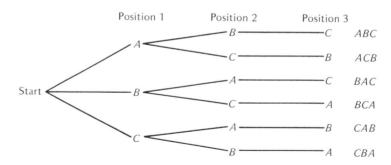

Factorial Notation It is easy to write $3 \times 2 \times 1$, but suppose that there had been 32 contestants and not just three. You would find it tiresome to write $32 \times 31 \times 30 \times 29 \times 28 \times \cdots \times 3 \times 2 \times 1$. Mathematicians have devised a compact notation for such expressions called *factorial notation* denoted "!". By this shortcut, $3 \times 2 \times 1 = 3!$ and $32 \times 31 \times 30 \times 29 \times 28 \times \cdots \times 3 \times 2 \times 1 = 32!$ The expression 32! is read "32 factorial."

Definition 4-1	If n is a natural number, $n!$ or n factorial $= (n)(n-1)(n-2)\cdots(3)(2)(1)$.
Example 4-4	$1! = 1$
Example 4-5	$5! = 5 \times 4 \times 3 \times 2 \times 1 = 120$
Example 4-6	$7! = 7 \times 6! = 7 \times 6 \times 5! = 5040$

Four distinct objects A, B, C, and D are to be arranged in a straight line without repetition. How many possible permutations are there? For each of the four choices for position one, there are three choices for the second position. For each of the 12 choices (4×3) for the first two positions, there are two choices for the third position. For each of the 24(12×2) choices for the first three positions, only one object remains to fill the last position. There are then 24(24×1) possible permutations of these four distinct objects using all four of the objects exactly once. They are

$$
\begin{array}{llll}
ABCD & BCAD & CABD & DABC \\
ABDC & BCDA & CADB & DACB \\
ACDB & BADC & CDBA & DBCA \\
ACBD & BACD & CDAB & DBAC \\
ADBC & BDAC & CBAD & DCAB \\
ADCB & BDCA & CBDA & DCBA
\end{array}
$$

What if we have eight distinct objects and we wish to select only three objects at a time, how many possible permutations are there? There are 8 (for the first position) $\times$ 7 (for the second position) $\times$ 6 (for the third position). Our answer then is $8 \times 7 \times 6$. Since we can always multiply by one, our answer may also be expressed as:

$$\frac{8 \times 7 \times 6 \times \left(5 \times 4 \times 3 \times 2 \times 1\right)}{\left(5 \times 4 \times 3 \times 2 \times 1\right)} = \frac{8!}{5!}$$

You can assume, unless otherwise stated, that all arrangements of objects are straight-line arrangements without repetition.

4-2 PERMUTATIONS

Example 4-7 Four automobile makers—Toyota, Ford, Honda, and Fiat—are interested in their display position at an auto show. The display positions lie in a straight line and each auto maker desires to be as close as possible to the entrance. State three possible permutations.

Solution Toyota, Ford, Honda, Fiat
Ford, Honda, Toyota, Fiat
Fiat, Toyota, Ford, Honda

Example 4-8 Find the number of permutations of eight objects taken four at a time ($_8P_4$).

Solution $_8P_4 = 8 \times 7 \times 6 \times 5 = 1680$

Example 4-9 Find the number of permutations of five objects taken five at a time ($_5P_5$).

Solution $_5P_5 = 5 \times 4 \times 3 \times 2 \times 1 = 5! = 120$

Definition 4-2 An ordered arrangement of r objects taken from a collection of n distinct objects is called a *permutation* of n objects taken r at a time. It is denoted by $_nP_r$; $r \leq n$.

The formula for computing $_nP_r$ is

$$_nP_r = \frac{n!}{(n-r)!} \quad r \leq n$$

Example 4-10 Use the permutation formula to evaluate $_8P_4$.

Solution $n = 8 \quad r = 4 \quad n - r = 4$

$$_8P_4 = \frac{8!}{(8-4)!} = \frac{8 \times 7 \times 6 \times 5 \times \cancel{4} \times \cancel{3} \times \cancel{2} \times \cancel{1}}{\cancel{4} \times \cancel{3} \times \cancel{2} \times \cancel{1}} = 1680$$

Example 4-11 Use the permutation formula to evaluate $_{10}P_3$.

Solution $n = 10 \quad r = 3 \quad n - r = 7$

$$_{10}P_3 = \frac{10!}{7!}$$
$$= \frac{10 \times 9 \times 8 \times \cancel{7!}}{\cancel{7!}}$$
$$= 10 \times 9 \times 8 = 720$$

84   PROBABILITY

Our formula for $_nP_r$ states that r can equal n. What if this happens? We have $_nP_n = n!/0!$. For convenience, we define $0!$ to be 1. Thus $_nP_n = n!/0! = n!$ See Example 4-9.

Observe that $_5P_5 = {_5P_4}$, $_7P_7 = {_7P_6}$ and, in general, $_nP_n = {_nP_{n-1}}$. With this one exception, for a fixed value of n as r increases, $_nP_r$ increases. We illustrate this for the case $n = 5$.

$_5P_5 = 5 \times 4 \times 3 \times 2 \times 1$ equals
$_5P_4 = 5 \times 4 \times 3 \times 2$ which is greater than
$_5P_3 = 5 \times 4 \times 3$ which is greater than
$_5P_2 = 5 \times 4$ which is greater than
$_5P_1 = 5$

EXERCISES A

4-18. Evaluate: (a) $4!/2!$ (b) $[(12 - 7)!]/6!$ (c) $6!3!/4!$ (d) $5!0!/5!$ (Examples 4-4, 4-5, 4-6)

4-19. Evaluate:
(a) $_6P_6$ (b) $_5P_3$ (c) $_8P_2$ (d) $_{12}P_4$ (e) $_{100}P_{97}$
(Example 4-8)

4-20. Find the number of possible arrangements of letters in the word, "police," if they are taken (Examples 4-10, 4-11): (a) six at a time (b) two at a time (c) five at a time (d) three at a time

4-21. How many different telephone exchange symbols, each consisting of two letters, can be formed from the letters of the English alphabet? Repetition is not allowed.

4-22. The president and the eight directors of the executive board of a large company are lined up in a row. In how many ways can they be lined up if the president is to stand in the center of the group?

B

4-23. In how many ways can six people stand in a row if two of them, A and B, must stand together? If A and B never stand together?

4-24. In how many ways can the manager of a softball team arrange her batting order if the pitcher must bat last and the catcher must bat first?

4-25. A man has two sets of books. One consists of six volumes; the other of three volumes. In how many ways can the books be arranged on a shelf if all of the volumes of each set are kept together?

4-26. How many different four-digit numbers can be formed with the symbols 0, 1, 2, and 3? How many of these numbers are odd? How many of these numbers are divisible by 5?

4-27. Twelve girls try out for a basketball team. Two can play only at center, four only as right or left guard, and the rest can play as right or left forward. In how many ways can the coach make up a team?

4-28. Recently, a wealthy man donated nine famous paintings to a museum. In how many different ways can three of these paintings be exhibited on the first floor of the museum if one is to be placed by the main entrance, one by the elevator, and one by the information desk?

4-29. In how many ways can four boys and three girls sit in a row if the boys and the girls must alternate?

4-30. In how many ways can five girls and five boys sit in a row if the sexes must alternate?

4-31. In how many ways can a convention schedule three speakers to speak simultaneously if there are five ballrooms?

4-32. A student has four examinations to write and there are 10 examination periods. How many possible arrangements are there of her exam schedule? How many arrangements are possible if her English exam must follow her mathematics exam? How many are possible if her English exam must *not* follow her mathematics exam?

4-3 COMBINATIONS

What if the order of arrangement of our objects is not important? In the five cards drawn for a poker hand, order does not matter. In choosing a committee of three from a group of three people, the order is not important. Five chores remain to ready a house for an 8 o'clock party, but the order in which the chores are done is immaterial as long as all get done. A problem of selecting objects where the order, or arrangement, is immaterial is called a problem of *combinations*.

In how many ways can we select three objects from four objects if order is unimportant? Let's call the objects *A*, *B*, *C*, and *D*. By inspection we find that there are exactly four combinations of these four objects taken three at a time. They are

$$ABC, ABD, ACD, DAB$$

The combination *ABC* is the same as the combination of *BAC* or *CAB*, and so forth. Thus, to be different, two combinations of *r* objects must differ by at least one object.

In how many ways can two objects be selected, without order, from the five objects *A*, *B*, *C*, *D*, and *E*? By inspection, there are 10 ways. They are

$$AB, AC, AD, AE, BC, BD, BE, CD, CE, DE$$

Definition 4-3 A *combination* is a selection of objects considered without regard to order. A selection of *r* objects from *n* distinct objects is called a *combination* of *n* objects taken *r* at a time. It is denoted, $_nC_r$ or $\binom{n}{r}$, where $r \leq n$.

The formula for calculating $_nC_r$ is

$$_nC_r = \frac{n!}{r!(n-r)!} \qquad r \leq n$$

Example 4-12 Evaluate $_5C_3$

Solution $n = 5$ $r = 3$ $n - r = 5 - 3 = 2$

$$_5C_3 = \frac{5!}{3!(5-3)!} = \frac{5!}{3!2!}$$
$$= \frac{5 \times 4 \times \cancel{3} \times \cancel{2} \times \cancel{1}}{\cancel{3} \times \cancel{2} \times \cancel{1} \times 2 \times 1} = 10$$

Example 4-13 Evaluate $_6C_4$.

Solution $_6C_4 = 6!/4!2! = 15$

Example 4-14 Find the number of four-member committees that can be chosen from eight people.

Solution   Since a committee consisting of A, B, C, D is the same as a committee consisting of B, A, D, C, and so forth, we are looking for the number of possible combinations of eight people taken four at a time.

$$_8C_4 = \frac{8!}{4!4!} = 70$$

Example 4-15 Evaluate $_5C_5$.

Solution $_5C_5 = 5!/5!0! = 1$

If $r > 1$, $_nP_r$ is always greater than $_nC_r$. That is, there are always more permutations than there are combinations. For example, $_5P_2 = 20 > {_5C_2} = 10$. Intuitively, this is because "order" is being considered in permutations. To see this mathematically we have but to look at the denominators of the two formulas for the numerators are equal.

EXERCISES A

4-33. Evaluate: (a) $_4C_2$ (b) $_7C_3$ (c) $_8C_2$ (d) $_{12}C_9$
(e) $_{12}C_3$ (f) $_{10}C_5$ (g) $_{100}C_2$ (h) $_5C_0$
(Examples 4-12, 4-13)

4-34. In how many ways can a committee of eight women be chosen from a group of 15 women? (Example 4-14)

4-35. A television dealer has 12 television sets to display. If he has room for only three at a time in his display window, in how many different ways can he select three of the television sets?

  4-36. Ten people have entered a beauty contest from which three winners will be selected. In how many ways can the winners be selected? If the order of selection of the three winners does not count, in how many ways can we select three winners?

4-37. A disc jockey has been given 12 records to play on the air. However, she has time to play only eight. In how many different ways can she select the eight records to be played?

4-38. How many five-card poker hands can be dealt from a deck of 52 playing cards?

  4-39. The personnel manager of Finast Sugar Company wishes to fill nine positions with five men and four women. In how many different ways can she fill these positions if 12 men and 10 women who are equally qualified apply for the positions?

4-40. Find the number of ways that seven bulbs, three green and four red, can be selected from a box containing 10 green bulbs and 8 red bulbs.

4-41. Find n if $_nC_2 = 36$.

B   4-42. How many combinations of at least one letter can be made from the letters A, B, C, and D?

  4-43. Each year a state tax department audits the books of five of the 15 largest companies doing business within the state. In how many different ways can the tax department select the five companies to be audited?

  4-44. Tony's Pizza House has eight workers on its morning shift and six workers on its afternoon shift. As an economy move, the management decides to fire two workers. All the workers are equal.
   a. In how many ways can the selection be made if management fires morning staff?
   b. Only afternoon staff?
   c. One from the morning staff and one from the afternoon staff?

4-45. A carton of 12 eggs contains two bad ones. In how many ways can you choose three eggs for an omelette so that:
   a. None of the bad ones are included?
   b. Both of the bad ones are included?
   c. Only one of the bad ones is included?

4-46. Thirteen cards are to be selected from a 52-card playing deck. In how many ways can this be done if each selection must contain four spades, three hearts, five clubs, and one diamond?

4-47. How many six-letter words (not necessarily in the dictionary), each consisting of three consonants and three vowels, can be formed from the letters of the word *equations*?

At this point, let's briefly review our chapter progress. We began with counting problems. After this came permutations (order) and combinations (no order), which are specialized counting topics. Why bother with these topics? In the preceding chapter we restricted ourselves to determining the probability of an event A where the process of counting the number of sample points in A and in S was relatively simple. As you now know, counting is not always simple. We need more sophisticated counting techniques to find many probabilities. For example, in the game of Poker, what is the probability of being dealt a straight flush on the initial deal? Here is the solution. For a single suit, considering ace low and high, there are 10 possible straight flushes (see below). For four suits there are 40 favorable cases.

$$A 2 3 4 5 | 6 | 7 | 8 | 9 | 10 | J | Q | K | A |$$

There are $_{52}C_5 = 2{,}598{,}960$ possible initial poker hands. Thus, P (of a straight flush) $= 40/2{,}598{,}960 = .0000153$. Can you imagine solving this problem in the last chapter?

4-4 MORE ABOUT PROBABILITY

In this section we use intersection and union of events to extend the probability theory of the last chapter.

What does it mean to say that two events have no sample points in common? Should this happen we say the two events are *mutually exclusive*.

Definition 4-4 Two events A and B are said to be *mutually exclusive* events if $A \cap B = \emptyset$.

If $S = \{1, 2, 3, 4, 5, 6\}$, $A = \{1, 3, 5\}$ and $B = \{2, 4, 6\}$ are mutually exclusive events. In addition, notice that the union of these two events is S. $C = \{1, 6\}$ and $D = \{2, 3\}$ are also mutually exclusive events, however, notice that the union of these two events is not S.

If two mutually exclusive events possess the additional property that their union is S, we say the events are *complementary*. In the preceding illustration A and B are complementary events and C and D are not complementary events.

Definition 4-5 Two events A and B are said to be complementary events if:

1. $A \cap B = \emptyset$ (they are mutually exclusive).
2. $A \cup B = S$ (their union is S).

If A and B are complementary events, $P(A) + P(B) = 1$ (or $P(B) = 1 - P(A)$. For example, if A corresponds to the event of drawing a club from an ordinary deck of cards and B corresponds to not drawing a club from the same deck of cards, then $P(A) + P(B) = 1/4 + 3/4 = 1$.

The concept of two independent events is extremely important. You may feel that two events are independent if they have nothing to do with each other (are mutually exclusive), however, this would only be true for the case where the events are $\emptyset$ and S. Our next definition formalizes the concept of two independent events.

Definition 4-6 Two events A and B are said to be *independent* if $P(A \cap B) = P(A)P(B)$.

If $S = \{1, 2, 3, 4, 5, 6\}$, $A = \{1, 2, 3\}$ and $B = \{1, 4\}$; A and B are independent because $P(A \cap B) = P(1) = 1/6 = P(A)P(B) = (1/2)(1/3)$. The events A and $C = \{1, 3, 5\}$ are not independent because $P(A \cap C) = P(1) = 1/6 \neq P(A)P(C) = (1/2)(1/2)$. Two events that are not independent are said to be *dependent*.

If you find the last definition confusing, you are probably mixing the probability definition of independence with its real-life definition. When we call two events, A and B, independent we mean that they have nothing to do with each other. That is, the occurrence of one event in no way influences the occurrence of the other. A Chicagoan's choice of "Mighty Crunch" breakfast cereal will not keep a Bostonian from choosing "Power Crisp" instead, or vice versa. And in coin tossing, a first throw of heads will not make tails any more likely on the second toss; the probability is still 1/2. The two throws are independent. Likewise, to the best of our knowledge, the sex of a child on two separate births, from the same mother, is independent.

Example 4-16 For the experiment of tossing two pennies and recording heads or tails:
a. State two events that are mutually exclusive but not complementary.
b. State two complementary events. c. State two independent events.
d. State two events that are dependent (not independent).

Solution
a. $A = \{HH\}$ $\quad B = \{TT\}$
b. $A = \{HH, HT\}$ $\quad B = \{TH, TT\}$
c. $A = \{HH, TT\}$ $\quad B = \{HH, HT\}$, where $A \cap B = \{HH\}$
because $P(A \cap B) = P(A)P(B)$
$1/4 = (1/2)(1/2)$.

d. One correct answer would be to use the events of part (a). Another answer is:

Let $A = \{HH\}$ $B = \{HH, HT\}$
$P(A \cap B) \neq P(A)P(B)$
$1/4 \neq (1/4)(1/2)$

Example 4-17 The experiment is to roll a die. A is the event the die shows 4 or less. B is the event the die shows an even number. Show that events A and B are independent.

Solution Since $A \cap B = \{2, 4\}$, we have

$$P(A \cap B) = P(A)P(B)$$
$$1/3 = (2/3)(1/2)$$

Therefore, we have independent events.

In practice, common sense tells you when events are independent. Only when an experience goes against common sense do we return to the sample space to try to prove independence. In practice, it may not be easy to show that two events are independent. For example, car A and car B sit in your driveway. Let A be the event that car A fails to start on a given morning. Let B be the event that car B fails to start on the same morning. It is not realistic to try to list all of the possible ways in which each of the two cars could fail to start: a moisture problem, dead battery, no gas, poor battery cable connection, and so forth. It is, at best, a chore to check two such events, A and B, against the formal definition of independent events. Our approach to such an applied problem goes more like this. Let's assume that the starting (or nonstarting) of car A in no way influences the starting (or nonstarting) of car B. This seems to be a reasonable assumption and it immediately leads us to the fact:

$$P(A \text{ fails} \cap B \text{ fails}) = P(A \text{ fails})P(B \text{ fails})$$

We can assume the starting of A and B is independent until such time as we discover a common denominator relative to the starting of the two cars. For example, such a common denominator might be the fact that the same mechanic has just tuned up both cars.

For the experiment of rolling a die, let A be the event that a prime number is rolled ($A = \{2, 3, 5\}$) and B be the event that an even number is rolled ($B = \{2, 4, 6\}$); find $P(A \cup B)$. $A \cup B = \{2, 3, 4, 5, 6\}$ and thus

$P(A \cup B) = 5/6$. Clearly, we cannot determine $P(A \cup B)$ by merely adding $P(A)$ and $P(B)$ for this yields $1/2 + 1/2 = 1$, which is incorrect. The following rule tells us how we may determine the $P(A \cup B)$ for any two events A and B.

$$P(A \cup B) = P(A) + P(B) - P(A \cap B)$$

In the event that A and B are mutually exclusive, the rule reduces to $P(A \cup B) = P(A) + P(B)$.

Example 4-18 Choose an integer from the first 200 positive integers. What is the probability that you chose an integer divisible by 6 or 8?

Solution

$$P(6 \text{ or } 8) = P(6) + P(8) - P(6 \text{ and } 8)$$
$$= \frac{25}{200} + \frac{33}{200} - \frac{8}{200}$$
$$= \frac{50}{200} = \frac{1}{4}$$

EXERCISES A

4-48. We are given the sample space $S = \{1, 2, 3, 4\}$. (Example 4-16)
  a. State two complementary events.
  b. State two mutually exclusive events that are not complementary.
  c. State two events that are independent.
  d. State two events that are not independent.

4-49. If the probability that North High will win its next football game is .82, what is the probability that it will lose?

4-50. State two events in everyday life that your common sense tells you are independent.

4-51. The experiment is to draw a card from an ordinary bridge deck. Find:
  a. $P(10)$.
  b. $P(\text{red card})$.
  c. $P(10 \text{ or a red card})$.
  d. The probability of drawing a card suitable to a given event A is 3/4; what is the probability of drawing a card that is not in this event?

4-52. For the experiment of tossing three coins and recording H or T, find:
  a. $P(\text{the coins fall alike})$.
  b. $P(\text{the first coin is an } H)$.
  c. $P(\text{the coins fall alike or the first coin is an } H)$.
  d. $P(\text{the first coin is a } T)$.
  e. $P(\text{the first coin is an } H \text{ and the first coin is a } T)$.
  f. $P(\text{the coins show at least one } H)$.
  g. $P(\text{the coins show no } H\text{'s})$.

B

4-53. The probability that Mr. A will live 10 more years is .75, Mr. B .50, and Mr. C .95. (Example 4-17)
  a. What is the probability that all three men will be alive in 10 years?
  b. What is the probabability that A and C alone will be alive in 10 years?
  c. What is the probability that all three men will not be alive?

4-54. The experiment is to draw a card with replacement from an ordinary bridge deck. Find the value of:
  a. Drawing a spade, a diamond, and then a club in that order.
  b. Drawing a red card, a heart, and then a deuce in that order.

4-55. If two coins are flipped, what is the probability of getting exactly two heads or exactly two tails?

4-56. If the probability of Liz's winning a race is 1/4 and the probability of Maria's winning the race is 1/3, what is the probability that either Liz or Maria will win should they both enter the race?

4-57. Your probability of hitting a dart board with a dart is .7. What is your chance of missing the board? What is the probability that you will hit the board on your first throw, miss on your second throw, and hit on your third throw? What is your chance of hitting the dart board in exactly two of three throws?

4-58. A, B, and C work individually on a problem. If the respective probabilities that each will solve the problem are 1/4, 1/3, and 1/5, find the probability that at least one of the three will solve the problem.

4-59. The probability that Joseph will make the track team is 1/4, and the probability that he will make the basketball team is 5/8. If the probability that he makes both teams is 5/24, what is the probability that he makes the track team or the basketball team?

4-60. Carmen and her college friends hope to visit Florida during the spring college vacation. The probability that they will go by car is 2/3 and the probability that they will go by plane is 1/5. What is the probability that they travel to Florida by car or plane?

4-61. Assume for a particular state that the probability of one spouse's voting in a national election is 3/11 and the probability that the other spouse votes is 4/7. If the probability that both vote is 5/77, what is the probability that either votes?

4-62. Two dice are rolled. What is the probability of rolling an even sum and a sum that is also less than 5? A sum that is even or less than 5?

4-63. The experiment is to choose a letter from the word *interval*. Let A be the event of choosing a consonant. Let B be the event of choosing a letter before r. Find: $P(A)$, $P(B)$, $P(A \cup B)$, $P(A \cap B)$.

4-64. The sex of the first three children born to a family is recorded. What is the probability of their having exactly one boy or exactly one girl?

4-5 CONDITIONAL PROBABILITY

The problem of choosing a left-handed person from your statistics class is not the same as selecting a left-handed female from your class. When we wish to determine a probability based on given information (a restriction on S) we say we are dealing with a *conditional probability*.

If we roll a die, $P(2) = 1/6$. If we roll a die and prior to stating the probability of obtaining a "2" we learn that the die has landed "even," we must think of our solution in terms of the given information. The given information eliminates the sample points 1, 3, and 5 from S. The sample points 2, 4, and 6 that remain are referred to as the *reduced sample space* of the experiment. Since the sample points were equally likely in the original sample space, we continue to assume that they are equally likely in the reduced sample space. Thus, the probability of obtaining a 2 given that the die has landed even is 1/3. See Fig. 4-2.

Figure 4-2
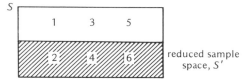
reduced sample space, S'

It is customary to denote the phrase "given that" by a vertical slash "|". The given information (this is really our reduced sample space) is written after the slash. Thus, $P(2|$ the die landed even$) = 1/3$. For this illustration, $S = \{1, 2, 3, 4, 5, 6\}$, and $S' = \{2, 4, 6\}$. Let $A = \{2\}$. In our illustration we considered $S' \cap A = \{2\}$.

Observe:

$$P(A \cap S') = 1/6 \quad \text{and} \quad P(S')P(A|S') = (1/2)(1/3) = 1/6$$

Thus,

$$P(A \cap S') = P(S')P(A|S')$$

The numbers 1, 2, 3, . . . , 20 are placed in a bag and thoroughly mixed. What is the probability of drawing a number less than 5? P(less than 5) $= 4/20 = 1/5$. Assume that before giving your answer you learn that a one-digit number is drawn. What is the probability of getting a number less than 5 based on this given information? Our sample space is now reduced to $S' = \{1, 2, \ldots 9\}$. P (less than 5| we have a one-digit number) $= 4/9$. Let $A = \{1, 2, 3, 4\}$.

Observe:

$$P(A \cap S') = 1/5 \quad \text{and} \quad P(S')P(A|S') = (9/20)(4/9) = 1/5$$

Thus,

$$P(A \cap S') = P(S')P(A|S')$$

Definition 4-7 The *conditional probability* of A, given B (B is the reduced sample space previously denoted S') is denoted $P(A|B)$ and $P(A|B) = P(A \cap B)/P(B)$ where $P(B) \neq 0$.

Example 4-19 A bag has 10 sealed envelopes. Five of these identical envelopes have a $1 bill inside, three have a $5 bill, and the last two have a $10 bill. Given that an envelope is drawn that does not contain a $1 bill, what is the probability that it contains $10?

Solution 1 Using the reduced sample space we are left with five envelopes—three with $5 and two with $10. Thus, our answer is 2/5.

Solution 2 Using Definition 4-7:

$$P(\$10 | \text{ a \$1 envelope was not selected}) = \frac{2/10}{5/20} = 2/5$$

We may solve a conditional probability problem by using either the reduced sample space or Definition 4-7.

Example 4-20 Given that a spade has been drawn from an ordinary deck of playing cards, what is the probability of drawing a face card (jack, queen, or king) on the same draw?

Solution

$$P(\text{spade}) = 13/52, \; P(\text{face card and spade}) = 3/52$$

$$P(\text{face card} | \text{a spade is drawn}) = \frac{3/52}{13/52} = 3/13$$

EXERCISES A

4-65. In a single roll of a die, what is the probability of obtaining a 2 or a 3 given that the die lands between 1 and 5 inclusive? (Example 4-19)

4-66. The numbers 1, 2, 3, . . . 20 are placed in a bag, thoroughly mixed, and one number is drawn. Given that a single-digit number is drawn, what is the probability that it is divisible by 3?

4-67. The sex of the first two children born to a family is recorded. Given that the first child born to the family is a girl, what is the probability that both children will be girls?

4-68. The sex of the first three children born to a family is recorded. If the first two children born are girls, what is the probability that the third child is a boy?

4-69. What is the probability of drawing a 10 from an ordinary deck of playing cards given that a red card is drawn? (Example 4-20)

4-70. A coin is tossed and a die is rolled. If we learn that the coin has landed H, what is the probability of obtaining a head and a number greater than 2?

4-71. Two dice are rolled. (a) What is the probability of obtaining a sum less than 5 given that the sum is less than 8 to begin with? (b) On the basis of the given information, what is the probability of obtaining a sum greater than 2?

4-72. Two balls are randomly selected from an urn that contains three white balls and two black balls. Given that one of the balls selected is black, what is the probability that both balls selected are black?

B

4-73. Use Definition 4-7 and show that $P(A) = P(A|S)$.

4-74. Consider three throws of an ordinary coin. Let A be "a head shows on the first throw." Let B be "a tail shows on the second throw." (a) Find $P(A|B)$. (b) Find $P(A)$. Observe that the two answers are identical. That is, the probability of event A is not influenced by the occurrence of event B. They are independent.

4-6 MATHEMATICAL EXPECTATION

Suppose that as a party stunt you have to throw a dart at the square dart board in Figure 4-3 without looking. The probability of hitting region I for one point is .50, region II for two points is .20, region III for three points is .20, and region IV for four points is .10.

Figure 4-3

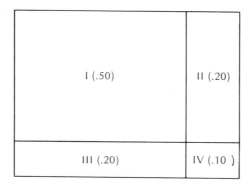

With only pure chance operating, what is the number of points you may reasonably expect to average? We compute this theoretical average, called the *mathematical expectation*, by summing the products of point values and their corresponding probabilities. In this case the mathematical expectation is $(1)(.50) + (2)(.20) + (3)(.20) + (4)(.10)$, which equals 1.90. This tells us that if we were to throw dart after dart at this target we would "average" 1.90 points.

For the case of discrete data we now introduce probability-based definitions for the population mean, μ, and the population variance, σ^2. Similar definitions for the continuous case require calculus and as such are beyond the scope of this text.

Definition 4-8 For a discrete population A of size N:
$$\mu = \sum_{i=1}^{N} a_i p_i, \text{ where } \sum_{i=1}^{N} p_i = 1 \text{ and } a_i \text{ is the } i\text{th data value whose probability of occurrence is } p_i.$$
We call μ the *mathematical expectation*.

To see that this definition is intuitively reasonable, consider the special case where each $p_i = 1/N$. $\mu = \frac{1}{N} \sum_{i=1}^{N} a_i$. This is our previous definition of μ.

Example 4-21 Compute μ for the experiment of rolling a die.

Solution

p_i	1/6	1/6	1/6	1/6	1/6	1/6
a_i	1	2	3	4	5	6

$\mu = 1/6 + 2/6 + 3/6 + 4/6 + 5/6 + 6/6 = 3.5$

The following is a pictorial sketch of S (Example 4-21) with μ

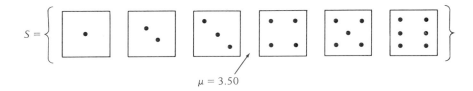

$\mu = 3.50$

Although no die could ever roll exactly 3.5, you can see that 3.5 is exactly halfway between 1 and 6, so this theoretical value for μ seems intuitively reasonable.

4-6 MATHEMATICAL EXPECTATION

Example 4-22 Finast Sugar Company is selling an obsolete piece of equipment. This piece of equipment has a scrap value, and the investment manager believes that with proper advertising she has a 22% probability of earning an additional $2500, a 36% probability of earning an additional $1000, a 28% probability of obtaining the scrap value ($0.00 improvement), and a 14% chance that she will have to accept $1500 less than the scrap value. What is the expected increase in revenue, μ, from advertising?

Solution $\mu = (.22)(\$2500) + (.36)(\$1000) + (.28)(\$0.00) + (.14)(-\$1500)$
$= \$700.00$. Theoretically, she will make money for the company by advertising.

Example 4-23 We are interested in recording the number of heads that occur when two pennies are tossed. Evaluate μ.

Solution

p_i	1/4	1/2	1/4
a_i	0	1	2

$$\mu = 1/2 + (2)(1/4) = 1$$

Example 4-24 Alicia and Miguel have never calculated this, but their probability of receiving one dime as a tip on their paper route is .6; two dimes, .3; three dimes, .07; four dimes, .03. What is the expected number of dimes in a customer's tip, μ?

Solution $\mu = (1)(.6) + (2)(.3) + (3)(.07) + (4)(.03)$
$= .6 + .6 + .21 + .12$
$= 1.53$

Definition 4-9 For a discrete population A of size N:

$$\sigma^2 = \sum_{i=1}^{N} (a_i - \mu)^2 p_i$$

where $\sum_{i=1}^{N} p_i = 1$, and a_i is the ith data value whose probability of occurrence is p_i.

Example 4-25 Compute σ^2 for the experiment of rolling a die.

Solution

$$\sigma^2 = \frac{1}{6}[(1 - 3.5)^2 + (2 - 3.5)^2 + (3 - 3.5)^2 \\ + (4 - 3.5)^2 + (5 - 3.5)^2 + (6 - 3.5)^2]$$

$$= 2.92$$

Example 4-26 Compute the value of σ^2 for the dart-throwing experiment that opened this section.

Solution

$$\sigma^2 = (1 - 1.9)^2(.5) + (2 - 1.9)^2(.2) + (3 - 1.9)^2(.2) + (4 - 1.9)^2(.1)$$
$$= 1.09$$

Example 4-27 Compute σ^2 for the experiment of recording the number of heads when two pennies are tossed.

Solution

$$\sigma^2 = 1/4 + 1/4 = 1/2$$

EXERCISES A

For Exercises 4-75 through 4-78 compute: μ, σ^2, and σ.

4-75.

P_i	1/2	1/2
x_i	0	1

(Examples 4-21, 4-23)

4-76.

P_i	1/4	1/2	1/4
x_i	1	2	3

4-77.

P_i	1/8	3/8	3/8	1/8
x_i	1	2	3	4

4-78.

P_i	0	1/6	2/6	3/6
x_i	1	2	3	4

4-7 THE BIRTHDAY PROBLEM AND THE WORLD SERIES PROBLEM (optional)

B

4-79. Suppose that Florida's weather bureau records show the probabilities for zero, one, two, three, four, five, six, or seven hurricanes per year as, respectively, .09, .22, .26, .21, .13, .06, .02, and .01. (Examples 4-22, 4-25)
   a. Theoretically, then, how many hurricanes can Floridians expect in any given year?
   b. Compute σ^2.

4-80. An importer is offered a shipment of rare jewelry for $5500. The probabilities that she will be able to sell it for $8000, $7500, $7000, or $6500 are, respectively, .25, .46, .19, and .10.
   a. How much can she expect to sell this shipment of jewelry for?
   b. What is her expected gross profit?
   c. What is the value of σ^2?

4-81. The probability that Mr. Jones will sell his house at a profit of $3000 is 3/20, the probability that he will sell it at a profit of $1000 is 7/20, the probability that he will break even is 9/20, and the probability that he will lose $1000 is 1/20. What is his expected profit? What is the value of σ^2? What is the value of σ?

4-82. The experiment is to record the number of heads in three tosses of a penny. Find the value of μ, the mathematical expectation, and σ^2.

4-83. A bag contains 8 nickels, 12 dimes, 7 quarters, 5 half dollars, and 3 silver dollars. You choose a coin from the bag without looking. What is your expected profit from the experiment?

4-84. Brian, a corporate executive, is in line for a promotion. Three positions will soon open and Brian is eligible for each of them. The position of controller will mean a salary increase of $2500 a year, and Brian feels the probability of his appointment to this position is .15. The position of chief auditor would mean a salary increase of $1800 a year, and the probability of this appointment is .30. Finally, the position of accounting supervisor would increase Brian's yearly salary by $900, and the probability of this appointment is .55. What salary increase can Brian write into his budget for next year?

4-7 THE BIRTHDAY PROBLEM AND THE WORLD SERIES PROBLEM (optional)

In this section, we introduce you to two classic probability problems: (1) the birthday problem, and (2) the World Series problem.

The Birthday Problem

There are k people in a room. What is the probability that at least two of these people have the same birthday (same day and month)?

Solution Forget about February 29 and deal with a 365-day year. Thus, there are 365 possibilities for each person's birthday, and 365^k possibilities for the birthday of k people. Therefore, our sample space, S, has 365^k sample points, each of which is an equally likely k-tuple of the form:

$$(x_1, x_2, x_3, \ldots, x_k)$$

where x_1 represents the birthday of the first person, x_2 represents the birthday of the second person, and so forth. Next consider the event A, "no two of the k people have the same birthday." Under this restriction, the first person's birthday has 365 possible values, the second person's birthday has 364 possible values, and so forth. The kth person's birthday has $365 - (k - 1) = 365 - k + 1$ possible values. Therefore,

$$P(A) = \frac{365 \times 364 \times 363 \times \cdots \times (365 - k + 1)}{365^k}$$

Finally, considering the complement of this event, we get:

$$P(\text{at least two birthdays are the same}) = 1 - P(A)$$

With as few as 23 people in the room there is a better than even chance that two people have identical birthdays. See the following table.

Number of people in the room	5	10	20	23	30	40	60
Probability that at least two birthdays are the same	.027	.117	.441	.507	.706	.891	.994

The World Series Problem

Our favorite baseball team is in the World Series. What is the probability that it will win?

Solution We assume that the probability of our team's winning each game is a constant value p. Let q or $(1 - p)$ represent the probability that our team loses (the other team wins). Next, we assume that the team performance in any game is not influenced by another game (the game performances are independent). Now, we think, how can our team win the World Series?

a. It can win by winning the first four games. This can happen in exactly one way with a probability of p^4.
b. It can win by winning four of the first five games. This can happen in four ways, each with a probability of $p^4(1 - p)$.
c. It can win by winning four of the first six games. This can happen in 10 ways, each with a probability of $p^4(1 - p)^2$.
d. Finally, it can win in seven games. This can happen in exactly 20 ways, each with a probability of $p^4(1 - p)^3$.

Thus, our answer is

$$p^4 + 4p^4(1 - p) + 10p^4(1 - p)^2 + 20p^4(1 - p)^3.$$

If you have trouble seeing this solution, construct the win (W)/loss (L) tree diagram.

Can you figure out the probability that the series ends in four games? (*Answer:* $p^4 + q^4$.)

SUMMARY

This chapter is optional since it is a continuation of the previous chapter. It is designed to provide maximum exposure to the important subject of probability. Problems of counting and in particular with the *multiplication principle* were dealt with in this chapter. *Permutations* and *Combinations* are specialized counting techniques. The former deals with counting where order is important; the latter deals with counting where order is not important. A formula for computing permutations of n objects taken r at a time, $_nP_r$, and a formula for computing combinations of n objects taken r at a time, $_nC_r$, were stated.

Factorial notation offers a convenient way of expressing answers to counting problems. For $n = 1, 2, 3, \ldots, n! = (n)(n-1)(n-2) \ldots 3, 2, 1$. A special case of factorial notation is 0! and $0! = 1$.

The chapter took a deeper look at probability. A definition of two *independent events* $[P(A \cap B) = P(A)P(B)]$ was given. *Mutually exclusive events* $(A \cap B = \emptyset)$ were discussed, and *complementary events* (1. $A \cap B = \emptyset$, 2. $A \cup B = S$) were discussed. Complementary events are always mutually exclusive, but mutually exclusive events are not necessarily complementary.

A rule for determining the probability of the union of two events was stated. It is: $P(A \cup B) = P(A) + P(B) - P(A \cap B)$. The important topic of *conditional probability* was discussed. This is the probability of an event A occurring given that an event B (the *reduced sample space*) has occurred. It is denoted $P(A|B)$ and $P(A|B) = P(A \cap B)/P(B)$, where $P(B) \neq 0$.

In section 4-6 *Mathematical Expectation* was discussed which deals with computing a theoretical average or μ. Based on mathematical expectation a probability explanation for σ^2 for the case of discrete data was able to be offered. The optional section dealt with two classical probability problems that I hope you had time to enjoy.

Can You Explain the Following?

1. multiplication principle
2. factorial notation, $n!$
3. permutation, $_nP_r$
4. combination, $_nC_r$
5. 0!
6. mutually exclusive events
7. complementary events
8. independent events
9. formula for $P(A \cup B)$
10. conditional probability, $P(A|B)$
11. reduced sample space
12. mathematical expectation, μ

MISCELLANEOUS EXERCISES

4-85. A market analyst claims that the probabilities that stock in the U.S. Can Company will go up more than three points, stay the same (within three points), or go down more than three points this year are .45, .24, and .31, respectively. A second analyst claims these probabilities are .47, .27, and .29. Comment on these claims.

4-86. The experiment is to select a prime number less than 25 from a bag in which all the prime numbers from 2 to 23 inclusive are written on slips of paper. (a) Write the sample space, S, for the experiment. (b) State the event that corresponds to the prime number's being less than 12. (c) State the event that corresponds to the prime number being less than 20 and greater than 15.

4-87. A man has purchased a ticket on every horse in a particular horse race. What is the probability that he holds the ticket on the winning horse (we did not say that he wins money)?

4-88. If the probability of catching a rare disease is .015, what is the probability of not catching this disease?

4-89. The population of Jasper County, Iowa, is 35,010 and of Lucas County is 10,031 (1970 census). If one inhabitant of these two counties is chosen at random, what is the probability she lives in Lucas County? What is the probability she doesn't live in Lucas County?

4-90. A card is randomly selected from an ordinary deck of playing cards. What is the probability that: (a) it is a 5? (b) it is a jack, queen, or king? (c) it has a value less than 7 (ace is high)?

4-91. The sex of the first three children born to a family is recorded.
  a. What is the probability of having at least one girl?
  b. What is the probability of having at most one girl?
  c. What is the probability of not having any girls?

4-92. In a local courtroom, 200 defendants are brought in to face misdemeanor charges. Of these 200, 110 are fined (event A), 60 go to jail (event B), and 30 are given a warning (event C). Assuming that these values are representative find: (a) $P(A)$; (b) $P(B)$; (c) $P(A \cap B)$; (d) $P(A \cup C)$; (e) $P(A \cup B \cup C)$

4-93. A child is selected from a fifth grade class in which 7 read at grade level, 11 read below grade level, and 7 read above grade level. Call these events A, B, and C, respectively. Find: (a) $P(A)$; (b) $P(A \cap B)$; (c) $P(A \cup C)$; (d) $P(A \cup B)$

4-94. At an inspection station, 1% of cars tested have bad brakes, bad headlights and cause too much pollution; 19% cause too much pollution; 14% have bad brakes; 5% have bad brakes and headlights; 3% have bad headlights and cause pollution; 16% have bad headlights; and 3% have bad brakes and cause pollution. [Hint: Draw a diagram.]
  a. What percentage have either bad brakes or bad headlights (or both)?
  b. What percentage have had brakes or bad headlights but not both?

4-95. A fair die is to be rolled once. You win $10 if the outcome is either even or divisible by 3. What is your probability of winning?

4-96. A customer enters a supermarket. The probability that the customer buys bread is .60, milk is .50, and both bread and milk is .30. What is the probability that a customer buys either bread or milk or both?

4-97. Fifty-two men and 48 women climb Mount Washington, and 14 of them get blisters; of those with blisters, 6 are women. If one group member is chosen at random, what is the probability that the person chosen either has blisters or is a woman (or both)?

4-98. Evaluate: a. $_{50}P_2$ b. $_6P_6$ c. $_{10}P_3$

4-99. Another symbol for $_nC_r$ is $\binom{n}{r}$. Evaluate:
(a) $\binom{5}{3}$ (b) $\binom{100}{2}$ (c) $\binom{8}{4}$

4-100. In how many ways can three of the four numbers 1, 3, 5, 7 be selected without replacement and placed in a straight line? Answer this question if numbers may be replaced as they are drawn.

4-101. If the same coin is tossed four times, how many different sequences of heads and tails are there that include three heads?

4-102. After a party, eight cigarettes are left on a table. Of these eight, five have filters and three do not have filters. Assume two cigarettes are chosen, without replacement, from the table. (a) What is the probability of getting two filter cigarettes? (b) What is the probability that exactly one has a filter?

4-103. The experiment is to toss a coin three times and record H or T each time. (a) Find the probability of $0H$, $1H$, $2H$, and $3H$. (b) Find the expected number (mathematical expectation) of heads.

4-104. The New York Lottery awarded, for each one million tickets sold at 50 cents:

   1   $50,000 prize
   9   5000 prizes
  90   500   prizes
 900   50   prizes

Find the expected value of each ticket.

5 SAMPLES AND SAMPLING

5-1 WHY SAMPLE

Suppose the leaders of a religious denomination are planning a recruitment drive and want to know how many adults in the United States claim membership in their church. How do they get this information? Suppose psychologists want to learn the relationship between drives and learning in rats. Specifically, how does hunger affect the number of trials a rat needs to learn a T-maze? Suppose sociologists want to study the differences in child-rearing practices between parents of delinquent and of nondelinquent children. Market researchers want to know what proportion of individuals prefer different car colors. Park attendants want to determine whether the ice is thick enough for safe public skating. Medical researchers wish to know the effects of a new flu shot on the general public.

What do all of these problems have in common? Someone is trying to get general knowledge about a population. As we said in the first chapter, because it is often impractical or impossible to analyze a whole population we must select a subset of this population, called a sample, and from this sample infer what is going on in the population.

5-2 SAMPLING WITH REPLACEMENT VERSUS SAMPLING WITHOUT REPLACEMENT

When we select a sample we generally return it to the population. But we have a choice as to when to return it, and that choice gives us two different ways of sampling. If we return each individual or unit of the sample to the population as soon as we record the desired information, and before choosing the next unit, we are *sampling with replacement*. If, on the other hand, we record information from the entire sample before replacing it, we are *sampling without replacement*.

5-2 SAMPLING WITH REPLACEMENT VERSUS SAMPLING WITHOUT REPLACEMENT

To illustrate the difference in effect between these two techniques, consider taking a sample of five students from your class for a weight study.

First, consider sampling with replacement. We select a first student at random and she gives us her weight. She promptly rejoins her class before we choose a second student. We now randomly choose a second student and he gives us his weight. Now this student rejoins the class before a third student is selected. We continue in this way until the fifth student has been chosen, recorded, and returned to the class. The key point to realize here is that the second student could conceivably have been the same as the first. In fact, it is possible, though highly unlikely, that we could have chosen that same student all five times.

Second, we consider sampling without replacement. Here we randomly select the set of five students. We record all five weights and only then do all five students return to the class.

Example 5-1 By sampling with replacement, record all the possible different samples of size $2(n = 2)$ from the population $A = \{a, b, c\}$. Here our population size $N = 3$.

Solution We use a little *tree diagram*.

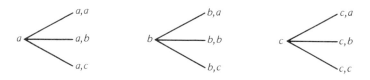

In the previous example, 3^2 or 9 different samples of size two can be drawn with replacement. In creating our sample there are three pieces of data—a, b, c,—available for selection as our first sample piece of data. Because we replace, there are again the same three pieces of data available for selection as our second sample piece of data.

Example 5-2 By sampling without replacement, record all the possible different samples of size $2(n = 2)$ from the population $A = \{a, b, c\}$. Here our population size $N = 3$.

Solution Get your solution from Example 5-1 by correcting for what doesn't apply. Specifically, because order does not count, a,b represents the same sample as b,a; a,c as c,a; and b,c as c,b. Thus, remove all samples that occur because of order in the selection.

Finally, since replacement is not allowed, there could be no such sample as b,b. Eliminating those, we find only three different samples possible: a,b; a,c; b,c.

It is a fact that:

1. In sampling with replacement, order of selection counts, but in sampling without replacement, it does not.
2. In sampling with replacement, a finite population acts as though it is infinite. Specifically, by sampling with replacement we could get a sample of 50 student weights from a class of 40 students. Obviously, we are speaking theoretically here for it would be silly to consider analyzing a sample bigger than the finite population from which it was drawn. It would be easier and more to the point simply to analyze the entire population itself.
3. Sampling with replacement yields many more different possible samples than does sampling without replacement. This follows immediately from the fact that we consider *order* of selection when sampling with replacement. Is it readily apparent that when we are sampling with replacement a sample of five students can be selected from a class of 40 students in 40^5 ways? There are 40 ways of selecting student one, 40 ways for student two, and so forth—a very large number indeed. A trivial sample of only two students can be sampled from a population of 5000 students with replacement in 5000^2 ways.

Practically speaking, sampling with replacement is sometimes impossible. For example, as an inspector for the U.S. Department of Weights and Measures hired to inspect gallons of milk, you cannot put a gallon of milk you have analyzed back on a store shelf. As another example, a marine biologist must select a fish from the ocean and test it for a possible lethal element. In order to conduct her test she may have to sacrifice the fish.

Example 5-3 By sampling with replacement, record all the possible different samples of size 3($n = 3$) from the population $B = \{x, y, z, w\}$. Here the population size $N = 4$.

Solution There are 4^3 different samples possible.

5-2 SAMPLING WITH REPLACEMENT VERSUS SAMPLING WITHOUT REPLACEMENT   107

xxx	yxx	zxx	wxx
xxy	yxy	zxy	wxy
xxz	yxz	zxz	wxz
xxw	yxw	zxw	wxw
xyx	yyx	zyx	wyx
xyy	yyy	zyy	wyy
xyz	yyz	zyz	wyz
xyw	yyw	zyw	wyw
xzx	yzx	zzx	wzx
xzy	yzy	zzy	wzy
xzz	yzz	zzz	wzz
xzw	yzw	zzw	wzw
xwx	ywx	zwx	wwx
xwy	ywy	zwy	wwy
xwz	ywz	zwz	wwz
xww	yww	zww	www

Example 5-4 By sampling without replacement, record all possible different samples of size $3(n = 3)$ from the population $B = \{x, y, z, w\}$.

Solution Again, start with Example 5-3 and eliminate choices that differ only by order and choices that repeat any element. Observe that there are only four different samples, far fewer than in Example 5-3: xyz; xyw; xzw; wyz.

In random sampling with replacement from a finite population each element has exactly the same probability of being drawn at every draw. In random sampling without replacement from a finite population, however, the probabilities differ with each draw because the composition of the population changes with each element drawn. To illustrate this problem, let's consider an ordinary 52-card bridge deck as a population. You want a random sample of five cards without replacement. You draw a first card and do not replace it. Now your probability for drawing any other card in the deck has risen from 1/52 to 1/51.

Observe again how many more different samples are possible in Example 5-3 than in 5-4.

Definition 5-1 A *sampling distribution* is a collection of the same statistic recorded from samples of equal size.

In taking a sample of 31 5-lb packages of sugar and recording the mean weight each week, the Finast Sugar Company (Chapter 1) is creating a sampling distribution of sample means. This is a very important sampling distribution that we deal with in detail in Chapter 7. Finast Sugar also records the sample variance (Chapter 2) each week, and in so doing creates a sampling distribution of sample variances.

The PVC Plastic Pipe Company (Chapter 1) is worried about the accuracy of the diameter measurement of its 4-in. pipe. The pipe's diameter must be accurate to within .001 (one-thousandth) of the advertised 4 in. A sample piece within those limits is called a "success." Each week the company takes a sample of 50 pieces and records the number of successes. PVC is creating a sampling distribution of success numbers.

5-3   THE RANDOM SAMPLE

How can we be sure that a sample is drawn in a random manner? How can we be sure that draft numbers, door prizes, or lottery numbers are drawn in a random manner? These questions deserve a little more thought than the average layperson gives them.

Definition 5-2   A sample is *randomly selected* if each element in the population has precisely the same probability of being included in the sample.

Statisticians think a lot about randomness, and have created the machinery to make sure that a selection process is random. They use a table of randomly generated digits, called a table of *random numbers,* in making selections from large populations. We have included a very brief table of randomly generated digits at the end of this text. Let us see how such a table might work.

You are visiting Truskin City, Calif., to select a random sample of 100 adults for making a political projection. Begin by assigning every one of the 50,000 adults in the city a natural number starting with 1. In other words, create a one-to-one (1–1) correspondence between a 50,000-member subset of $\{1, 2, 3, 4, 5, \ldots\}$ and the adults of Truskin City. Next, turn to any row of any page in your table of random numbers and read off five-digit numbers in a consistent manner, either vertically or horizontally. Assume you read: 24,639; 19,275; 63,921; 00,201; etc. The first set of five digits corresponds to a certain adult in Truskin City. This person is now the first element in your sample. The second set of five digits also corresponds to an adult in Truskin City. This person is now the second element in your sample. The third set of five digits does not correspond to a person in Truskin City, so pass this set of digits by. The fourth set of digits again corresponds. Continue this process until you have the desired 100 adults. In the event that a pertinent five-digit number repeats, omit it after its first appearance. When you've drawn your 100 adults, mark your stopping place in the table. If you need to draw more Truskin City adults for this same question, you can go back to the table and pick up where you stopped.

Random selection is a useful and necessary technique in statistics, but it does have some problems: (1) It is not an easy chore to get randomly selected people to cooperate. Just trying to get people to show up or let interviewers into their homes can be time-consuming. (2) We may want more than random selection in our sample. For example, we may also want to have a proportion factor operating. We may, for instance, want our sample to be 60% female and 40% male. (3) Finally, we may be dealing with a population that doesn't readily lend itself to a tagging process. Such populations would be all the trees in a forest or all the fish in the Atlantic Ocean.

At such times statisticians turn to nonrandom sampling techniques. Three of the commonest nonrandom sampling procedures are stratified, cluster, and systematic sampling. This is our next section.

5-4 STRATIFIED, CLUSTER, AND SYSTEMATIC SAMPLING

Suppose you want a sample of houses in Boston for a study on house maintenance. In order to use a table of random numbers you would have to number every house in Boston. This would be a staggering job. For the many cases where using a table of random digits is impractical, we have the following three proven alternatives.

1. *Stratified sampling:* Here we divide the population into subpopulations (strata) and draw equal numbers from each stratum. For example, dividing Boston into 25 neighborhoods and choosing 40 houses from each yields a sample of 1000 houses. Stratified sampling is sometimes modified by proportional sampling within strata. For example, we might want twice as many women as men selected from a stratum because we know there are twice as many women in the stratum to begin with. The key to success in this sampling method is to strive to get data within each stratum as *homogeneous* (as much alike) as possible.

2. *Cluster sampling:* Here we again divide the population into subpopulations, but this time we only use some of them. The usefulness of this technique is that it saves money. For example, if we want a sample of American university students in order to estimate financial need, random sampling would require numbering every student in the United States in one listing followed by a random drawing. This process would be extremely difficult and costly. Cluster sampling lets us use a limited number of universities as well as existing lists such as registration lists. Expenses are thus kept to a minimum. The key to success in this method of sampling is making the data within each cluster as *heterogeneous* (varied) as possible.

3. *Systematic sampling:* This is an easy method to use. Starting from a randomly chosen point on an existing list of a population, such as a telephone book, we select every nth element. For example, the Deluxe Car Company can systematically sample cars from its production line by marking every third car for inspection.

Example 5-5 You are conducting a public opinion poll before an election. Experience tells you that income level is an important factor in your poll. Since 30% of all eligible voters have low incomes, 60% have medium incomes, and 10% have high incomes, you want a sample of size 400 with 120 voters from the low-income group, 240 from the medium-income group, and 40 from the high-income group. The voters in each group are chosen at random. This is an example of stratified sampling.

Example 5-6 Atlantic Electric Inc. wishes to select a sample of 50 customers to check on their billing. It seems reasonable to select every twentieth customer's bill from current files. The first customer's bill is chosen from among the first 20 on the list by a table of random digits. Now the clerk continues down the list, taking every twentieth name. This is an example of systematic sampling.

Example 5-7 The N and C Can Company wishes to sample a trainload of its tin containers. The train is already loaded, but the intended customer has heard about a possible defect. One of the workers selects several containers from the top layer of the shipment for analysis. This is an example of cluster sampling.

Although results based on cluster or systematic sampling are generally less reliable than results based on simple random sampling of the same size sample, they can give good reliability at a more reasonable cost.

Thus, the four types of sampling are

1. Random
2. Stratified
3. Cluster
4. Systematic

(Note that, in actual practice and in all types of sampling, we frequently sample without replacement.)

In some situations, none of the four methods will work. At such times selecting a sample may be no more than using the available data. This often happens in the medical area where researchers must take the cases at hand or nothing.

In summary, then, statistical theory is primarily designed for and based on random sampling. If time and cost make random sampling impossible, we must choose another sampling scheme and proceed with caution, hoping that a theory intended for random sampling will still apply. Basically, this should be the case if we have used common sense in our sampling. Random sampling of data often leads to a famous family of probability curves called *normal curves*. These curves are the topic of the next chapter.

5-5 BIAS

One thing a sample should not have, and random sampling is designed to avoid, is *bias*.

Definition 5-3 A sample is said to be biased (a *biased sample*) when one or more pieces of data from our population fail to receive an equal chance of being included in our sample.

Intuitively speaking, a sample is biased if it is in some way prejudiced; it favors one kind of element over another. Perhaps you were supposed to interview shoppers at random in a shopping center, and noticed about noon that most of your interviewers were young and attractive. Perhaps you were supposed to get a definite ratio of males to females, and failed to do so. A sample chosen by a biased method tends systematically to underselect or overselect different pieces of data from the population. In such circumstances, clearly each element in the population did not have an equal chance of being selected.

Following are some examples of bias in a sample.

Example 5-8 When a coin is tossed we assume H and T are equally likely to appear. However, if the coin happens to be heavier on the H side, a sample of 100 tosses is likely to contain more than 50 H's.

Example 5-9 A machine working on an assembly line makes flaws in bottles in a rhythmic manner, every 10 bottles. Any systematic sampling along the belt following this machine's output could easily yield a biased sample. Not only every tenth bottle on the right rhythm, but multiplies of 10, such as every twentieth, or hundredth bottle would also yield a much higher percentage of flaws than the population (the output of several machines) actually contains.

Example 5-10 Suppose the city of Denver pays a major portion of the debt of the local rapid transit system. A polling firm hired to learn how people could feel if the state took over this burdensome debt interviewed mainly people from the greater Denver area. The poll showed the public highly in favor of Colorado's assuming the Denver debt. However, if the poll had sampled the population of the entire state the result might have been different. Do you see why?

Bias Outside the Sample

Just as a sample of data may be biased, so a statistic being calculated from a sample may also be biased—perhaps because it is coming from a biased sample. Begin with the sampling distribution of a statistic such as the mean. Intuitively speaking, if the average of values in this sampling distribution yields the parameter value (μ for the distribution of means) corresponding to our statistic, we say that our statistic is an *unbiased estimator* of the parameter.

The sample mean is an unbiased estimator of the population mean. This will be proved in the optional section of this chapter. It can be shown that s^2 is an unbiased estimator of σ^2. However, interestingly s is a biased estimator of σ.

Is it possible to have more than one unbiased estimator for a population parameter? Yes. For example, the median and the mean drawn from a normal population are both unbiased estimators of the population mean, μ. In such a case we use the estimator whose sampling distribution possesses the smallest standard deviation. This is the mean.

5-6 MORE ABOUT BIAS (optional) 113

EXERCISES A

5-1. What does it mean to say you have a biased sample?

5-2.
a. By sampling with replacement from the population $A = \{a, b, c, d, e\}$, record all possible different samples of size 2. (Example 5-1)
b. Record all possible samples of size 2 for sampling without replacement. (Example 5-2)

5-3.
a. By sampling with replacement from the population $B = \{x, y, z, u, v\}$ record all possible different samples of size 3. Just state the total number and start the listing.
b. Record all possible samples of size 3 for sampling without replacement.

5-4. State an example of a population that would be hard to sample randomly using a table of random digits.

5-5. Assume that your college has 5000 students. You want to find out whether they prefer pass–fail grading or the traditional A, B, C, D, and E. Explain, in detail, how to use a random-number table to select a random sample of 50 students for your poll.

5-6. Devise a method for selecting a sample of five rabbits from a population of 100.

5-7. A large metropolitan newspaper serving 500,000 people asked its readers to express their opinion on a major issue by filling out a response form in the newspaper. The newspaper received 1500 responses, of which 950 were for the issue and 550 were against it. On the basis of these data, the newspaper concluded that the people of the area were for the issue. Discuss the validity of this conclusion.

5-8. Listed here are various ways of getting a sample of families needed for a town health survey.
a. Discuss the validity of the sample scheme.
b. Discuss whether or not each method would yield a truly random sample.
   i. Visit a collection of families selected from the telephone book.
   ii. Use pure chance to open the telephone book and select families to call on the telephone.
   iii. Select every tenth house or every tenth apartment in an apartment building.
   iv. Send questionnaires to all families on a given postal route with return postage included.
   v. Interview families as they pass the busiest street corner in the town.
   vi. Use a random-digit table to select families from the town school list.
   vii. Send a questionnaire to all people in the town who have visited a doctor or hospital in the past year.

5-6 MORE ABOUT BIAS (optional)

When we have to use a statistic to estimate a parameter it is only human to wonder how accurate that estimate is. If μ is known, we may compare the value of $\bar{x}$ with it.

For the experiment of rolling a die, we have recorded the following sample of 50 rolls.

SAMPLES AND SAMPLING

```
6 6 2 4 6 3 5 2 4 3
5 2 3 4 1 2 2 6 6 2
4 3 5 1 5 5 5 1 3 4
2 6 5 4 1 3 6 4 5 5
1 5 3 5 2 5 2 2 3 1
```

Here our sample mean is 3.60, which is just a little over the true mean value of $\mu = 3.50$. Thus, we assume that the mean is a good (accurate) estimator for this sample. In statistical terms, when the expected value of the estimator approaches the parameter it is supposed to estimate, the estimator is said to be *unbiased*. If it is not close or equal to that value it is said to be *biased*.

Definition 5-4 A statistic $\hat{\theta}$ is an *unbiased estimator* of a parameter θ if $\mu_{\hat{\theta}} = \theta$.

Theorem 5-1 $\bar{x}$ is an unbiased estimator of μ.

Proof It will be helpful to imagine that each x_i, $1 \leq i \leq n$ is a box through which all N elements of our finite population flow.
Now:

$$\mu_{\bar{x}} = \sum_{i=1}^{N} \bar{x} p_i = \sum_{i=1}^{N} \left[1/n \sum_{j=1}^{n} x_j \right] p_i$$

$$= 1/n \left[\sum_{i=1}^{N} (x_1 + x_2 + x_3 + \cdots + x_n) p_i \right]$$

$$= 1/n \left[\sum_{i=1}^{N} x_1 p_i + \sum_{i=1}^{N} x_2 p_i + \cdots + \sum_{i=1}^{N} x_n p_i \right]$$

$$= 1/n [\underbrace{\mu + \mu + \mu + \cdots + \mu}_{n \text{ mu's}}]$$

$$= \mu$$

SUMMARY

The problems of sampling; and, in particular, how time and money force us to sample from a population were discussed in this chapter. We discussed *sampling with replacement* (datum is selected, recorded and then returned to the population one piece at a time), and *sampling without replacement* (the entire sample of data are selected, recorded, and then the sample is returned). Thus, the question of sampling is not, "do we return the data to the population?" for we always return it, but "when do we return the data?".

Random sampling was discussed and you were introduced to a table of randomly generated digits that is often helpful in selecting a random sample. Statistical theory is based on random sampling. What happens if, because of time or money, you cannot select a random sample? If this is the case, resort to one of three basic sampling procedures: (1) *stratified sampling,* (2) *cluster sampling,* or (3) *systematic sampling.* Each of these sampling procedures, when employed with common sense, provides reliable results in a cost efficient manner. The exact method of sampling to be used depends on the nature of your research project, and your research project may employ a combination of sampling procedures.

The chapter dealt at length with the subject of *bias.* There are two meanings to the word bias. It is possible to be dealing with a *biased sample;* this was the subject of section 5-5. On the other hand, it is possible to be dealing with *a statistic that is biased.* Because this topic is a bit more sophisticated it was left to the optional section.

Can You Explain the Following?

1. sampling with replacement
2. sampling without replacement
3. sampling distribution
4. tree diagram
5. random sampling
6. stratified sampling
7. cluster sampling
8. systematic sampling
9. biased sample
10. biased statistic

MISCELLANEOUS EXERCISES

Exercises 5-9 through 5-12 contain a serious source(s) of probable bias. In each case discuss the reason you suspect bias may occur.

5-9. The Detroit Police Department wants to know how black residents of Detroit feel about police service and a questionnaire is prepared. A sample of 300 mailing addresses in predominantly black neighborhoods is chosen, and a police officer is sent to each house to administer the questionnaire to an adult living there.

5-10. A large television rating service selects its sample from households at random using telephone directories.

5-11. A bread company wants to know what fraction of Toledo households bake some or all of their own bread. A sample of 500 residential addresses is taken, and interviewers are sent to these addresses. The interviewers are employed during regular working hours on weekdays, and interview only during those hours.

5-12. A behavioral researcher working on sexual habits uses subjects who volunteer to discuss their behavior or otherwise participate in the research.

5-13. How many possible samples of size 50 drawn with replacement can be selected from a population of 8000 Republicans? (Leave the answer in exponential form.)

5-14. An "ideal cluster" is one that acts as an exact miniature of the population. Discuss the money-saving features of such an ideal cluster.

5-15. The owner of an automobile dealership decides to sample 750 customers who received new-car repair service last year. Because of time limitations, he decides to sample only 20 customers. Explain how he may use a table of random numbers to do this.

5-16. Suppose a business population consists of 1000 accounts receivable and a sample of 50 such accounts is to be obtained by systematic sampling of the accounts receivable files. Explain how this sample may be collected.

5-17. A company with four subdivisions (A, B, C, and D) wants to solicit employee opinion. Divisions A, B, C, and D comprise 40%, 30%, 20%, and 10% of the company workforce. The company decides to select a sample of 50 employees, viewing each subdivision as a stratum. How many employees should be selected from each stratum? What method can be used actually to obtain the employees needed from within each stratum?

5-18. For the sampling distribution of means only, and when our population is finite and we are sampling without replacement, a correction factor is inserted in σ^2 means.

$$\text{Factor: } \frac{N-n}{N-1} \qquad \sigma^2 \text{ means} = \frac{\sigma^2}{n}\left[\frac{N-n}{N-1}\right]$$

a. If $N = 5000$ and $n = 50$, find the value of the correction factor.
b. If $N = 500$ and $n = 50$, find the value of the correction factor.

5-19. As a rule of thumb, the correction factor of the last exercise is dropped when n constitutes 5% or less of N. First observe the two answers of the last exercise, and then explain why this is a reasonable rule.

5-20. What is the probability that two adjacent digits in a table of random numbers are identical?

6 THE NORMAL DISTRIBUTION

6-1 THE NORMAL DISTRIBUTION

The most important probability curves in applied statistics are the normal probability curves. These curves are extremely important because they describe the distribution of many sets of data in nature, industry, and research. In 1733 Abraham DeMoivre developed the mathematical equation of the normal probability curve. This provided the basis for much of the theory of applied statistics.

HISTORICAL NOTES

DeMoivre (1667–1754) was born in France. In 1697 he was elected to the Royal Society and later to the Academies of Paris and Berlin. As an adult, he traveled to England where he met Newton. His mathematical interest extended through many branches of mathematics. His famous text in probability is entitled *Doctrine of Chance*. It is interesting to discover that he derived the theory of permutations and combinations from the principles of probability, whereas now we do the reverse.

The Granger Collection
ABRAHAM DEMOIVRE 1667–1754
French mathematician

The standard normal curve is shaped like a cross section of the Liberty Bell. Hence some people call it a *bell curve*. Others call it a *Gaussian probability curve* in honor of Carl Gauss (1777–1855), who derived its equation from a study of errors in repeated measurements of the same quantity. Thus, two talented men whose life spans did not overlap independently derived the same important probability curve.

There is really a family of these probability curves. The family is infinite in number and a particular probability curve is determined by specifying values for μ and σ that appear in the normal curve equation. Here we will simply state the equation and leave further observations for the optional section of this chapter.

Culver Pictures Inc
CARL FRIEDRICH GAUSS

The normal probability curve:

$$n(x; \mu, \sigma) = \frac{1}{\sigma \sqrt{2\pi}} e^{-(1/2)\left(\frac{x-\mu}{\sigma}\right)^2}$$

where x is defined over the whole real-number line.

Figure 6-1

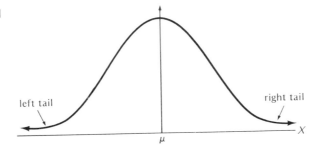

Figure 6-1 is a sketch of a normal probability curve. The word normal *here* simply means common. Normal should not be thought of in the sense that distributions producing other shapes are abnormal. Figures 6-2 and 6-3 are the result of first holding μ constant while σ varies and then holding σ constant while μ varies.

Figure 6-2

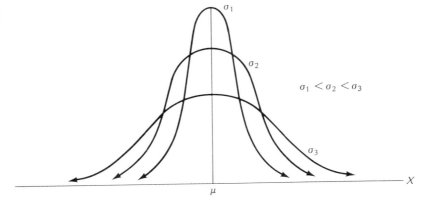

Figure 6-3

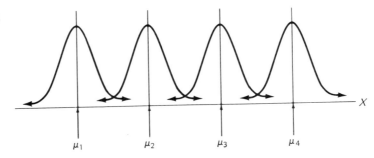

6-1 THE NORMAL DISTRIBUTION

Characteristics of Normal Curves

1. Normal curves satisfy all the properties of probability curves. See Section 3-3.
2. The modal value is $x = \mu$.
3. Each curve is symmetric to the line $x = \mu$.
4. The extreme left of the normal probability curve is called the *left tail* and the extreme right is called the *right tail*. As the value of x increases (decreases), the right tail (left tail) moves closer to the X-axis but never touches it. (See Fig. 6-1.)
5. Each curve changes from concave down (or up) at the points $x = \mu \pm \sigma$.

How can we ever hope to analyze an entire (infinite) family of probability curves? We don't even try. We settle for analyzing one particular member of this family of curves. This one curve will serve as the standard for all members of this family of curves. As such it is called the *standard normal probability curve*. This curve is completely tabulated in a table (Table I) at the end of this text. Our interest will be solely in how to read this important table and not in how the table was derived. At this point, two important questions should come to mind.

1. How do we read this table?
2. If we are to use only one curve of the family of curves, how do we transform other members of this family of curves into that one?

The first of these questions is the easiest and the one that will be attended to first.

Definition 6-1 The *standard normal probability curve* is that particular normal curve whose mean is 0 and whose standard deviation is 1.

Figure 6-4 is a geometric sketch of the standard normal probability curve. Table I (page 368) is an accurate tabulation of areas (probabilities) below this curve over intervals of finite length.

Figure 6-4

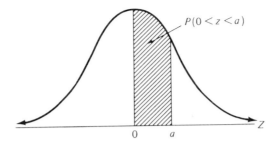

Table I records areas (probabilities) for intervals that start at $z = 0$ and end at $z = a > 0$. This is the only type of interval that is tabulated. With the curve's symmetry, plus maybe a little ingenuity, we can find areas corresponding to any interval of finite length along the Z-axis. The horizontal axis is traditionally labeled Z. The points of this real number line are traditionally called Z scores. The unit of measure on this real-number line is the standard deviation, which we have already stated is equal to 1.

In Fig. 6-4 we shaded a typical interval that would be tabulated in the table. Now turn to Table I on page 368. The leftmost column, under the title z, records whole standard deviation units and tenths of standard deviation units. The row at the top of the table records hundredths of standard deviation units. As we move along the Z-axis from $z = 0$ to $z = 1$, we travel 1 standard deviation unit. As we move from $z = 1$ to $z = 2$, etc., we travel 1 standard deviation unit.

The body of the table records, in decimal form, the probability (area) of obtaining a value in the finite interval from $z = 0$ to $z = a > 0$. Finally, the table records no probabilities beyond $z = 3.+$ —because to the right of $z = 0$ the total area (probability) is exactly .50 to begin with, and by the time we have traveled from $z = 0$ to $z = 3.+$, we have already accumulated .499+ of this area. Here the table stops recording area (probability) because there is so little area (probability) left.

You may find it hard to believe that the total area below any normal probability curve is 1, even though the curve goes on forever in both directions. Yet that is true! However, a formal proof of the fact is beyond the scope of our work here.

The following examples will help in learning to use the standard normal table.

Example 6-1 What is the probability of getting a z score from the standard normal distribution between the values $z = 1$ and $z = 2$? This is expressed: $P(1 \leq z \leq 2)$.

Solution

.4772
−.3413
─────
.1359

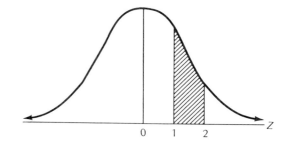

Note that because the area above a point is zero we could have asked for $P(1 < z < 2)$ in the previous example. Thus, the equal signs may be included or omitted.

Example 6-2 What is the probability of obtaining a z score greater than 1? This is expressed: $P(z > 1)$.

Solution

.5000
−.3413
─────
.1587

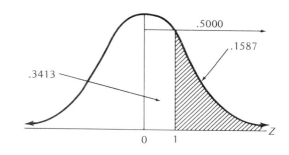

Example 6-3 Find $P(-1 < z < 2)$.

Solution

.4772
+.3413
─────
.8185

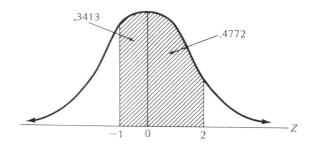

Example 6-4 What is $(1.35 < z < 2.14)$?

Solution

$$\begin{array}{r} .4838 \\ -.4115 \\ \hline .0723 \end{array}$$

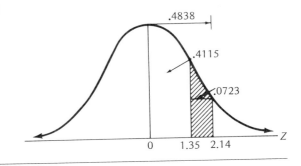

Example 6-5 What is $P(z < -3 \text{ or } z > +3)$?

Solution

$$2(.4987) = .4974$$

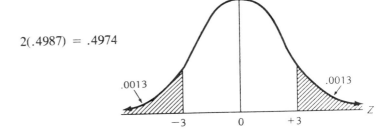

Example 6-6 Find $P(-2 < z < -1)$.

Solution

$$\begin{array}{r} .4772 \\ -.3413 \\ \hline .1359 \end{array}$$

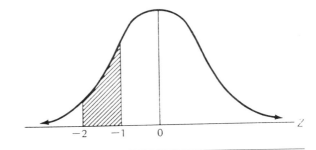

Example 6-7 Evaluate the probability that $P(-.82 < z < .27)$.

Solution

$$\begin{array}{r} .2939 \\ +.1064 \\ \hline .4003 \end{array}$$

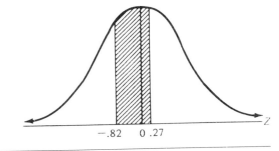

Note that none of the standard normal distribution examples had the flavor of actual applications. This is because although collections of data in nature, industry, and research are often normally distributed, they are not standard normally ($\mu = 0$ and $\sigma = 1$) distributed. To illustrate, consider the case of the distribution of the weights of people in Oregon. This distribution is approximately normally distributed but not standard normally distributed. If this distribution were ever to be standard normally distributed, the mean weight of people in Oregon would be 0. Even more interesting, some people in Oregon would have negative weights.

At this time, we offer an empirical rule that works for normal distributions only. For the applied case substitute $\bar{x}$ for μ and s for σ.

Empirical Rule

1. $\mu - \sigma$ to $\mu + \sigma$ will contain about .68 of our data.
2. $\mu - 2\sigma$ to $\mu + 2\sigma$ will contain about .95 of our data.
3. $\mu - 3\sigma$ to $\mu + 3\sigma$ will contain about 99.7 of our data.

A major appliance company keeps records on the life spans of its appliances. The company has thousands of such records and is confident that the distribution of life spans is approximately normal with a mean of 36 months and a standard deviation of 6 months. The company wants to determine the probability that an appliance has a life span that ranges from 27 months to 42 months. The company statistician realizes that her normal distribution is not the one tabulated in Table I of this text. How does she go about evaluating her normal distribution?

We can solve this problem directly in terms of the given normal distribution but instead we will opt for a more uniform approach that will serve for one and all normal distributions. *Given a normal distribution that is not standard normal, we can always transform it into the standard normal distribution. This will be our consistent approach.* We will use the standardizing formula of Section 2-2 of Chapter 2. This formula, as you recall, lets us change values from any distribution into a new distribution whose mean is 0 and whose standard deviation is 1. Here it lets us transform any normal distribution into the standard normal distribution (Table I). The object then is first to determine the interval(s) in the normal distribution that interests us and observe where the data within this interval(s) are transformed in the standard normal distribution. Once we have determined the corresponding interval(s) in the standard normal table, we look them up. It's that simple.

Our problem is somewhat analogous to that of the lion hunter who felt that searching out a lion in the wilds of Africa was an overwhelming task. Because of this, he went to a clearing through which flowed a sparkling stream and there he constructed a sturdy cage on 25 square feet of land. As he relaxed beneath a tree he realized one other problem remained—the lion he sought was in some other 25 square feet of land in Africa. Being clever as well as famous, our hunter then devised an equation that would transform any 25 square feet of land in Africa into the plot of land upon which he had constructed his cage. In fact, his equation was so clever that not only could he move any 25 square feet of land into the very clearing before him, but he could also move it back. This was very helpful for it meant that to capture any lion whatsoever he had merely to move it from its existing location to the interior of his cage. Then he could remove the lion, at his convenience, and return the land to its original state.

Just as our hunter could transform any 25-square-foot parcel of land that interested him into the interior of his cage, we can transform any normal probability curve that interests us into the standard normal probability curve at the end of the text (and back again). Let us illustrate this for the appliance company. The z score corresponding to 27 months is -1.50 ($z = (27 - 36)/6 = -1.50$). The z score corresponding to 42 months is 1 ($z = (42 - 36)/6 = +1$). Thus, the interval in the standard normal distribution that corresponds to the original interval of 27–42 months is $(-1.50, 1)$. This interval in the standard normal distribution has a probability of .7745 associated with it. Therefore, the probability of our appliance having a life span from 27 months to 42 months is .7745. Figure 6-5 is a geometric sketch of this.

Figure 6-5

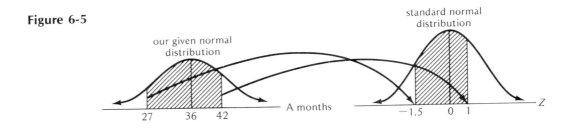

Example 6-8 We have a normal distribution A with $\mu = 2$ and $\sigma = 3$. What is $P(0 < a < 4)$?

Solution

a_i	z_i	Desired a Interval	Corresponding z Interval
0	$-2/3$	$0 < a < 4$	$-2/3 < z < 2/3$
4	$2/3$		

Figure 6-6 is a geometric sketch of this example.

Figure 6-6

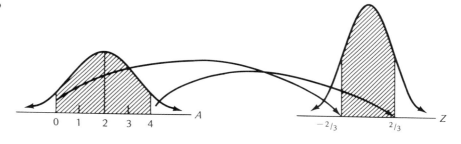

$$P(0 < a < 4) = P(-2/3 < z < 2/3) = 2(.2486) = .4972$$

For the previous example, let's observe the effects of the standardizing formula. The given normal distribution was symmetric about the line $a = 2$. We had to move the entire normal distribution two units to its left. The given normal distribution had a data deviation from its mean that was three times that of the standard normal distribution. We had to contract this normal distribution by a factor of $1/3$. Observe that the length of the original interval was four units and the length of the corresponding interval in the standard normal distribution is $4(1/3) = 1.33$ units long. All the area is accommodated in a smaller horizontal interval because the standard normal distribution rises to a greater vertical height at its mean value than does our initial normal distribution.

Example 6-9 In any given year, the U.S. Army has a very large sample of American youth on active military duty. The Army knows from its records that the mean height of a soldier is 69 in. Further, this distribution of data is approximately normal with a standard deviation of 3 in. The Army would like to know the percentage of its soldiers with heights from 63 in. to 75 in. inclusive.

Example 6-10 The Acqua Net Company manufactures nets to catch salmon. Biologists who work for the company estimate the mean length of the salmon population at 650 mm and the standard deviation at 25 mm. Experiments lead the company to believe that its nets can catch any salmon whose length is at least 635 mm. Assuming that the company experiments are correct and the population of lengths is normally distributed, what percentage of the salmon can be expected to escape through the net?

Solution Our z score = $(635 - 650)/25 = -.6$
$P(\text{length} < 635 \text{ mm}) = P(z < -.6) = .2743$

Example 6-11 The cost of an average toll call for 100,000 overseas telephone calls is $20 and the standard deviation is $5. Assuming the cost of toll calls is approximately normally distributed, find:
a. The number of tolls between $5 and $25.
b. The number of tolls greater than $30.

Solution
a. $P(5 < x < 25) = P(-3 < z < 1) = .8400$

$$\begin{array}{r} .4987 \\ +.3413 \\ \hline .8400 \end{array}$$

The answer is $(.84)(100,000) = 84,000$ toll calls
b. $P(x > 30) = P(z > 2) = .0228$
The answer is $(.0228)(100,000) = 2280$ toll calls

Example 6-12 Psychologists have determined that the distribution of intelligence quotients obtained by applying a particular test is approximately normally distributed with a mean of 100 and a standard deviation of 15. Estimate the percentage of the population that has an intelligence score of 120 or above.

Solution Our z score = $(120 - 100)/15 = 1.33$
$P(z > 1.33) = .0918$
Thus, we estimate that, based upon this particular test, approximately 9% of the population has an intelligence score in excess of 120.

6-1 THE NORMAL DISTRIBUTION

Another strength of the standardizing formula is the fact that it is reversible. This feature allows us to solve the formula for a in terms of z. What does this mean? It means that we can go backward if we want to. For a specified value of z, we can determine the value of a that would transform into it.

1. Using parameters: $a_i = \mu + \sigma z_i$ $1 \le i \le N$
2. Using statistics: $x_i = \bar{x} + s z_i$ $1 \le i \le n$

Example 6-13 We are given a normal distribution whose mean is -3 and whose standard deviation is $1/2$. What value transforms into the z value of 1?

Solution $a = -3 + (1/2)(1) = -2.5$

We note that we could get the same answer by first substituting the given values into the standardizing formula and then solving for a.

Example 6-14 In Example 6-9, what height value will transform into a z score of 2.5?

Solution $a = 69 + (3)(2.5) = 76.5$ in.

Example 6-15 Suppose that the life span of a type of light bulb is normally distributed with a mean of 100 hours and a standard deviation of 10 hours.
  a. Find the probability that a randomly selected bulb lasts between 92 and 124 hours.
  b. What length-of-life value produces a z score of 2?

Solution
  a. $P(-.8 < z < 2.4) = .4918 + .2881 = .7799$
  b. A life span of 120 hours.

Example 6-16 Find a value of z such that the probability of a larger z value is exactly .025.

Solution   The probability of a larger z value is the area to the right of the desired z value. The area between zero and our desired z value must be $.5000 - .0250 = .4750$. We now search the body of the standard normal table to locate the value .4750. By chance, this exact value appears. Our desired z value is 1.96.

Example 6-17 A national brand of automobile battery has a life span that is normally distributed with a mean of three years and a standard deviation of .5 year.

How long a full-warranty period should the company give its customers if it wants to be sure that less than 1% of the batteries are ever returned?

Solution We are interested in the 1% under the left tail of this normal distribution. Our z value is -2.33.

$$-2.33 = \frac{a - 3}{.5}$$
$$a = (.5)(-2.33) + 3$$
$$a = 1.835 \text{ years}$$

First, notice that this is nearly a 2-year warranty. Second, the actual answer is too awkward to put on a warranty. Why not round it down to 1.5 years? It looks good, sounds good, and hardly anyone will be back to use it.

EXERCISES A

6-1. Define a median value for the standard normal distribution.
6-2. Use the standard normal table to find: (Examples 6-1, 6-4, 6-5)
   a. $P(1.5 < z < 2.0)$
   b. $P(z < 1.75)$
   c. $P(z < -1.5 \text{ or } z > 1.5)$
6-3. Use the standard normal table to find:
   a. $P(z < 1.96)$
   b. $P(.38 < z < 1.24)$
   c. $P(-.43 < z < 1.76)$
   d. $P(-1.96 < z < 1.05)$
6-4. We have a standard normal distribution, find c such that: (Example 6-16)
   a. $P(z < c) = .025$
   b. $P(-c < z < c) = .95$
   c. $P(z < c) = .028$
6-5. We have a standard normal distribution, find c such that:
   a. $P(z < c) = .31$
   b. $P(-c < z < c) = .45$
6-6. If $\mu = 2$ and $\sigma = 3$, what value of a will yield a z value of -2? (Example 6-13)
6-7. If $\mu = 3$ and $\sigma = 1/2$, what value of a will yield a z value of 2?
6-8. Assume that you have a normal population such that $\mu = 2$ and $\sigma = 3$, what is the probability of obtaining a value in the interval (0, 1)? (Example 6-8)
6-9. Assuming that you have a normal population such that $\mu = 4$ and $\sigma = 1/2$, what is the probability of getting a value greater than 6?

6-10. A national standardized test has a mean score of 500 points and a standard deviation of 100 points. What percentage of the students receive a score between 400 and 600 points? Between 453 and 672 points? (Examples 6-9, 6-10)
6-11. In Exercise 6-10, we want to pass the top 80% of the students taking this national exam. What is the lowest passing grade?

6-12. The Army wishes to form a company of soldiers with heights above 6 ft. If the mean height of the soldiers is 68 in. with a standard deviation of 2 in. and the distribution is normal, how many soldiers out of 1200 will have heights above 6 ft?

6-13. A certain scale makes measurement errors that are normally distributed with a mean of 0 and a standard deviation of .1 oz. If we weigh an object on this scale, what is the probability that the weight will be correct to within .3 oz?

 6-14. The incomes of textile workers in the northeast region of our country are normally distributed with a mean of $11,000 and a standard deviation of $250. What percent of the incomes are between $10,500 and $11,500? Between $10,700 and $11,600?

 6-15. The lengths of the fish in your favorite lake are normally distributed with a mean of 8 in. and a standard deviation of 2 in. For breeding purposes, the fish and game department will let you catch and keep only the biggest 25% of the fish in the lake. What is the smallest fish that you can keep?

 6-16. The life span of a washing machine produced by a major company is normally distributed with a mean life span of 3.5 years and a standard deviation of 1.39 years. What is the probability that a randomly selected machine will last more than 4 years?

 6-17. Mortgage statistics collected by a bank indicate that the number of years that the average homeowner will occupy a house before selling is normally distributed with a mean of 6.3 years and a standard deviation of 2.31 years. What is the probability that a randomly chosen homeowner will sell before 3 years are up?

B **6-18.** We have a normal population with a mean of 100 and a standard deviation of 10. If there are 1359 pieces of data in the interval from 110 to 120, how many pieces of data are there in the entire population?

 6-19. If the service lives of electron tubes in a particular appliance are normally distributed, and if 92.5% of the tubes have lives greater than 2160 hours and 3.92% have lives greater than 17,040 hours, what are the mean and the standard deviation of the service lives?

 6-20. Students in schools A and B took the same national exam. In school A the mean was 48 and the standard deviation was 8. In school B the mean was 56 and the standard deviation was 12. We may assume that the grades in both schools are approximately normally distributed.
  a. What percentage of the students in school A were better than the average student in school B?
  b. What percentage of the students in school B were worse than the average student in school A?

 6-21. The average rainfall for the month of June in Arlington, Va., recorded to the nearest hundredth of an inch, is 3.63 in. Assuming a normal distribution with a standard deviation of 1.03 in., find the probability that next June Arlington receives:
  a. Less than .72 in. of rain.
  b. More than 2 in. but not more than 3 in.
  c. More than 5.3 in.

 6-22. The Super Tire Company knows that the life span of its top line of tires is normally distributed with a mean of 1.81 years and a standard deviation of .20 year. For sales purposes the company's dealers want an attractive warranty. They want to offer a guarantee of total replacement. However, they don't really want to replace many tires. How long should the warranty period be if they hope to replace at most .1% of their tires?

6-23. Start with the formula $z = (a - \mu)/\sigma$ and derive the formula $a = \mu + \sigma z$.

6-2 A DEEPER LOOK AT THE STANDARD NORMAL DISTRIBUTION (optional)

At this time, we would like to take a more penetrating look at the standard normal distribution (probability curve). Our equation is that of the normal probability curve with $\mu = 0$ and $\sigma = 1$. Figure 6-7 is a sketch of $n(x; 0, 1) = (1/\sqrt{2\pi})e^{(-1/2)x^2}$.

Figure 6-7

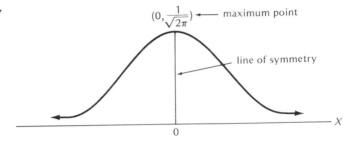

1. Because $e^{(-1/2)x^2} > 0$ for all x, it follows that $n(x; 0, 1) > 0$. Thus our standard normal probability curve lies entirely above the X-axis. (We left the equation in x rather than in z to match the defining equation.)

2. Observe! $e^{(-1/2)x^2}$ takes on its largest value when $x = 0$. So, $n(x; 0, 1)$ reaches its maximum value when $x = 0$, and $(0, 1/\sqrt{2\pi})$ is the highest point on the curve.

3. For $x = \pm a$, $n(x; 0, 1)$ has the same value. This shows that the curve is symmetric about the line $x = 0$.

4. Finally, as $x \to \infty (-x \to -\infty)$ $n(x; 0, 1) \to 0$, which implies that $n(x; 0, 1)$ is asymptotic to both extremes of the X-axis. This means that as x grows steadily larger $n(x; 0, 1)$ grows steadily smaller but remains greater than zero.

SUMMARY

There is no probability curve (distribution) in applied statistics that is more important than the *normal distribution*. There are an infinite number of normal distributions. Each normal distribution is entirely determined by specifying two parameters, μ and σ. Because there are an infinite number of normal distributions, we can never hope to study them. We study, in detail, one particular normal distribution called the *standard normal distribution*. This distribution has two important features: (1) $\mu = 0$, and (2) $\sigma = 1$. Table I allows us to quickly determine probabilities (areas) below specific portions of our standard normal distribution. Mathematicians have used calculus and computer facilities to derive this table for us. The probability of our raw data ever being standard normally distributed is nearly zero. Our job is to transform any normal distribution whatsoever into the standard normal distribution. It is possible to accomplish this by using the standardizing formula, $z = (a - \mu)/\sigma$ where $\sigma \neq 0$. Here a is the value in the normal distribution that we happen to be dealing with and z is the corresponding value in the standard normal distribution. The value of z is commonly referred to as the *z score* (meaning the z score of a). The standardizing formula is especially powerful because we may solve it for a in terms of z. That is, we are able to determine the original value of a when we are given its z score.

Can You Explain the Following?

1. normal probability curve (distribution)
2. standard normal probability curve (distribution)
3. empirical rule
4. z score
5. tail (left or right)

MISCELLANEOUS EXERCISES

6-24. Assuming that your data are standard normally distributed, what percentage of the data lies in the following intervals?
   a. $z > 1.26$
   b. $-2 < z < 1.30$
   c. $-3 < z < -2$

6-25. How much probability lies within an interval of 1.23σ about the mean of a standard normal distribution?

6-26. When dealing with any normal distribution the empirical rule tells us that $\mu - 2\sigma$ to $\mu + 2\sigma$ will contain about .95 of our data. Compare this fact with Chebyshev's Theorem for the same interval, which is true for any distribution whatsoever.

6-27. Find the initial value (a_i) in a normal distribution if $\mu = 2$, $\sigma = .5$, and $z_i = 1$.

6-28. Find the initial value (a_i) in a normal distribution if $\mu = 1.45$, $\sigma = .20$, and $z_i = .68$.

6-29. If x is normally distributed with $\mu = 100$ and $\sigma = 10$, find:
   (a) $P(x > 115)$ (b) $P(95 < x < 110)$ (c) $P(85 < x < 95)$

6-30. If the ages of heads of households in a particular planned community have a normal distribution with a mean of 30 years and a standard deviation of four years, find the probability that the head of a randomly selected household is:
   (a) over 40 years old   (b) between 25 and 40 years old

6-31. In a certain high-rent district, the monthly rental for apartments is approximately normally distributed with a mean of $390 and a standard deviation of

$60. Above what value is the highest 25% of the monthly rentals in this district?

 6-32. Error scores on a maze test for a particular strain of rats have an approximately normal distribution with a mean of 32 and a standard deviation of 8. In one experiment dealing with a control sample of six rats, one animal makes 78 errors. What argument can be advanced for discarding the results of this animal?

6-33. A food processor packages 12-oz containers of iced tea. The population of all 12-oz containers is normally distributed with a mean of 12 oz and a standard deviation of .02. What percentage of this population of containers will have weights ranging from 11.90 to 12.10 oz?

7 THE SAMPLING DISTRIBUTION OF MEANS FOR LARGE SAMPLES

7-1 INTRODUCTION

In Chapter 1 we introduced you to the Finast Sugar Company and its efforts at quality control. We now assume that Finast has been collecting its weekly sample of 31 5-lb packages of sugar for some time, and that the quality control department now wants to evaluate the collection of sample means it has obtained to study the long-run behavior of its product. Toward this end, a histogram was constructed from the sample means obtained during the first 25 weeks of sampling. In an effort to create the most informative histogram possible, the mean of these 25 means was calculated for use as the midpoint of the middle class in the histogram. The process was refined even further by calculating the standard deviation of these 25 sample means as an aid in selecting the most desirable class range. But when all the work was done (representing nearly a half year's worth of data), it was found that the 25 sample means did not yield an informative histogram.

The process was repeated using the first 50 sample means and the mean and standard deviations of these 50 means were calculated. In an effort to obtain even better results, it was decided to increase the sample size to the first 100 weekly samples of 5-lb packages of sugar. At this point, it was felt that the inclusion of additional sample means would not appreciably affect the histogram. Table 7-1 shows, in condensed form, the results. Those based on only 10 weeks have been included as an aid in trying to determine a long-run pattern.

	Mean of Sample Means	Standard Deviation of Sample Means
For 10 weeks	4.999 lb	.1000
For 25 weeks	5.001	.0100
For 50 weeks	5.000	.0010
For 100 weeks	5.000	.0001

TABLE 7-1

The following is the frequency distribution table for the first 100 sample means. The class range is .0005. Figure 7-1 shows the department histogram.

Classes	Frequency (f)
5.00125–5.00175	1
5.00075–5.00125	8
5.00025–5.00075	22
4.99975–5.00025	38
4.99925–4.99975	20
4.99875–4.99925	10
4.99825–4.99875	1
	100

What did the Finast quality control department learn from this process? For one thing, as the sample size increases (as more sample means are included), the mean of the means ceases to fluctuate and settles at 5.000 lb. Interestingly, this is the mean of the entire population of 5-lb packages of sugar. Further, as the sample size is increased, the standard deviation of this collection of means decreases toward zero. Finally, the symmetry of this histogram is noted. The department correctly projects that increasing the number of means beyond 100 and further refining the interval width will create a sequence of histograms that will steadily approach a normal distribution.

The most important sampling distribution we have is the sampling distribution of means. In practice, we learn all about the nature of the sampling distribution of sample means for large-size samples ($n \geq 30$), and once we understand this distribution we collect only one sample and hence only one sample mean. To analyze this one sample mean we use our knowledge of the distribution of means.

Figure 7-1

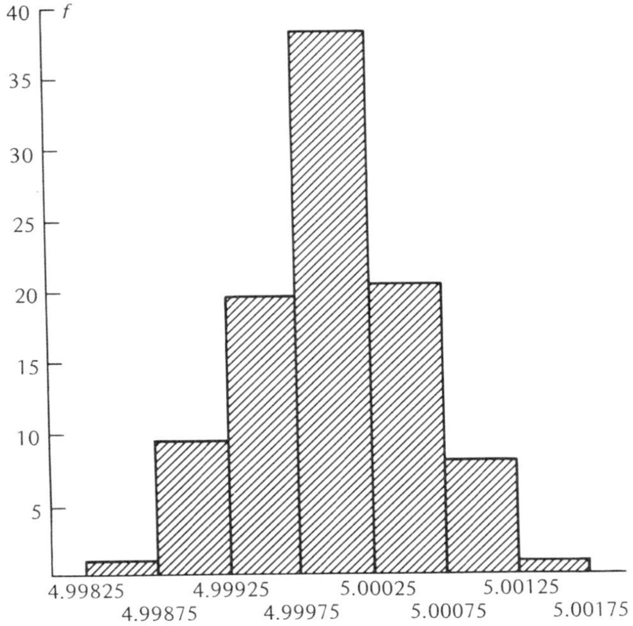

[Histogram with x-axis labels: 4.99825, 4.99875, 4.99925, 4.99975, 5.00025, 5.00075, 5.00125, 5.00175; y-axis f from 0 to 40]

Theorem 7-1 relates the mean and the standard deviation of the sampling distribution of means (denoted $\mu_{\bar{x}}$, $\sigma_{\bar{x}}$, respectively) to the mean and the standard deviation of the original population. The theorem is stated without proof.

Theorem 7-1 When we take a sample of size $n \geq 30$ from an infinite population or from a finite population with replacement:

a. $\mu_{\bar{x}} = \mu$ b. $\sigma_{\bar{x}} = \dfrac{\sigma}{\sqrt{n}}$

This theorem is both important and remarkable. Specifically, it tells us that the mean of all of the sample means is exactly the mean of the population, and it relates the standard deviation of the set of means to the standard deviation of the original population. Observe that n will never be zero. Because the mean of the sample means approaches a constant value as n increases in size, there must be very little fluctuation within the sample mean data itself. Thus, it is not surprising to observe that as n becomes infinite, the standard deviation of the sample means tends toward zero. Example 7-1 illustrates this theorem. For purposes of illustration we have kept all the sample sizes equal.

Example 7-1 From the population $A = \{1, 2, 3\}$ select all possible samples of size $2(n = 2)$ with replacement and verify Theorem 7-1.

Solution

Set of Samples	Corresponding Means
1, 1	1.0
1, 2	1.5
1, 3	2.0
2, 1	1.5
2, 2	2.0
2, 3	2.5
3, 1	2.0
3, 2	2.5
3, 3	3.0
	18.0

$\mu_{\bar{x}} = (1/9)(18) = 2$; $\mu_A = (1/3)[1 + 2 + 3] = 2$ also.

$\sigma_{\bar{x}}^2 = (1/9)(3) = 1/3$; $\dfrac{\sigma_A^2}{2} = \dfrac{(1/3)[1+1]}{2} = 1/3$ also.

(Recall that σ^2, the population variance, was defined in Chapter 2.)

$$\therefore \sigma_{\bar{x}} = \dfrac{\sigma_A}{\sqrt{2}}$$

Example 7-2 The distribution of weights of ball bearings at the Aconda Manufacturing Plant has a mean of 22.40 oz and a standard deviation of .048 oz. For samples of size 36, drawn with replacement, what are the mean and standard deviation of the sampling distribution of means?

Solution

$$\mu_{\bar{x}} = 22.40 \text{ oz}, \sigma_{\bar{x}} = \dfrac{.048}{6} = .008 \text{ oz}$$

Table 7-2 is an abbreviated record of sample data representing 260 weekly samples (five years) of 31 5-lb bags obtained by the Finast Sugar Company. All measurements have been recorded to the nearest .001 of a pound.

(1) Sample Number	(2) Sample Mean	(3) Sample Standard Deviation
1	5.000	.000
2	5.001	.001
3	5.000	.000
4	5.023	.002
5	5.001	.001
.	.	.
.	.	.
.	.	.
256	5.001	.001
257	5.034	.037
258	5.000	.000
259	5.001	.001
260	5.000	.000

TABLE 7-2

Column (2) of Table 7-2 represents a sampling distribution of means and column (3) represents a sampling distribution of standard deviations. The curve in Fig. 7-2 is the result of smoothing out the frequency polygon of the sampling distribution of 260 means. Observe that it is a normal distribution with mean = 5.000 lb.

Figure 7-2

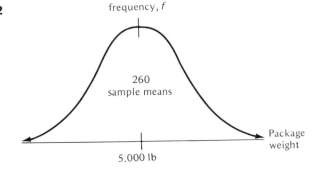

138 THE SAMPLING DISTRIBUTION OF MEANS FOR LARGE SAMPLES

Definition 7-1 When we are dealing with the sampling distribution of means a sample of 30 or more pieces of data ($n \geq 30$) is considered a *large sample*.

Theorem 7-2 (The central limit theorem for means) If n is large, the sampling distribution of means can be approximated closely by a normal distribution with mean $\mu_{\bar{x}} = \mu$ and $\sigma_{\bar{x}} = \sigma/\sqrt{n}$.

This theorem represents one of the cornerstones of applied statistics. Because its proof requires calculus, it is beyond the scope of our work here. The following observations are related to this theorem.

1. Three points should be noted: (a) $n \geq 30$; (b) we are dealing with the sampling distribution of means; and (c) this distribution is approximately normally distributed. It is difficult to say precisely how large n must be before the theorem applies, but unless the population has a very unusual shape, the approximation will be good when $n \geq 30$.

2. By introducing our standardizing formula, which allows us to transform any normal distribution into a standard normal distribution, we may obtain an equivalent version of this theorem. That is, for large sample sizes the sampling distribution of the statistic

$$\frac{\bar{x} - \mu}{\sigma/\sqrt{n}}$$

is approximately standard normally distributed.

3. It is a mathematical fact, which can be derived, that should our population be normally distributed, the theorem applies regardless of the size of the sample.

Consider the following applied example.

Example 7-3 T.W.E. Airlines wishes to analyze the cost of an inflight movie on its new U.S. air route. On the basis of its international air routes, it knows that a fully loaded plane will average 50 people who are willing to pay to see an inflight movie, with a standard deviation of 8. These international parameters are assumed valid for the new U.S. air route.

T.W.E.'s cost analyst has a problem: In order for the company to break even financially on inflight movie expenses, 46 paying customers are needed on any fully loaded plane. The analyst will check to see what happens on the first 36 fully loaded planes on this new U.S. air route. Assuming that the international parameters are valid, what is the probability that for a sample of 36 planes our sample mean will be less than 46 customers? Discuss the problem.

Solution $\mu = 50$, $\sigma = 8$, $n = 36$
The z test statistic $= (46 - 50)/(8/\sqrt{36}) = -3$
Therefore, $P(\text{sample mean} < 46) = P(z < -3) = .0013$

Conclusion There appears to be very little chance that the company will lose money on the movie.
a. What if the sample mean exceeds 46? We will believe that the movie is a potential money maker.
b. What if the sample mean exceeds 46 and the planes aren't fully loaded? This is also fine.
c. What if the sample mean is less than 46? We must take stock of the situation. Just how severe is the financial loss? Then, again, we may wish to take a loss here because of increased revenue from another source, such as increased beverage sales.

Definition 7-2 The *standard error* of a statistic is the standard deviation of the sampling distribution of that statistic.

In this chapter, we have been discussing the standard error of the means, $\sigma_{\bar{x}}$. When the standard error of a statistic is small we feel that our sampling is *precise*. We say we are sampling with *precision*.

7-2 TESTING STATISTICAL HYPOTHESES ABOUT μ

Statistical decision making plays a very important role in our lives. Each of us is faced daily with personal decisions that require predictions about the future. A metallurgist seeks to use the results of experiments to test the tensile strength of a new metal alloy. A stockbroker would like to predict the future behavior of the stock market. The government is concerned with predicting the national debt or with making projections about trade balances with other countries. A housekeeper wishes to know whether detergent A will be more effective than detergent B in the family washing machine. The inferences made in statistics are based on the collection and testing of relevant data.

Most of the time, statistical hypotheses will be conjectures that deal with population parameters. The number of statistical hypotheses that we can actually test is finite. Fortunately, they cover most of the key questions that we would like answered in our applied work. In this section we deal with statistical tests of hypotheses about the mean of a population.

The object of statistical testing is to try to negate the statistical hypothesis (conjecture). The researcher tests a statistical hypothesis by testing a sample. It is a process of indirect reasoning. For example, if a coin lands and we wish to know which side is facing downward, we reason that it must be the opposite of the side facing upward. Should the coin land with the tail side facing upward, then the side we cannot see must be the head side. The fact that we are trying to nullify our statistical hypothesis motivates the following definitions.

Definition 7-3   A statistical hypothesis called a *null hypothesis* and denoted H_0 (H nought) is an assumption we are testing.

Definition 7-4   The hypothesis that we accept if the null hypothesis is rejected is called the *alternative hypothesis*. It is denoted H_a (H sub a).

Frequently, the alternative hypothesis is the negation of the null hypothesis; however, this is not always true, for we may have H_0: $\mu = 50$ and H_a: $\mu < 50$.

Often a researcher will design an experiment in such a way that the alternative hypothesis represents the researcher's personal belief. This means that if the researcher should secure a sample that is significantly different from the null hypothesis, it will lend credence to that personal belief.

The reader should be aware that although just one example can prove something to be false, any number of examples will not prove a statement true.

Example 7-4 H_0: $\mu = 5$ oz H_a: $\mu \neq 5$ oz

Example 7-5 H_0: Paul Williams who plays guard for our football team weighs 200 lb.
               H_a: Paul Williams doesn't weigh 200 lb.

Example 7-6 H_0: We will receive a grade of B in economics.
               H_a: We will not receive a grade of B in economics.

Example 7-7 H_0: Blondes have more fun.
H_a: Blondes don't have more fun.

Note that a colon is used in the foregoing examples. This is the form we employ to state our null and alternative hypotheses.

Definition 7-5 A *test statistic* is a real number computed from a sample(s) that is used to reject (or not to reject) our null hypothesis.

We are going to make an assumption about the value of μ. This assumption represents a statistical hypothesis (H_0). We will then select a sample and use the sample mean to determine the accuracy of our assumption. Our test statistic here is the sample mean.

Now, evaluating an assumption about μ on the basis of one sample is at best a concept that should be treated cautiously. *If our sample mean agrees with the statistical hypothesis about μ—that is, it matches H_0—the testing procedure is over.* However, accepting a statistical hypothesis based on only one sample is a risky business. We do not say, "Accept H_0." Instead we say, "There is insufficient evidence to reject H_0." Memorize this statement for it is what we write every time that our sample fails to nullify H_0, that is, when the test statistic matches the conjecture stated in H_0.

If we had to characterize the entire statistical testing procedure with one word, we would pick "reject." To reject H_0 is equivalent to accepting H_a.

Example 7-8 We are economists working for the U.S. government and we wish to test the null hypothesis that the mean salary of teenage males is $6000 (yearly).

H_0: $\mu = \$6000$ H_a: $\mu \neq \$6000$

W-2 forms of 36 teenage males are randomly selected from our files. This sample has an average of $6000, and a standard deviation of $50. Notice that the sample mean matches the population mean assumed in H_0.

Conclusion There is insufficient evidence to reject H_0.

The rejection of a null hypothesis is a question of probability. We never reject a null hypothesis with 100% certainty. Statistical hypothesis testing is a form of applied gambling.

Definition 7-6 The *level of significance* corresponds to the probability with which we are willing to risk rejecting a true null hypothesis. The level of significance is denoted by the Greek letter alpha (α).

Although one might test at various levels, our task has been simplified. Years of experience have proved that the best applied results are obtained by using one of two specific values for α: .05 and .01. These are the conventional statistical testing levels, and special tables are available to facilitate their use.

What does it mean to say that we reject a null hypothesis at the .05 level of significance? We are saying that, based on our sample results, H_0 can be true only in the most unbelievable .05 or less of all possible cases. Our answer of "reject" is a probability gamble based on a sample assumed to be representative of the population. Remember our null hypothesis about μ? Here, when we say reject H_0 at the .05 level of significance, we mean that the difference between the value of the sample mean and the value of μ in H_0, with which we are comparing it, is so great that it could happen only in the most unbelievable .05 or less of all possible cases.

Definition 7-7 When a sample leads to the rejection of a null hypothesis we say that the test results are *significant*.

Note that this definition is just the opposite of what our intuition might lead us to believe.

Once we have selected a level of significance (either .05 or .01), we use this choice of probability to set up a critical region below our standard normal distribution. (In this chapter we use the standard normal probability curve for testing hypotheses.) Because, for convenience we are adopting the fixed procedure of always transferring from a normal distribution to the standard normal distribution, our test statistic will always be a z score of a sample mean, not the mean of the sample itself. The critical region is placed below the standard normal distribution in such a manner that should the z score of our sample mean land here it would represent an extreme difference between the value of the sample mean and the value of μ with which we are comparing it in H_0. When the value of the z score of the sample mean lands within the critical region, we conclude that H_0 should be rejected at this previously determined level of significance.

Definition 7-8 A *two-tail critical region* is equal in size to the level of significance and is located in two equal parts below the right tail and the left tail of the probability curve.

It is common practice for some texts to refer to the critical region as the *region of rejection. For this chapter, the probability curve is the standard normal distribution*. In future chapters we will meet and use other probability curves.

Definition 7-9 The points that border the critical region are called *critical points*.

When we conduct a test of H_0 using a two-tail critical region, we say we are conducting a *two-tail test;* see Fig. 7-3. Thus, when we run a two-tailed test at the .05 level of significance, the critical region consists of the extreme .025 beneath the left tail of the standard normal distribution and the extreme .025 beneath the right tail of the standard normal distribution. A .01 level of significance would be split into .005 and .005.

Figure 7-3

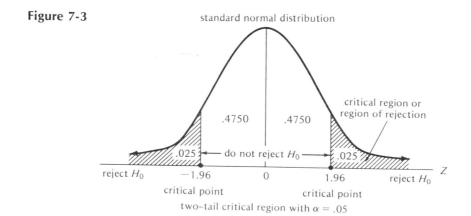

two-tail critical region with $\alpha = .05$

Example 7-9 Let us return to our economist of Example 7-8 who is testing the null hypothesis that the average teenage male worker in the U.S. earns exactly $6000. Assume our sample data are different and we conduct a statistical test at the .05 level of significance. Because we are testing the exact value of $6000, we will execute a two-tailed test. The positive critical point value of 1.96 is obtained by looking up .4750 in the standard normal table. From the symmetry of the standard normal distribution we determine that -1.96 is the other critical point.

Assume that a different sample of 36 teenage male W-2 forms had been randomly selected. This time the sample mean is $7000 and the sample standard deviation is $1000. The value of the sample mean doesn't match the value of μ conjectured in our null hypothesis. There is a chance for rejection! However, is the difference between the two values extreme enough? It looks extreme: $7000 − $6000 = $1000, which is a lot of money for most of us. But is the difference significant? We obtain a concrete answer by calculating the z-score of the sample mean and observing whether or not it lies within our two-part critical region. Our sample mean corresponds to one value in a sampling distribution of means that is approximately normally distributed (the central limit theorem). Look at the standardizing formula now:

$$z = \frac{\bar{x} - \mu_{\bar{x}}}{\sigma_{\bar{x}}}$$

We have the value of the sample mean. What about the value of $\mu_{\bar{x}}$? Fortunately, Theorem 7-1 tells us that this is equivalent to μ. The value of μ has just been given (conjectured) in our null hypothesis. How about the value of $\sigma_{\bar{x}}$ (which equals $\sigma/\sqrt{n}$)? Now we have a real problem. We need to know the value of σ in order to test a value of μ. Recall there is an order involved in calculating the standard deviation. We must calculate μ prior to calculating σ^2. We are in a bind.

Our predicament is not unlike that of a runner who wishes to compete in a 100-yd dash and discovers that the starter refuses to shoot the starting gun until he is told who will win the race. If our runner knew who was going to win, he probably wouldn't bother to run. What does our runner do? He looks over the field of athletes, and since he knows their previous performance times, he can make an approximation. He tells the starter who he believes will win. The gun goes off. The race is run. If our runner has correctly predicted the winner, fine. If his approximation was wrong, he simply apologizes to the starter and tells him who actually won.

7-2 TESTING STATISTICAL HYPOTHESES ABOUT μ

Well, we don't know σ and we can't calculate it. Further, the only approximation available to us is the sample standard deviation, and so we plug it into the standardizing formula and keep going. This may throw our answer off, but not by much, if we have a large-size sample.

When we use s as an approximation for σ in our problem, we obtain:

$$z = \frac{\$7000 - \$6000}{\$1000/\sqrt{36}} = 6 > 1.96, \text{ our critical point}$$

Our conclusion is to reject H_0. There is a significant difference between the value of the sample mean and the conjectured value of μ in H_0 and the difference was extreme at the .05 level of significance (see Fig. 7-4). Our z test statistic is located six standard deviations to the right of zero in our standard normal distribution.

Figure 7-4

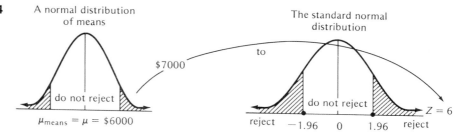

Whether we say, "reject H_0" or find "there is insufficient evidence to reject H_0," we are in danger of making a possible error. And the key word here is "possible." In our previous problem we were not worried that our conclusion might be in error because our z test statistic was so far beyond the critical point value of 1.96 standard deviations. We do worry about making an error in stating our conclusion when our test statistic is close to the critical point value. We will not attempt to answer the classic student question of how close is close because we haven't a simple, convincing answer.

Definition 7-10   A *type I error* is committed if the null hypothesis is rejected when it is true. The probability of a type I error is alpha (α).

Definition 7-11   A *type II error* is committed if the null hypothesis is not rejected when it is false.

The probability of making a type II error is denoted by the Greek letter beta (β). Determining the probability value α in a statistical test is quite easy, as we control this value. Evaluating β is much more difficult and is beyond the scope of this text. When the value of the test statistic does not fall in the rejection region we withhold judgment and say, "There is insufficient evidence to reject H_0." This eliminates the possibility of making a type II error. (Note that we use Roman numerals—the usual convention.) When we select a .05 level of significance, we are limiting the probability of a type I error to at most .05.

Definition 7-12 A *one-tail critical region* is equal in size to the level of significance, α, and is located under the right tail (or left tail) of the probability curve.

When we conduct a test of H_0 using a one-tail critical region, we say we are conducting a *one-tail test*. Figure 7-5 illustrates a one-tail critical region based on the value of $\alpha = .01$. The value of this critical point is found by locating .4900 in Table I. We see that it lies between .4898 and .4901. We select the z score in the table closest to our desired value (in this case, it is .4900) and we come up with a critical point value of 2.33.

Figure 7-5

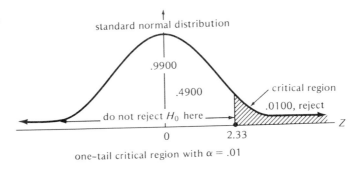

one-tail critical region with $\alpha = .01$

In the event that we are trying to locate a value in the body of our table that is equidistant between two table values, we consider either table value as applicable. This will be the case when we conduct a one-tail test (assume right-tail) and the area in the critical region is $\alpha = .05$. We now check Table I and see that the desired value of .4500 is equidistant between the table values .4495 and .4505. Again, we will assume that a z score of either 1.64 or 1.65 is applicable.

7-2 TESTING STATISTICAL HYPOTHESES ABOUT μ

How do we know whether we wish to conduct a one-tail test or a two-tail test? This is an important question, but it is not always easy to answer. Here is what to do. Read the problem carefully to see if it is asking us to determine if something is "less than" a specific value or "greater than" a specific value (this may be implied, not actually stated). If the answer is yes, we conduct a one-tail test with this less-than or greater-than statement as our alternative hypothesis. If the answer is no, we conduct a two-tail test.

An educator decides to select a random sample of 100 U.S. college students to test their IQ level. From her experience, she believes that 110 is a separating value, and she desires to test this at a .05 level of significance.

Examples 7-10a–e illustrate by comparison different situations that might confront our educator. You should not infer that the testing process is a repetitive process; that is, if we don't like the results the first time, we repeat the testing process in the hope that we will find something more suitable to our taste on the next try.

Example 7-10a The educator decides to test the null hypothesis that $\mu = 110$.

$$H_0: \mu = 110 \qquad H_a: \mu \neq 110$$

A random sample is obtained and it is found to have a mean of 110 and a standard deviation of 5.

Solution Stop the testing process. The sample mean agrees with the value of μ in H_0. It provides no evidence whatsoever to disbelieve the conjecture of $\mu = 110$ in H_0.

Example 7-10b The educator again decides to test the null hypothesis that $\mu = 110$.

$$H_0: \mu = 110 \qquad H_a: \mu \neq 110$$

A random sample is obtained and it is found to have a mean of 120 and a standard deviation of 30.

Solution Because our sample mean differs from the value of μ in H_0, there is a possibility of rejection. Our educator is testing the null hypothesis that $\mu = $ exactly 110. A sample can fail to support this (cause rejection) by being either significantly less than 110 or significantly greater than 110. A two-tail test is called for. The critical points are ± 1.96.

The z test statistic $= (120 - 110)/(30/\sqrt{100}) = 3.33 > 1.96$

Conclusion Reject H_0.

Let's comment on this example. The conclusion is understood to mean reject H_0 at the .05 level of significance. An alternative way of stating the conclusion would be to say, "The test results are significant at the .05 level." If we were to rewrite this example using the .01 level, we would be unable to state a conclusion without first looking up the critical point value in our standard normal table. At the .01 level the critical point value is 2.57 (or 2.58). Thus, our conclusion would be to reject H_0 at the .01 level as well.

Example 7-10c Our educator believes that the average IQ of American college students is greater than 110. Because she would like to have her personal feelings accepted, she makes them the alternative hypothesis.

$$H_0: \mu = 110 \qquad H_a: \mu > 110$$

A random sample is selected and it is found to have a mean of 120 and a standard deviation of 30.

Solution Because our sample mean is greater than 110, there is a possibility of rejection. In essence, our educator is running the test with a stipulation that only sample mean values in excess of 110 can fail to support H_0. This is a one-tail test situation with the entire critical region below the right tail. The single critical point is 1.64 (or 1.65).

Figure 7-6

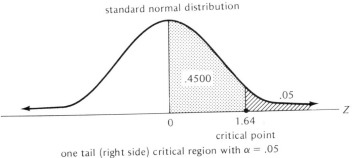

one tail (right side) critical region with $\alpha = .05$

The z test statistic $= 3.33 > 1.64$

Conclusion Reject H_0 (accept the educator's feelings).

Let's discuss this example. H_0 is always a statement of strict equality, regardless of whether we are running a one-tail or a two-tail test. Why is this true in the one-tail test? We cannot substitute an infinite number of values for μ into our standardizing formula for the computation of the z test statistic. We settle for the one value of μ that is most revealing—the boundary value of $\mu \leq 110$. *Thus, our thinking may be H_0: $\mu \leq 110$, but what we actually write is H_0: $\mu = 110$.*

Example 7-10d Our educator believes that the average IQ of American college students is greater than 110. Because she would like to have her personal feelings accepted, she makes them the alternative hypothesis.

$$H_0: \mu = 110 \qquad H_a: \mu > 110$$

A random sample is selected and it is found to have a mean of 105 and a standard deviation of 5.

Solution Stop the testing process! All sample means less than or equal to 110 support H_0. That is, they provide no sample evidence to cause us to doubt the truth of H_0. There is insufficient evidence to reject H_0.

Example 7-10e Our educator believes that the average IQ of American college students is less than 110. Because she would like to have her personal feelings accepted, she makes them the alternative hypothesis.

$$H_0: \mu = 110 \qquad H_a: \mu < 110$$

A random sample is selected and it is found to have a mean of 105 and a standard deviation of 30.

Solution Those sample mean values that match H_0 are 110 or greater. Because our sample mean is not one of these values, there is a chance for rejection to take place. Our critical point value is -1.64.

The z test statistic $= (105 - 110)/(30/\sqrt{100}) = -1.67 < -1.64$

Conclusion Reject H_0.

Let's talk about this example. Reject is a correct conclusion, but our test statistic is close to the critical point value. We should worry about a type I error. If we had any reservations, it would be best to retest.

> *Five Points About Hypothesis Testing*
>
> 1. H_0 is always a statement of strict equality, because:
>   a. A two-tail test is a test of exactly one conjectured value.
>   b. A one-tail test is a test of an infinite number of conjectured values. However, we can never hope to test an infinite number of values and so we settle for testing the conjectured value in H_0 that is most revealing. This is the lower (or upper) boundary value of all values in H_0.
> 2. H_a is always $>$, $<$, or $\neq$.
> 3. A *rule of thumb:* To locate the critical region in a one-tail test, look to see where the greater-than or less-than sign points in the alternative hypothesis.
> 4. What do we do if the value of our test statistic equals the value of the critical point? We have never seen this happen in a research situation. Assume it does happen. An elementary, but artificial, solution would be to define the critical point value as being within (or not within) the critical region.
> 5. We should observe that there is an element of "role playing" involved in setting up a null hypothesis.

Consumer complaints are pouring into the U.S. Department of Weights and Measures office in Omaha. The people are complaining that a large regional milk distributor is underfilling his gallon containers of milk. Because there are so many documented claims, the department decides to test their validity. At the same time, the milk distributor decides to begin a testing process to exonerate himself from consumer allegations. In the first of the next two examples we play the role of the U.S. government conducting the test. In the second example, we play the role of the milk distributor. It is interesting to see the contrasting conclusions based on the same sample.

Example 7-11a As testers for the U.S. government, we will see if we can accept the consumer claims on the basis of a random sample of 100 gallon containers of milk. A gallon of milk contains 3.785 liters. We will test at the .01 level of significance.

$$H_0: \mu = 3.785 \text{ liters} \qquad H_a: \mu < 3.785 \text{ liters} \text{ (consumer)}$$

A random sample of 100 gallon containers is selected and found to have a mean of 3.475 liters with a standard deviation of 1.5 liters.

7-2 TESTING STATISTICAL HYPOTHESES ABOUT μ 151

Solution We are running a one-tail test. The critical point value is -2.33. The sample means that match H_0 are those values greater than or equal to 3.785 liters. We have a chance for rejection; see Fig. 7-7.

Figure 7-7

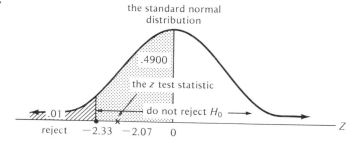

The z test statistic $(3.475 - 3.785)/(1.5/\sqrt{100}) = -3.1/1.5 = -2.07 - 2.07$
$\not< -2.33$

Conclusion There is insufficient evidence to reject H_0.

Now let's make some comments. Our sample mean was below average but not significantly below average. The difference between the sample mean and the value conjectured as μ was not so great as to place us in the most unbelievable .01 of all possible cases. If the test were conducted at the .05 level of significance, the critical point value would have been -1.64 and the test conclusion would be to reject.

We must note in passing that a sample of size 100 may be too expensive to collect. If money is a research problem, and it frequently is, we may have to settle for a more practical sample size of, say, $n = 40$. Most likely, this value of n will not appreciably alter our test result.

Example 7-11b The milk distributor asks us to conduct a test to exonerate him. He maintains that if he is guilty of anything it is that he overfills his gallon containers. In this case, also, we will test at the .01 level of significance.

$$H_0: \mu = 3.785 \text{ liters} \qquad H_a: \mu > 3.785 \text{ (milk company)}$$

A random sample of 100 gallon containers is selected and it is found to have a mean of 3.475 liters with a standard deviation of 1.5 liters.

Solution We are running a one-tail test. The critical point value is 2.33. Because the sample mean matches H_0, it offers no evidence to make us doubt the conjecture stated in H_0.

Conclusion There is insufficient evidence to reject H_0.

It is true that the sample does show some short filling. However, at the .01 level, the government can't support the consumer claim of short filling, nor can the distributor support his feelings of generosity. At the .05 level, things are different. The government can support the claim of short filling and the distributor cannot support his claim of generosity.

For these examples, then, it is clear that the use of a two-tail test is inappropriate.

We shall present, and answer, three interesting questions about hypothesis testing. First, if we conduct a statistical test at the .01 level and the conclusion is to reject H_0, what, if anything, can be said at the .05 level? Reject H_0 immediately at the .05 level. The .01 critical region is always contained within the .05 critical region. Second, if we conduct a statistical test at the .05 level and our conclusion is that there is insufficient evidence to reject H_0 at this level, what, if anything, can be said at the .01 level? There is insufficient evidence to reject H_0 at the .01 level. Finally, if we conduct a statistical test at the .01 level and our conclusion is that there is insufficient evidence to reject H_0 at this level, what, if anything, can be said at the .05 level? We cannot determine the answer without knowing the value of the test statistic.

Testing a Null Hypothesis About μ

Tools: (1) The sampling distribution of means, (2) the central limit theorem, (3) one sample, and (4) the standardizing formula.

Steps:
1. Specify a level of significance (.05 or .01).
2. Decide whether it is to be a one-tail or a two-tail test. (State the null hypothesis.)
3. Remember that to reject H_0 is to accept H_a.
   (Recall that H_0 is a statement of strict equality.)
4. Obtain a large sample.
5. Evaluate your test statistic. (It is the z score of the sample mean.)
6. Compare your test statistic with the value of the critical point in question.
7. State a test conclusion.

Example 7-12 Engineers at the Delux Motor Company have developed a new braking system for the company's entire line of cars. They believe that, on the average, this new braking system will bring a car traveling at 40 mph to a full stop in 25 feet. They would like to test their belief on a sample at the .01 level. If the average stopping distance of the cars in the sample being tested proves anything less than 25 feet, the engineers will, of course, be delighted. Common sense dictates that they run a one-tail test. Rejection should occur only for exceptionally small z test (negative) values (indirect reasoning). The entire region of rejection will be placed under the left tail of the standard normal distribution; see Fig. 7-8.

Figure 7-8

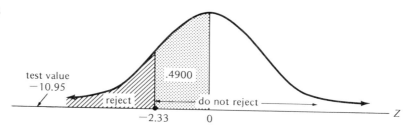

Solution

$$H_0: \mu = 25 \text{ feet} \qquad H_a: \mu < 25 \text{ feet (Delux Motor Company)}$$

A random sample of 100 cars is tested and it is found that the mean stopping distance is 22.7 feet with a standard deviation of 2.10 feet. The critical point value is -2.33.

The z test statistic $= (22.7 - 25.0)/(2.10/\sqrt{100}) = -10.95 < -2.33$

Conclusion Reject H_0. Statistically, the engineers can feel 99% confident that the new braking system is performing correctly. Perhaps they should feel even more confident. After all, the test value lies deep inside the critical region. The sample result actually promotes great confidence in the new braking system.

Example 7-13 A psychologist believes that the average person sleeps 7.50 hours per night. She decides to test this null hypothesis at the .05 level of significance. For testing purposes, a random sample of 100 college students is selected and it is found that the mean sleeping time is 7.40 hours, with a standard deviation of 1.05 hours.

Solution H_0: $\mu = 7.50$ hours (psychologist) H_a: $\mu \neq 7.50$ hours.
The critical points are ± 1.96.

The z test statistic = $(7.40 \text{ hours} - 7.50 \text{ hours})/(1.05 \text{ hours}/10) = -.95 > -1.96$

Conclusion There is insufficient evidence to reject H_0. The sample matched the psychologist's feelings, which were stated in the null hypothesis. Do college students and their sleeping habits really portray the "average person"? If this is not the case, there is a defect in the design of this experiment.

The concepts we have presented in this chapter will appear again and again in future chapters. It is hoped that a careful look at these concepts now will result in a more economical use of time later.

Among those concepts that will reappear in later chapters are null hypothesis, alternative hypothesis, test statistic, a level of significance (.05 or .01), critical region, critical point, two-tailed test, one-tailed test, type I error, and type II error.

EXERCISES A

7-1. In each part of this exercise we will be given H_0. Write the corresponding H_a. (Examples 7-4 and 7-5)
   a. Our planet is the result of an exploding star.
   b. Retarded youngsters perform better in integrated classes.
   c. $x = 2$. We are executing a two-tail test.

7-2. State the three symbols that we may find in H_a.

7-3. We will be given H_a. Write the corresponding H_0.
   a. $\mu \neq 25$ lb
   b. $\mu < 25$ lb

7-4. We will be given H_a. Write the corresponding H_0.
   a. $\mu > 2.78$ in.
   b. $\mu \neq 2.78$ in.

7-5. If $\mu = 40$, $\sigma = 100$, and $n = 100$, find $\mu_{\bar{x}}$ and $\sigma_{\bar{x}}$. (Example 7-2)

7-6. If $\mu = 6.85$, $\sigma^2 = 100$, and $n = 25$, find $\mu_{\bar{x}}$ and $\sigma_{\bar{x}}$.

7-7. We are dealing with the population of 10-year-olds in New York City. Assume that the mean IQ of this population is 100 and the standard deviation is 15. We draw samples of 36 with replacement.
   a. What is the mean of the sampling distribution of means?
   b. What is the standard error of the sampling distribution of means?

7-8. On the basis of years of testing, owners of the Micro Manufacturing Plant, which produces steel wire, believe that the mean strength of the wire is 200 lb and the standard deviation is 10 lb. If they are using samples of 49, what is the mean of the sampling distribution of means? What is the standard error of the means?

7-9. If the standard error of the means is 20 and our sample size is 25, what is the value of the standard deviation of the population?

7-2 TESTING STATISTICAL HYPOTHESES ABOUT μ

 7-10. To test the null hypothesis that the average student height at Southwestern University is 68 in., we selected a random sample of 49 students. It was found that this sample has a mean of 67 in. and a standard deviation of 3 in. Test this null hypothesis at the .05 level.

 7-11. Because we have just finished looking for and purchasing a new home, we believe we know from experience that the amount the average American pays for a first home is $50,000. The state assessment bureau lets us test this hypothesis by allowing us to obtain a random sample of 100 recently purchased homes. The sample has a mean of $50,050 with a standard deviation of $200. Test this null hypothesis at the 1% level. (Example 7-8)

 7-12. Because of consumer complaints, the better business bureau in a suburban community selects several jars of a popular brand of jelly for a check on the advertised net weight. The label advertises the net weight as 12 oz. A sample of 36 jars shows the weight as 11.92 oz with a standard deviation of .30 oz. Test the consumer complaints at the .01 level. (Examples 7-10e and 7-12)

 7-13. Brian and Steven have a paper route. During the January and February snows their father drives them in the family car. For each of these winter weeks they average $5.00 in tips. They both believe that as soon as they do the route entirely by bike their weekly tips will increase substantially. For the first 36 weeks after they start to "bike it," they average $5.06 in tips with a standard deviation of .10. At the .05 level, were the boys correct in their judgment? (Example 7-10c)

 7-14. Assume that a sociologist examines a large apartment complex to see if the number of persons per family unit differs significantly from the national mean of 4.80 persons. From interviews with 100 of the 1500 families in the complex, it is found that the average is 5.30 persons with a standard deviation of .80. At the .05 level, is there a significant difference between the sample mean and the national mean?

 7-15. We believe that, on the average, a name brand of fuse must carry over 20 amperes before it will burn out. To test this belief, we obtain a random sample of 64 fuses and carefully test them. It is found that the average number of amperes needed to burn out these fuses is 20.25 with a standard deviation of .75. Test this belief at the .05 level.

 7-16. A random sample of $n = 36$ persons is subjected to a stimulus and observed for the time delay in a specified reaction. If the sample mean and standard deviation are 2.2 seconds and .057 second, respectively, test the null hypothesis that the mean time delay is $\mu = 1.6$ seconds against the alternative hypothesis that the mean time delay is smaller than 1.6 seconds. Test at the .01 level.

 7-17. A hospital claims that the average length of patient confinement is five days. A study of the length of patient confinements on $n = 36$ people shows a mean of 4.2 days with a standard deviation of 5.2 days. Do these data present sufficient evidence to support the hospital claim? $\alpha = .01$.

CUMULATIVE REVIEW

 7-18. A dealer in wholesale fruit lots wishes to determine whether the mean sugar content per orange shipped from a certain grove is less than .025 lb. A random sample of 49 oranges is found to have a mean of .023 with a standard deviation of .003 lb. At the 1% level, do the data present sufficient evidence to indicate that the mean is less than .025?

7-19. Welfare applicants in a major U.S. city complain that it takes too long to process their claims. The city offers them a new streamlined application process. To see if this new process is working, a sample of 100 applications is followed from beginning to end. The sample average is 42 days with a standard deviation of 20 days. Previously, the average was 48 days. Test at the .05 level to see if there is evidence that this new streamlined process is really working better.

  7-20. The Hecht Vacuum Cleaner Manufacturing Company claims that the average noise level of its latest model is at most 75 decibels. We don't believe their claim, and we would like to substantiate our belief. A random sample of size 48 is obtained and the mean of this sample is 76.4 decibels with a standard deviation of 3.6 decibels. Test at a .01 level of significance to see if we can verify our personal feelings.

7-3   THE TWO TYPES OF PARAMETER ESTIMATES

Definition 7-13   For a given parameter the corresponding sample statistic is called a *point estimate* of that parameter: (a) $\bar{x}$ is a point estimate of μ; (b) s is a point estimate of σ.

Geometrically, both the population parameter, of which we don't know the value, and the corresponding statistic, which we have evaluated, may be represented by points on the number line. We use the point on the line which we can locate (the statistic) to approximate the point on the line whose location we don't know (the parameter).

Point estimates of parameters are important, and are used every day in statistical analysis. They may be thought of as our best guess. Their weakness lies in the fact that we are unable to determine the relative position of the point estimate (the statistic) to the point being estimated (the parameter); see Fig. 7-9.

Figure 7-9

7-3 THE TWO TYPES OF PARAMETER ESTIMATES

A second type of parameter estimate is an interval estimate. For a specified probability, we calculate a finite interval in which we believe the parameter point lies. The specified probability represents how much confidence we have of finding our parameter point within the estimating interval, and so the interval estimate is called a *confidence interval* estimate. Thus, we will calculate finite interval estimates of parameters such that we are 95% or 99% confident of finding the parameter point within.

Definition 7-14 An estimate of a population parameter formed by trapping this estimate between two points on the real number line is called an *interval estimate* of a parameter.

Typical examples of Definition 7-14 include the biologist who would like to have an interval estimate of the mean diameter of skulls for a particular species of bird, the sociologist who wants an interval estimate of the mean family size for a specific population of people, the anthropologist who would like an interval estimate of the mean number of calories consumed daily by an indian tribe in Colorado, or the insurance company that needs an interval estimate of the average life span of 20-year-old males in the United States.

In yet another example, a construction executive is planning a multimillion dollar development and is trying to estimate the average number of weeks needed to complete this development. He would be happy to have a finite interval estimate of this number in which he is 95% confident. Or a major food chain would find it of great value to have an interval estimate of the average weekly gross sales for a planned store obtained from the data of previously opened stores. Imagine how it would help the owners to plan (or to drop the plan) for this new store by having an interval estimate of the weekly gross sales in which they are 99% confident. They could plan to absorb the potential 1% loss by spreading it throughout the entire chain of stores.

Theorem 7-3 The end points of a $100(1 - \alpha)\%$ confidence interval estimate of μ are $\bar{x} \pm z\, s/\sqrt{n}$, where $\bar{x}$, s, and n are determined by a sample and where z is a positive z score determined by specifying a probability value for this interval.

Proof We are dealing with the sampling distribution of means and $n \geq 30$. Begin with a symmetric interval (zero is the midpoint) on the Z axis (stan-

158 THE SAMPLING DISTRIBUTION OF MEANS FOR LARGE SAMPLES

dard normal distribution) of the desired probability and evaluate the corresponding interval in our population.

$$-z \leq \frac{\bar{x} - \mu}{s/\sqrt{n}} \leq z, \, z > 0$$

$$-zs/\sqrt{n} - \bar{x} \leq -\mu \leq zs/\sqrt{n} - \bar{x}$$

$$\bar{x} + zs/\sqrt{n} \geq \mu \geq \bar{x} - zs/\sqrt{n}$$

Therefore,

$$\boxed{\bar{x} - zs/\sqrt{n} \leq \mu \leq \bar{x} + zs/\sqrt{n}}$$

Example 7-14 We wish to obtain a 95% confidence interval estimate of the mean salary for the population of working American males. For this purpose, a random sample of size 36 is obtained and it is found to have a mean of $7000 and a standard deviation of $1000; see Fig. 7-10.

Solution The starting symmetric interval on the Z axis is (−1.96, 1.96).

$$\$7000.00 \pm \frac{(1.96)(\$1000.00)}{6}$$

$$\$6673.33 \leq \mu \leq \$7326.67$$

Figure 7-10

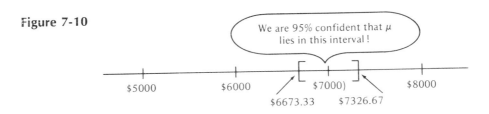

What we learn from this last example and Fig. 7-10 is that hypothesis testing and confidence intervals are related. If we were to test a conjecture about μ in H_0 to the effect that μ is a value from $6673.33 to $7326.67 inclusive, it would be supported by the sample at the .05 test level.

Example 7-15a The Mica Furniture Company wishes to estimate the amount that the average family spends to furnish a living room. The company would like a 95% confidence interval estimate of this mean. For their purpose, a sample of size 36 is obtained and is found to have a mean of $1975 and a standard deviation of $100.

Solution $1975.00 ± [(1.96)($100.00)]/6
Therefore,

$$\$1942.33 \leq \mu \leq \$2007.67$$

Example 7-15b Refer to the preceeding example 7-15a and calculate a 99% confidence interval for the Mica Furniture Company.

Solution $1975.00 ± [(2.58)($100.00)]/6
Therefore,

$$\$1932 \leq \mu \leq \$2018$$

Figure 7-11 shows both the 95% and 99% confidence intervals for the Mica Furniture Company. *Observe that the length of the 95% confidence interval is smaller than the length of the 99% confidence interval. This is always the case. Next observe that we used the value of the point estimate in our calculation of the interval estimate of our parameter. This too is always the case.*

Figure 7-11

Example 7-16 a. A clinic wishes to determine a 95% confidence interval for the mean weight loss, in months, for a new experimental diet. In order to do this, the clinic obtains a random sample of 64 people. These people have a mean weight loss of 12 lb with a standard deviation of 1 lb.
b. Find the 99% confidence interval.

Solution

a. $12 \pm \dfrac{(1.96)(1)}{8}$

   $12 \pm .245$

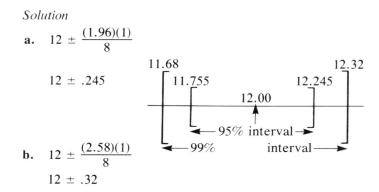

b. $12 \pm \dfrac{(2.58)(1)}{8}$

   $12 \pm .32$

Example 7-17 Find a 95% confidence interval for the mean load of a population of steel cables produced by the Armstrong Cable Corporation. Base your work on a random sample of 81 cables with a mean load value of 11.06 tons and a standard deviation of .71 ton.

Solution The end points of the interval are $11.06 \pm .15$ ton.

It should be kept in mind that it is the sample that varies in our present discussion and not the value of the population mean, μ—which is an unknown constant. As the mean and the standard deviation of a sample change, so too will the end points of the confidence interval determined by that sample change.

Figure 7-12

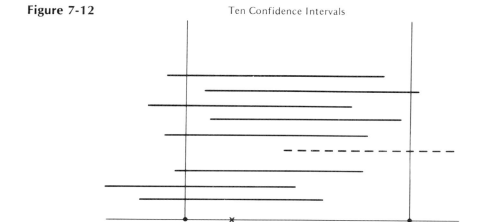

Ten Confidence Intervals

When we say that we are 95% confident we are saying (in the long run) that we expect collections of 100 samples to yield confidence interval estimates of μ that are accurate 95% of the time.

Figure 7-12 shows the confidence interval estimate of μ as obtained in Example 7-17, as well as nine similar interval estimates of the same value of μ. Note that the dotted interval does not contain the parameter point for μ.

Example 7-18 We wish to estimate the book value of the inventory of our client, Helex Corporation. We have been careful to select a random inventory sample, gathered during the pricing and extension of inventory. Our data are as follows:

Total number of items in the inventory—12,700
Sample size—400
Sample mean—$96
Sample standard deviation—$28

a. Estimate the total value of the inventory.
b. Compute a 95% confidence interval estimate of μ.
c. Use your solution to part (a) to obtain an interval estimate of the total value of the inventory in which we can be 95% confident.

Solution
a. ($96.00)(12,700) = $1,219,200.00
b. $96.00 \pm [($28.00)(1.96)]/\sqrt{400}$ (assume the distribution of item prices is normally distributed.)
$93.256 < \mu < $98.744
c. We will use the end points of the interval from part (b).
($93.256)(12,700) < total inventory value < ($98.744)(12,700)
$1,184,351.20 < total inventory value < $1,254,048.80

EXERCISES A

7-21. A machine company wishes to calculate a 95% confidence interval for the mean diameter of its ball bearings. For this purpose, a sample of $n = 36$ is drawn and is found to have a mean of .52 in. with a standard deviation of .01 in. Find this interval. (Example 7-14)

7-22. The Ace Brick Works would like a 99% confidence interval estimate of the mean weight of its bricks. A random sample of 36 bricks is selected from the ovens and is found to have a mean of 2.10 lb with a standard deviation of .2 lb. Find this confidence interval. (Example 7-15b)

7-23. A new drug is hailed as a fast treatment for an old disease. Forty-nine patients treated with this drug showed signs of full recovery in an average of nine days with a variance of four days. Find the 95% confidence interval for the average number of days required for full recovery under this new drug treatment.

7-24. It is desired to determine a 99% confidence interval for the mean weight of

3-month-old babies. In order to do this, a sample of 64 babies is selected and it is found that the mean weight is 14.25 lb with a standard deviation of .26 lb.

 7-25. An accountant working for a department store takes a random sample from its very extensive files, and finds that the amounts owed in 100 overdue accounts have a mean of $27.63 and a standard deviation of $6.74. Find a 95% confidence interval for the true mean of the amounts owed in all of the department store's accounts.

 7-26. Given that the lengths of the skulls from 100 fossil skeletons of an extinct species of birds have a mean of 5.73 cm and a standard deviation of .34 cm, find the 95% confidence interval for the mean length of the skulls of this species of birds.

7-4 A PARTIAL ANSWER TO THE QUESTION OF SAMPLE SIZE

What should we choose as our sample size? This is a perennial student question. What you are seeking is a formula that will yield an exact value of sample size, preferably under all conditions. We know of no such formula, and so this question is not easy to answer.

We begin by making our sample large. We talk to friends and researchers in our field. We consider copying sample sizes used in previous, successful research efforts. This may not be the best choice of sample size but it does represent a starting point.

Still the question is one of formula. We will offer some partial solutions. Each will represent "a give-and-take answer." The give will be in the form of precise restrictions upon what we are doing. The take will be a formula for determining the value of n under these precise restrictions. The question of determining sample size boils down to giving before receiving. Even after we have determined a value for n, it will be a value in which we are only 95% or 99% confident.

Here is our first partial solution.

The mechanics for determining an appropriate sample size for μ comes from our discussion of confidence interval estimates of μ. This is what motivated us to leave Theorem 7-3 in the main text rather than defer it to the optional section. From the first line of this theorem's proof, we proceed to obtain a distance measurement of $\bar{x}$ relative to μ.

1. $-z \leq (\bar{x} - \mu)/(s/\sqrt{n}) \leq z$, $z > 0$; to begin we assume $n \geq 30$.
2. $-z\, s/\sqrt{n} \leq \bar{x} - \mu \leq z\, s/\sqrt{n}$
3. $|\bar{x} - \mu| \leq z\, s/\sqrt{n}$ (Note that the bars denote absolute value.)

This last line tells us that the distance between $\bar{x}$ and μ is less than or equal to $z\,s/\sqrt{n}$. Now set $z\,s/\sqrt{n}$, in turn, less than a desired distance, d, of $\bar{x}$ to μ. Thus we are assured with $(1-\alpha)$ probability that $|\bar{x}-\mu| \le z\,s/\sqrt{n} \le d$, which implies

$$n \ge \frac{z^2 s^2}{d^2}$$

At this point, let's take stock of our situation. The object is to solve the inequality $z\,s/\sqrt{n} < d$ for n. We pick the tolerance distance d. We select a probability level that, in turn, determines the value of our critical point z. Now all we need to know is the value of s. To obtain a value for s, we need a previously obtained sample from our population.

Another method of estimating s is illustrated in Exercise 7-27.

Example 7-19 We are dealing with the average life span of a population of television picture tubes. How large a sample must we obtain to be 95% confident that the difference between our sample mean and μ will be less than 20 hours?

Solution A previously obtained sample from the same population yielded a standard deviation of 100 hours. Our critical point has a value of 1.96.

$$n \ge \frac{(1.96)^2 (100)^2}{20^2} = 96.04$$

Thus, the answer is $n = 97$ (or larger).

Figure 7-13

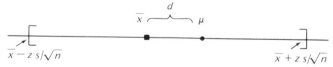

A point and an interval estimate of μ for a specified tolerance distance d.

Example 7-20 Solve Example 7-19 again for the new tolerance distance of $d = 10$ hours.

Solution $n \ge [(1.96)^2 (100)^2]/10^2 = 384.16$
Therefore, $n = 385$
Observe that cutting the tolerance distance in half does not double the value of n.

Example 7-21 A businessman is interested in the average price of copper in the United States. He estimates the standard deviation to be $.02. How large a sample should he obtain to be 95% confident that the sample mean estimates μ with an error (within a tolerance distance) of less than $.01?

Solution

$$n \geq \frac{(1.96)^2(\$.02)^2}{(\$.01)^2} = 15.37$$

Therefore,

$$n = 16$$

EXERCISES A

7-27. Assuming that we know that our population is normally distributed, explain why the range/4 would serve as an estimate of σ. This can be used as our estimate of s.

7-28. In Example 7-21, find a value of n in which we are 99% confident. Hint: Let $z = 2.58$.

7-29. Union officials are concerned about reports of inferior wages being paid to employees of a large corporation under their jurisdiction. They wish to be 99% confident that their selection of sample size will yield a mean value within $.50 of the true mean wage of employees of this corporation. They know from the data of a similar corporation that the standard deviation of employees wages is approximately $1.50. (Example 7-19)

7-30. A dermatologist wishes to learn the average number of minutes that a person spends taking a shower. How large a sample must be selected if the dermatologist wishes her sample mean estimate to be within 2 minutes of μ? We would like to give our answer with 99% confidence. The dermatologist estimates $\sigma = 7.39$ from a previous research project. (Hint: s is the point estimate of σ.)

7-31. For testing purposes, we have taken a sample of 36 drinks from a soft-drink machine. The sample mean is found to be 7.50 oz, the advertised value, with the sample standard deviation equal to .04 oz. Based upon the data from this initial sample, how large a sample must be selected if we wish to be 95% confident that our sample mean will be within .01 oz of the true mean?

B **7-32.** We are working on minimum federal wage legislation for restaurant employees. In order to evaluate this legislation intelligently, we want to obtain a sample of restaurant employees in which we are 99% confident that our sample mean will differ from the population mean by no more than $.05.

An initial sample of 100 employees, 10 from each of 10 large restaurant chains, is obtained. Of the 10 employees selected from each company, two are randomly selected from each of five job levels. The sample mean is $3.30 (hourly wage) and the sample standard deviation is $.30.

On the basis of these initial sample data, what should the sample size be in order to ensure that our sample mean will be within $.05 of the true mean?

SUMMARY

This chapter concerns large samples ($n \geq 30$). When dealing with the sampling distribution of means an amazing discovery is made called the *Central Limit Theorem:* that this sampling distribution is approximately normally distributed. Further, that $\mu_{\bar{x}} = \mu$, and $\sigma_{\bar{x}} = \sigma/\sqrt{n}$. These last two facts allow us to standardize any mean in this sampling distribution.

With the Central Limit Theorem as an indispensable aid it is possible to discuss: (1) a statistical hypothesis about μ, and (2) obtain an interval estimate of μ. Testing a statistical hypothesis is a process of indirect reasoning. We try to nullify the statistical hypothesis and hence it is called a *null hypothesis* (denoted H_0). The hypothesis that is accepted in the event the null hypothesis is rejected is called the *alternative hypothesis* (denoted H_a). Experience has shown that the best test results are obtained by using either a 5% test-level or a 1% test-level. These test levels are commonly spoken of as *alpha-levels* (Greek letter α). The alpha level corresponds to the probability of rejecting a null hypothesis that is true. Such an error is called a *type I error*. A *type II* error occurs if we accept a null hypothesis that is false. This text opts for a conservative approach throughout that prevents us from making a type II error. This is accomplished by the simple device of never accepting H_0. Should the sample data fail to cause a rejection of H_0 (it supports H_0), the phrase "there is insufficient evidence to reject H_0" is always written. Thus, we hesitate to accept a null hypothesis based on one sample, but do not hesitate to reject a null hypothesis. Exactly what does rejection of H_0 at the α-level mean? It means that the difference between the statistical conjecture in H_0 and the sample data is so great that it can only happen in the most unbelievable α%. Thus, rejecting a statistical hypothesis does not mean that it cannot happen; it means that it is highly unlikely that it can happen based on the sample data. Rejecting a null hypothesis is a form of applied gambling.

This chapter dealt with estimating the parameter μ. For every parameter, including μ, there are two types of estimates: (1) *a point estimate,* and (2) *an interval estimate.* The sample mean is the point estimate of μ (our sample being drawn from the population of which μ is the mean). In general, to obtain a point estimate for any population parameter, calculate the corresponding sample statistic. For example, s is the point estimate of σ. Armed with a sample mean and the Central Limit Theorem, it is possible to calculate a *confidence interval* estimate of μ. A confidence interval is always a finite interval in which I believe μ lies with $100(1 - \alpha)$% confidence. The two most common confidence intervals are 95% and 99%.

The chapter ends with a formula for determining sample size. This formula will keep the sample mean within a specified distance, d, of μ, with a probability of $(1 - \alpha)$. I found that to determine this sample size one must first

166 THE SAMPLING DISTRIBUTION OF MEANS FOR LARGE SAMPLES

specify: (1) the desired distance, d, (2) the desired probability, and (3) provide a sample variance. This last condition is especially interesting for it implies that one must first obtain a sample in order to determine a sample size.

Can You Explain the Following?

1. Central Limit Theorem
2. standard error
3. large sample
4. $\mu_{\bar{x}} = ?$
5. $\sigma_{\bar{x}} = ?$
6. null hypothesis (H_0)
7. alternative hypothesis (H_a)
8. critical region (or region of rejection)
9. the two α test levels
10. "significant result"
11. one-tail test
12. two-tail test
13. type I error
14. type II error
15. point estimate of μ
16. interval estimate of μ
17. critical point(s)
18. test statistic
19. formula for determining n

CUMULATIVE REVIEW

MISCELLANEOUS EXERCISES

7-33. Show that $\sigma^2 = n\sigma_{\bar{x}}^2$.
7-34. We always use a point estimate in the determination of an interval estimate. True or false?
7-35. Which is always larger—a 90% confidence interval for μ or a 95% confidence interval for μ?
7-36. State the central limit theorem in your own words.
7-37. Given H_a, write the corresponding H_0.
(a) $\mu > 50$ (b) $\mu \neq 50$
7-38. A null hypothesis (H_0) has been rejected at the 1% level. What, if anything, could be said if the testing were to be redone at the 5% level?

7-39. A clinic believes the mean weight loss for a new diet to be 12.5 lb. In order to test $\mu > 12.5$ lb, a random sample of size 64 is selected. This sample has a mean of 12.75 lb and a standard deviation of 1 lb. Test at the 5% level.
7-40. Use the sample data of Exercise 7 to compute the 95% confidence interval for μ.

7-41. The chief executive of Handsome Clothes, Inc., is required to make a decision regarding the opening of a new store. For a store to be successful in a particular area, the executive feels, the average family income in the area must exceed $18,000. It is decided to test at the 1% level. A sample of 225 families in the area reveals an average income of $18,400 with a standard deviation of $500. Will the executive open this new store?
7-42. Use the sample data of Exercise 7-41 to compute the 99% confidence interval for μ.

7-43. We believe that the average professional baseball player has a batting average of 225. To test this null hypothesis, we obtain a random sample of 36 players. The sample has a mean of 240 and a standard deviation of 10. Test the hypothesis at the 5% level.

 7-44. Use the sample data of Exercise 7-43 to compute the 95% confidence interval for μ.

 7-45. A market research firm wishes to estimate the mean income in a marketing area with 95% confidence. It is felt necessary to keep the sample mean estimate within $100 of μ. The market research firm knows from previous surveys of this marketing area that the standard deviation is about $10. Find the required sample size.

 7-46. The production manager of Hefty Bag Inc. wants to obtain a sufficiently large sample to estimate a critical measurement value to within .05 cm of the true mean value. The manager wishes to be 99% confident in using this sample size. On the basis of a pilot sample run, he estimates σ to be .25 cm. Find the required sample size.

8 THE SAMPLING DISTRIBUTION OF MEANS FOR SMALL SAMPLES

8-1 DEGREES OF FREEDOM

Three identical boxes are set before us. We are instructed that we are to place an integer into each of the three boxes. We have three completely free choices. We say that we have three *degrees of freedom*.

Next we are given the same three boxes. This time one of the boxes contains the integer 5. We are told that we may place the integer we desire in one of the remaining two boxes. We have two completely free choices. We say that we have two degrees of freedom.

We are given the same three boxes again. This time one box contains the integer 5 and we are told that the sum of the three integers which we pick must be 16. The result is that we have one free choice. Of course, we may decide to use this one free choice in either of the two open boxes. Once this decision is made, the third integer that must be placed in the last remaining box is determined. We have one degree of freedom.

Finally, we are again given the same three boxes. One box still has the integer 5 inside. The sum of the three integers selected must still be 16. In addition, we must now pick an integer twice as big as 5 and place it in one of the remaining boxes. Indeed, one box has not yet been assigned an integer value. However, the restrictions are so severe that we really have no freedom of choice for the remaining integer value. Do you see what the missing value is? We have zero degrees of freedom.

$$\boxed{10} + \boxed{5} + \boxed{?} = 16$$

Example 8-1 Mary and Jane enter a candy store and each girl buys a bar of candy. There are two degrees of freedom. Now we discover that Mary always makes the same choice of candy as Jane. The result is one degree of freedom.

8-1 DEGREES OF FREEDOM

Definition 8-1 The number of *degrees of freedom* that we have in a problem is the number of independent choices that we have. Degrees of freedom is commonly abbreviated as d.f. or ν (This is the Greek letter nu).

Example 8-2 We are in the U.S. Mint and we are told that we may select any four coins we like. Here, $\nu = 4$. Now we are told that one coin we select must be a dime and another coin must be all copper. Here, $\nu = 2$.

Example 8-3 We are given the slope-intercept form of a straight line. $y = mx + b$. Now we are restricted by being told that all straight lines must have a slope of 1. The letter b still remains independent. Here, $\nu = 1$.

Example 8-4 We are free to pick any three integers that we like and name them x, y, and z. Here, $\nu = 3$. Next, we are told that $y = 2x$. Here, $\nu = 2$. Finally, we are told that $y = 2x$ and $x = 3z$. Here, $\nu = 1$.

Example 8-5 A student has taken five history exams, computed his mean grade, and then misplaced his exams. What is the least number of exams that he must find in order to recall all five grades? The answer is four. The computed mean serves as a restriction.

Example 8-6 We have four coins in our pocket: a penny, a nickel, a dime, and a quarter. How many degrees of freedom do we have in tipping if we wish to tip exactly two coins in each of the following cases?
  a. We wish to give the penny in our tip. Here, $\nu = 3$.
  b. We wish to give the quarter in our tip. Here, $\nu = 3$.
  c. We wish to give the nickel and the dime in our tip. Here, $\nu = 0$.
  d. We wish to give either the dime or the quarter in our tip. Here, $\nu = 1$.

In Examples 1-23 and 1-24 we verified that

$$\sum_{i=1}^{n} (x_i - \bar{x}) = 0$$

For any collection of data, the sum of the differences from the mean is zero.
Assume we have three values: $x_1 = 2$, $x_2 = 4$, and $x_3 = ?$. In addition, the mean, $\bar{x} = 4$. What is the value of x_3?

$$\sum_{i=1}^{3}(x_i - 4) = 0$$

$$(x_1 - 4) + (x_2 - 4) + (x_3 - 4) = 0$$

$$(2 - 4) + (4 - 4) + (x_3 - 4) = 0$$

$$-2 + \quad 0 + \quad x_3 - 4 = 0$$

$$x_3 = 6$$

Assume we have four values: $x_1 = ?$, $x_2 = 7$, $x_3 = 13$, $x_4 = 3$, and $\bar{x} = 7$. Find x_1.

$$\sum_{i=1}^{4}(x_i - 7) = 0$$

$$(x_1 - 7) + (x_2 - 7) + (x_3 - 7) + (x_4 - 7) = 0$$

$$(x_1 - 7) + (7 - 7) + (13 - 7) + (3 - 7) = 0$$

$$x_1 - 7 + \quad 0 + \quad 6 - \quad 4 = 0$$

$$x_1 = 5$$

In general, given the value of $\bar{x}$,

$$\sum_{i=1}^{n}(x_i - \bar{x}) = 0$$

has $(n - 1)$ degrees of freedom. That is, we are free to assign any value we desire to any $(n - 1)$ of the x's with the result that the nth x value, the remaining one, is completely determined.

The number of degrees of freedom equals the number of choices less the number of restrictions imposed upon these choices. The computed statistics being utilized as estimates of parameters will be a main source of restrictions in our work. Degrees of freedom is an important concept and one that is not easily understood. We hope that the several preceding examples have helped to provide a degree of intuitive understanding.

8-2 THE t-DISTRIBUTION

We know from the last chapter that, for large samples, the statistic $z = (\bar{x} - \mu)/(\sigma/\sqrt{n})$ is approximately normally distributed. Seldom, if ever, are we so fortunate as to know the value of σ^2. For samples of size $n \geq 30$, s^2 provides a good estimate of σ^2 with little variation from sample to sample.

For large samples, the statistic $z = (\bar{x} - \mu)/(s/\sqrt{n})$ is approximately standard normally distributed. It is this fact that motivated our earlier definition of a large sample. When our sample size is small ($n < 30$), the value of s^2 tends to fluctuate from sample to sample. Because of this the statistic $(\bar{x} - \mu)/(s/\sqrt{n})$ can no longer be approximated by a standard normal distribution.

HISTORICAL NOTE

The Granger Collection
WILLIAM SEALY GOSSET

An investigation of just what happens to the sampling distribution of this statistic for $n < 30$ was undertaken by William S. Gosset (1867–1937). Gosset was educated at Oxford in chemistry and mathematics. In 1899 he went to work for the Arthur Guinness and Sons Company, a brewery in Dublin, Ireland, where the management had just begun a policy of employing scientists to make tests and recommendations. Confronted with the problem of gathering data about the brewing process, he recognized that small samples yielded a value of s that failed to serve as a good point estimate of σ, and that the normal-curve model was inappropriate for small samples.

Gosset obtained the records of the Bertillon measurements of thousands of British criminals. The Bertillon System, the first scientific method for criminal identification, was largely based on the classification of body measurements. Gosset recorded these measurements on cards, and in so doing built an approximately normal distribution from which he could conveniently draw samples of various sizes. By drawing large numbers of samples of a given size, he was able to establish empirically the sampling distribution of means for a particular sample size. He discovered that the distribution of means differed markedly from the normal distribution for samples of size less than 30. There was actually a family of such distributions with each sampling distribution directly dependent on the size of the sample, n.

Today, mathematicians are able to represent the family of empirical distributions of Gosset by a family of continuous probability curves called *t distributions*. We know of no particular significance attached to the choice of the letter t; it seems to have been selected because it was available.

In 1908, Gosset wanted to publish his work on the t distributions. However, because of a company rule that Guiness employees could not publish (which probably was established to protect brewing secrets), he had to publish his results in 1908 under the pseudonym of *student*. For this reason, t distributions are sometimes called student t distributions.

Gosset was highly regarded by the Guinness Company, and in 1937, just a few months before his death, he was appointed head brewer.

The continuous probability t distributions are based on the assumption that our samples are being selected from a *normal population*. This is so because only in the normal distribution is it true that the mean and variance are independent.

With each t distribution there is associated the number of degrees of freedom, $\nu = n - 1$. The loss of one degree of freedom (the minus 1) is because we know the value of the sample mean. (See Section 8-1.) The following lists the main characteristics of t distributions. Figure 8-1 compares the standard normal curve and a few standard t distributions.

Figure 8-1

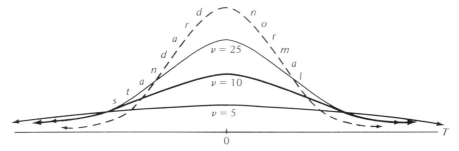

Basic Characteristics of t Distributions
1. The graph of each t distribution lies entirely above the horizontal axis.
2. Each t distribution is symmetric (bell-shaped) about line $t = 0$.
3. For each t distribution $\nu = n - 1$.
4. Each t distribution steadily approaches the positive and negative extremes of the horizontal axis.
5. The mean of each t distribution is 0.
6. When $\nu \geq 3$ the variance of each t distribution may be computed by the formula: $\sigma^2 = \nu/(\nu - 2)$.

Note that the last characteristic is especially interesting. The variance of a t distribution is given in terms of the number of degrees of freedom, ν. It is always greater than one, which means that there tends to be slightly more probability under the tails of the t distribution than under the corresponding tails of the standard normal distribution. As n gets larger, ν gets larger; as a result the variance of the t distribution (which becomes the normal distribution) approaches one. This is not an unexpected result.

Sample Size, n	ν	σ^2
4	3	3
5	4	2
10	9	1.29
15	14	1.17
20	19	1.12
30	29	1.07

Example 8-7 When $\nu \geq 3$, the formula for the variance is $\sigma^2 = \nu/(\nu - 2)$. Explain why this formula would not make good sense if we insist on using the value $\nu = 1$.

Solution Should we attempt to use this formula for the case $\nu = 1$, we would find $\sigma^2 = -1$. This is an unacceptable result. Because $\sigma = \sqrt{\sigma^2}$, we always want the value of the variance to be greater than or equal to zero.

Table II, at the end of this text, represents a tabulation of probabilities (areas) for the family of t distributions. At the top of this table is a sketch of a typical t distribution. The shaded region under the right tail represents the region (area) for which the probabilities are being tabulated. Those probabilities that are being quoted lie in a row across the top of the table. The particular t distribution we want is represented by exactly one row of this table, and this row is located by knowing that $\nu = n - 1$. We read the value of ν from the left column of the table. The body of the table is a set of positive critical points. Negative critical points will be determined by using the symmetry of the t distribution. For example, if we desire the positive critical point for a sample of size 21 and we wish to have .025 under the right tail, we look in the row $\nu = 20$ under the column $t_{.025}$ to obtain the value 2.086.

A close scrutiny of the critical points in this table will provide us with a better understanding of the family of t distributions and the closely related standard normal distribution. In fact, we may think of the standard normal distribution as a special t distribution. For all practical purposes, when $n \geq 30$, the t distribution in our table becomes the standard normal distribution.

Look at the column $t_{.025}$. Observe that the values of this column steadily decrease as we go down the column until we reach the bottom critical point value of 1.96. If we read down the column labeled $t_{.005}$, we observe the same decreasing trend and end with the critical point value of 2.576. These values are exactly the same critical point values that we have already obtained from our standard normal table. Because of the bottom line in this t distribution table, we could actually lose the standard normal table and carry on with business as usual.

A rather common custom is to name the critical points by using the double subscript notation, $t_{\alpha, \nu}$. Thus, the critical point that corresponds to a probability value of .01 under the right tail and $\nu = 15$ is $t_{.01, 15} = 2.602$. A probability value of .05 under the right tail and a sample size of 18 ($\nu = 17$) yields the critical point, $t_{.05, 17} = 1.740$. However, the format of the table is not meant to imply that only one-tail tests of hypotheses are possible.

The next sections deal with hypothesis testing and confidence intervals,

174 THE SAMPLING DISTRIBUTION OF MEANS FOR SMALL SAMPLES

relative to μ, for small samples. Because this work is similar to work covered in Chapter 7, these sections will be brief.

Example 8-8 Our population has 4875 pieces of data recorded less than the value $t_{.025,25}$. How many pieces of data are there in the entire population? See Fig. 8-2.

Figure 8-2

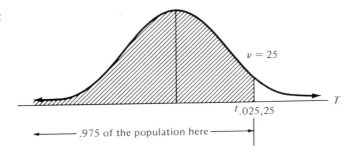

Solution There must be .975 of our population size (N) located below $t_{.025,25}$. Thus,

$$.975 N = 4875$$

$$N = 5000$$

EXERCISES A

8-1. Our population has 2670 pieces of data located on the t-axis greater than $-t_{.10,20}$ and less than $t_{.01,20}$. How many pieces of data are in the entire population? (Example 8-8)

8-2. Our population has 3400 pieces of data located on the t-axis greater than $-t_{.05,15}$ and less than $t_{.10,15}$. How many pieces of data are in the entire population?

8-3. Our population possesses a t distribution with $\nu = 18$. This population has 4875 pieces of data less than the t value, $t = 2.101$. How many pieces of data are in the entire population?

8-4. Our population possesses a t distribution with $\nu = 10$. This population has 7920 pieces of data greater than the t value, $t = -2.764$. How many pieces of data are in the entire population?

8-5. We are dealing with a t distribution where $n = 17$. Find:
  a. $P(t > 1.746)$
  b. $P(-1.746 < t < 2.120)$
  c. $P(t < 1.337)$
  d. $P(t > 2.583$ or $t < -2.583)$

8-6. We are dealing with a t distribution where $\nu = 20$. Find:
  a. $P(t < 1.325)$
  b. $P(t > -1.325)$
  c. $P(-1.725 < t < 2.528)$
  d. $P(t < -1.725$ or $t > 2.086)$

8-7. We are dealing with a t distribution where $\nu = 18$. Find:
   a. $P(t > 1.330)$
   b. $P(t < 1.330)$
   c. $P(t > -2.101)$
   d. $P(t < 2.101)$

8-8. The following represent critical points in the t-distribution table. Fill in the blanks (?).
   a. $t_{.05, 11} = \underline{\quad ? \quad}$
   b. $t_{?, 25} = 2.485$
   c. $t_{.1, ?} = 1.315$

8-9. The following represent critical points in the t distribution table. Fill in the blanks (?).
   a. $t_{.025, 19} = \underline{\quad ? \quad}$
   b. $t_{?, 23} = 1.714$
   c. $t_{.01, ?} = 2.681$

8-10. The following represent critical points in the t distribution table. Fill in the blanks (?).
   a. $t_{.01, 15} = \underline{\quad ? \quad}$
   b. $t_{?, 17} = 1.333$
   c. $t_{.05, ?} = 1.721$

8-11. a. What value should we define as the most suitable value for the median of a t distribution?
   b. What value should we define as the most suitable value for the mode of a t distribution?

8-12. When $\nu \geq 3$ the formula for the variance is $\sigma^2 = \nu/(\nu - 2)$. Explain why σ^2 would not make good sense should we insist upon using the value $\nu = 2$. (Example 8-7)

8-3 TESTING HYPOTHESES ABOUT μ

Hypothesis testing for small samples is exactly the same as hypothesis testing for large samples, with two slight changes: (1) Our test statistic is now a t score, instead of a z score. (2) Our critical point(s) are obtained from the t distribution table, instead of the standard normal table.

Suppose that we believe that people consider $3000 as the average value when they purchase a used car. In the next examples, we will obtain a random sample of 16 people who have just purchased a used car and test our feelings at the .05 level.

Example 8-9a We decide to test

$$H_0: \mu = \$3000 \qquad H_a: \mu \neq \$3000$$

A random sample is obtained and it is found to have a mean of $3000 with a standard deviation of $10.

Conclusion There is insufficient evidence to reject H_0. We should immediately stop the testing process. The sample mean is an exact match to the conjecture stated in H_0.

Example 8-9b We decide to test

$$H_0: \mu = \$3000 \qquad H_a: \mu \neq \$3000$$

A random sample is obtained and is found to have a mean of $3280 with a standard deviation of $400. Because the value of our sample mean doesn't match our conjecture in H_0 we have a chance for rejection.

$$t\text{-test statistic} = \frac{\$3280 - \$3000}{\$400/\sqrt{16}}$$
$$= 2.80 > t_{.025,15} = 2.131$$

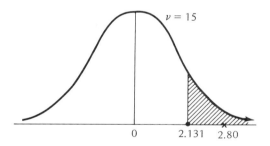

Conclusion Reject H_0.

Example 8-9c We decide to test our belief that $\mu < \$3000$

$$H_0: \mu = \$3000 \qquad H_a: \mu < \$3000 \text{ (our feelings)}$$

The sample is found to have a mean of $3280 with a standard deviation of $400.

Conclusion There is insufficient evidence to reject H_0. We should immediately stop the testing process. The sample mean matches our conjecture in H_0. (In essence, we are running a test with the stipulation that only sample means less than $3000 may fail to support H_0.)

Example 8-9d We decide to test our belief that $\mu > \$3000$.

$$H_0: \mu = \$3000 \qquad H_a: \mu > \$3000 \text{ (our feelings)}$$

A random sample is obtained and is found to have a mean of $3180 and a standard deviation of $400.

$$t\text{-test statistic} = \frac{\$3180 - \$3000}{\$400 \; \sqrt{16}}$$
$$= 1.80 > t_{.05,15} = 1.753$$

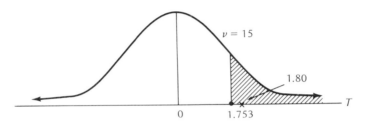

Conclusion Reject H_0. Our belief is accepted at the .05 level. Finally, because our test statistic is so close to the critical point value we should at least worry a little about the possibility of making a type I error.

Example 8-10 A church official is concerned about the amount of financial support he is getting from the people of his congregation. He believes that the average family income in his township is at least $16,000. He decides to test his feelings at the .05 level. For this purpose, he selects a random sample of 16 families and finds that their average yearly income is $16,100, with a standard deviation of $50. Does this sample provide sufficient evidence to support the church official's personal feelings?

Solution $H_0: \mu = \$16,000 \qquad H_a: \mu > \$16,000$ (church official)
Our test statistic equals 8, which is certainly greater than the critical point value.

Conclusion Reject H_0. We will accept H_a. The church official's feelings are strongly supported at the .05 level.

Example 8-11 A test of the breaking strengths of nine chains manufactured by a company showed a mean breaking strength of 805 lb with a standard deviation of 10 lb. The manufacturer claims that the breaking strength of this type of chain exceeds 800 lb. Test the manufacturer's claim at the .01 level.

Solution $H_0: \mu = 800$ lb $H_a: \mu > 800$ lb (manufacturer)

$$t\text{-test statistic} = \frac{805 \text{ lb} - 800 \text{ lb}}{10 \text{ lb}/3}$$
$$= 1.50 < t_{.01,8} = 2.896$$

Conclusion There is insufficient evidence to reject H_0. There is a difference between the value of the sample mean and the value of μ conjectured in our null hypothesis. The difference is in the manufacturer's favor; however, it is not a significant difference.

Example 8-12 A study of depth perception was done at a state college psychology department. Nine people were asked to estimate the distance between two markers. The true distance was 5.10 ft. The test was conducted at a .05 level of significance. The nine participants gave the following depth-perception estimate, in feet: 4.6, 4.0, 4.7, 5.1, 5.0, 4.2, 4.8, 4.7, 5.2

Test the ability of these people to judge depth correctly.

Solution $H_0: \mu = 5.10$ ft $H_a: \mu \neq 5.10$ ft
The sample mean $= 4.70$ ft and $s = .40$ ft

$$t\text{-test statistic} = \frac{4.70 \text{ ft} - 5.10 \text{ ft}}{.40 \text{ ft}/3}$$
$$= -3 < -t_{.025,8} = -2.306$$

Conclusion Reject H_0. On the basis of this small sample, we must conclude that people have difficulty in judging distance accurately.

Example 8-13 A large hospital decided to conduct a follow-up study to determine if its antismoking campaign had persuaded people to stop smoking, or at least cut down on smoking. The hospital knew, prior to its campaign, that a smoker smoked 11 cigarettes per day, on the average. After the campaign was over, it was found that the average was 10 cigarettes per day with a standard deviation of 3. The sample size was 100 (large). Observe that the drop in the average figure is only one cigarette. Should the hospital be discouraged by such an apparently small drop? Decide by testing at the .05 level.

Solution $H_0: \mu = 11$ $H_a: \mu < 11$ (the hospital)

$$t\text{-test statistic} = \frac{10 - 11}{3/10} = -3.33 < -1.64$$

8-3 TESTING HYPOTHESES ABOUT μ

Conclusion Reject H_0. Accept H_a. Indeed, the hospital should feel good. A decrease of a single cigarette drop in the average number smoked is significant.

EXERCISES A

8-13. Explain how a t score for a sample mean can be equal to zero. Explain how a t score for a sample mean can be negative.

8-14. An ornithologist believes that the number of fleas on a blue warbler is at most 20 at any one time. For research purposes, 16 blue warblers were studied in a large out-door enclosure. The mean number of fleas was found to be 23.40, with a standard deviation of 1.18. Test the ornithologist's belief at the .05 level. (Example 8-9c)

8-15. The police chief of a major city maintains that the mean age of rapists is 26.4 years. A women's liberation group obtains a sample of 25 rape cases from police files. This sample has a mean of 25.2 years with a standard deviation of two years. Are these data consistent with the police chief's claim at the .05 level? (Example 8-9b)

8-16. An automobile tire manufacturer advertises that, on the average, the company's tires are good for 15,000 miles. A sample of 25 tires is found to have a mean lifetime of 14,500 miles with a standard deviation of 1800 miles. Test H_a: $\mu < 15{,}000$ miles at the .05 level.

8-17. A vending-machine distributor knows that his machines will sell an average of 612 drinks per machine. He tests 16 new vending machines with neon-flashing lights to see if more customers will be attracted to his machines. He finds the sample mean to be 620 drinks per machine with $s = 50$. If we test at the .05 level, can we conclude that the flashing neon lights improved sales? (Example 8-10)

8-18. A chemist is testing the hypothesis that the boiling point of a certain substance is 846°C. He makes a determination based on a sample with size $n = 4$, a mean of 845°C, and $s = 1.40$. Test this hypothesis at the .05 level.

8-19. A disgruntled golfer claims that a new set of golf clubs has upset her game. She points to the fact that she used to average 93.4 strokes per round, but in the 36 rounds she has played since obtaining the new clubs, her average has jumped to 97.7, with a standard deviation of 8.1. Test this claim with $\alpha = .01$. Remember that in golf low scores are desirable.

B

The remaining exercises use *raw data*. This means that the data are being used exactly as initially recorded. For this type of exercise, we have restricted ourselves to small samples in order to keep the solution time reasonable.

CUMULATIVE REVIEW

8-20. Test H_a: $\mu \neq 7$ at the 5% level based on the sample $\{2, 4, 6, 8, 10\}$.

8-21. The following set of measurements represent the diameters of micrometeors taken from samples of moon soil: 1.00, 1.72, 1.66, 1.32, 1.61, 2.02, 1.21, 1.46. Test H_a: $\mu \neq 1.70$ at the 1% level.

8-22. The following six IQ scores of freshmen at Big U. were recorded: 110, 90, 100, 112, 108, 110. Using these raw data, test the alternative hypothesis H_a: $\mu > 95$ at the .05 level.

8-23. A research engineer is trying to create a new gasoline engine that will run for at least 20 minutes on a gallon of gasoline. Test runs of the experimental engine produced six trial values of 25, 29, 22, 24, 33, and 23 minutes, respectively. Test at the .05 level of significance to see if the engineer has found the desired engine.

8-4 THE TWO TYPES OF ESTIMATES OF μ

The sample mean remains the point estimate of μ. There is no change here, although the use of a small sample size does tend to make us feel a little less secure in our use of this point estimate.

Our second method of estimating μ is still a finite interval estimate. We still use the point estimate of μ in our calculation of the interval estimate. The end-points for the confidence interval estimate of μ, for $n < 30$, are given by Theorem 8-1.

Theorem 8-1 The end-points of a $100(1 - \alpha)\%$ confidence interval for μ are

$$\bar{x} \pm \frac{(t_{\alpha, n-1})s}{\sqrt{n}}$$

This result is exactly analogous to the result obtained in Chapter 7 (when $n \geq 30$).

Example 8-14 The fire department of Simpsonville wants to know the average amount of time needed to put out a fire there. For this purpose, pertinent data are gathered from the next 25 fires. The mean amount of time necessary to put out a fire is 1.60 hours, with a standard deviation of .40 hour. Compute the 99% confidence interval for μ.

Solution $1.60 \pm [(2.797)(.40)]/5$ hours

$1.60 \pm .224$ hours

Example 8-15 Sixteen fifth graders take a standardized achievement test in mathematics. The mean of the 16 scores is 75 and the standard deviation is 2.5. Find the 95% confidence interval for the true mean of the entire population of fifth graders.

8-4 THE TWO TYPES OF ESTIMATES OF μ 181

Solution $75 \pm [(2.131)(2.50)]/4$

$$75 \pm 1.33$$
or
$$73.67 < \mu < 76.33$$

EXERCISES A

8-24. It is desired to approximate a population mean by a 95% confidence interval. For this purpose, a sample of size 16 is obtained with a mean of 3 and a standard deviation of 1.23. Find this interval. (Example 8-15)

8-25. We wish to obtain a 99% confidence interval for the mean number of childhood diseases of children in the $8000 to $11,000 family income bracket. Twenty-five children selected at random are found to have an average of five childhood diseases with a standard deviation of .82. Compute this interval. (Example 8-14)

8-26. In 25 horse races the average distance between the winner and the runner-up is found to be 6.70 ft, with a standard deviation of 2 ft. Find the 99% confidence interval for the true value of μ.

8-27. A soup-filling machine is checked during a quality control test by selecting a sample of 16 cans filled at random. On the basis of this small sample, an entire shipment will be rejected if the 90% confidence interval for μ fails to contain the advertised weight of 10 oz. The 16 cans selected have a mean weight of 9.80 oz, with a standard deviation of .20 oz. Will the shipment be rejected?

B   The remaining exercises use *raw data*. This simply means that the data are being used exactly as initially recorded. For this type of exercise, we have restricted ourselves to small samples in order, to keep the solution time reasonable.

CUMULATIVE REVIEW

8-28. Five measurements of the reaction time, in seconds, of an individual to an electric stimulus were recorded: .28, .30, .27, .33, .32.
   a. Compute the 95% confidence interval for μ.
   b. Compute the 99% confidence interval for μ.
   c. Compare the lengths of these two intervals.

8-29. Use the data in Exercise 8-22 to calculate a 95% confidence interval for the value of μ.

8-30. The following set of measurements represent the batting average of 8 professional ball players: 310, 215, 288, 296, 244, 252, 300, and 255. Compute the 95% confidence interval for μ.

8-31. A candy bar manufacturer has machines that are supposed to produce 2.30-oz bars. It is known that the weights vary slightly; however, they are normally distributed, with the standard deviation varying slightly from machine to machine, and often from day to day. To test quality, a random sample of four bars is drawn from the day's output, and four bars are found to weigh: 2.24 oz, 2.34 oz, 2.28 oz, and 2.30 oz, respectively. Find the 98% confidence interval for the mean weight of the day's output.

SUMMARY

This chapter deals with small samples ($n < 30$). It was discovered by W. S. Gosset that the sampling distribution of means for small samples was not normally distributed. In fact there are a family of such distributions called *t distributions* (or student t distributions). Table II is a tabulation of probabilities (areas) below a specified portion of each standard t distribution. Each t distribution (row of Table II) depends on a quantity called *degree of freedom*. There are two common abbreviations for degrees of freedom: (1) d.f., and (2) ν, (This is the Greek letter nu.). Roughly speaking, the value of ν, in a problem, depends on the number of independent variables (choices) that you are dealing with. In this chapter $\nu = n - 1$ (the sample size less one). Using the standardizing formula, it is possible to standardize any piece of data from a given t distribution into one of the standard t distributions of Table II. When this is done we say we are obtaining a "t score" (versus a z score in the preceding chapter). For most standard t distributions of Table II: (1) $\mu = 0$ and $\sigma^2 = \nu/(\nu - 2)$ where $\nu > 2$. As soon as we become familiar with t distributions, our job of testing a statistical hypothesis (H_0) about μ, and our job of estimating μ is identical to the work done in the preceding chapter. When our sample size becomes large, our standard t distribution of Table II is sufficiently close to the standard normal distribution as to make further tabulation of t distributions unnecessary. This provides motivation for saying that large sample sizes begin at 30.

Can You Explain the Following?

1. small sample
2. degrees of freedom (d.f. or ν)
3. t distribution (or student-t distribution)
4. W. S. Gosset

CUMULATIVE REVIEW

MISCELLANEOUS EXERCISES

8-32. State another name for t distribution.

8-33. We are looking at our t distribution table. Our critical point is 2.080 and .975 of our data are to the left of this critical point value. What is the value of ν?

8-34. We are dealing with a t distribution where $n = 25$. Find:
   a. $P(t < 1.711)$
   b. $P(-1.711 < t < 2.064)$
   c. $P(2.064 < t < 2.492)$

8-35. The fire department of a large city believes that the time involved from initial response to full control of the fire situation is less than 1.75 hours. To test this belief at the 5% level, a sample of 25 fires of the past year are randomly selected. The mean for this sample is 1.60 hours and the standard deviation is .40 hour. Conduct the test for the fire department.

8-36. Use the sample data of Exercise 8-35 to compute the 99% confidence interval for μ.

 8-37. Sullivan Tire Company intends to base an advertising campaign on the claim that the average life of its new radial tire is more than 35,000 miles. Before proceeding, a life test is conducted on 20 tires. The tires are placed on a special machine and driven until they all fail. The average time to failure for this sample is 36,750 miles with $s = 600$ miles. Should the tire company continue with the planned advertising campaign? Testing is to be done at the 5% level.

 8-38. Use the sample data of Exercise 8-37 to compute the 95% confidence interval for μ.

9 NONPARAMETRIC STATISTICS

9-1 INTRODUCTION

We are supposed to select two independent samples from a population; however, we really don't feel certain about the independence of the two samples. What do we do? We are supposed to select samples from populations with the same variance, but we really aren't sure that the population variances are equal. What do we do? We need to draw a sample from a normal population, but we have reason to doubt the normality of the population. What do we do? A basic assumption in sampling is that our sample is randomly selected. However, in some instances the sample is collected over an interval of time and the sample selection process can't be rigidly controlled. Again, what do we do?

In this chapter we begin to discuss a solution—we can use *nonparametric procedures*.

Definition 9-1 Statistical procedures that do not depend upon knowledge of (1) the distribution or (2) the population parameters are called *nonparametric* or *distribution-free* procedures.

The disadvantages of such procedures include the following:

1. At times background work, such as ranking all the data in order, is very tedious.
2. Because nonparametric procedures are easily applied, they sometimes are used in cases where parametric procedures would be preferable. A parametric procedure will make more efficient use of the data than will a nonparametric procedure.

In the next section we meet our first nonparametric test: *the runs test*. This test was chosen for two reasons: (1) It deals with an important concept—the randomness of a sample. (2) The null hypothesis in this nonparametric test is clearly unrelated to any population parameter. H_0: The sample sequence of data is random.

For many parametric procedures there is a nonparametric procedure that does approximately the same job. However, in this chapter, we consider only the runs test, to give us time to assimilate the nonparametric concept. In future chapters, we meet other nonparametric tests: the sign test, the Mann-Whitney U-test, the Spearman rank correlation coefficient test, and the Kruskal-Wallis H-test.

9-2 THE SAMPLING DISTRIBUTION OF RUNS

In recent years, considerable attention has been given to testing a sample to ascertain its randomness or lack of randomness. Now, although more than one technique exists for testing the randomness of a sample, perhaps the most popular test today is the runs test. This nonparametric test is based on the order in which observations are made. For example, suppose a safety expert has been monitoring a radar set on a busy expressway. Each time a car passes that is exceeding the speed limit, he writes F, for fast. Each time a car passes that is doing less than the speed limit he writes S, for slow. The results after 40 cars are

$$\underbrace{SS}_{2} \ \underbrace{FFFF}_{4} \ \underbrace{SSSSSSSSS}_{9} \ \underbrace{F}_{1} \ \underbrace{SS}_{2} \ \underbrace{FFFF}_{4} \ \underbrace{SSSSSSS}_{7} \ \underbrace{FFFF}_{4} \ \underbrace{SSSS}_{4} \ \underbrace{F}_{1} \ \underbrace{SS}_{2}$$

Our safety expert is interested in the question: Do speeders and nonspeeders pass the monitor in a random manner?

Definition 9-2 A *run* may be a single letter or a succession of letters (symbols), which is followed and preceded by a different letter or by no letter at all.

Our sample sequence of F's and S's has 11 runs. Count them. We start by assuming that we have exactly two letters in our sample sequence of data. When we use the runs test, the length of each run is not important. We are interested in the number of runs and the total number of times that each letter appears in the entire sample sequence of data.

In our previous sample sequence of F's and S's, let n_1 denote the total number of S's and n_2 denote the total number of F's. Here, $n_1 = 26$ and $n_2 = 14$. It is immaterial which letter total is denoted n_1 and which letter total is denoted n_2. This follows immediately from the fact that n_1 and n_2, in Theorem 9-1, are commutative.

Definition 9-2 tells us that there are two sets of circumstances in which nonparametric procedures arise. The two types of circumstances are usually related.

When should we use a nonparametric procedure?

1. When the hypothesis to be tested does not involve a population parameter.
2. When we are not certain that we can meet the assumptions necessary for executing a parametric procedure. For example, we may feel that our sample size is too small to justify a parametric test.
3. When the results are needed in a hurry (use this reason judiciously).
4. As a check on a parametric procedure.

What are the advantages of a nonparametric procedure?

1. Most of the nonparametric procedures are based on a minimum set of assumptions, and this tends to reduce the chance of their being used improperly.
2. The arithmetic computations necessary for the application of most nonparametric procedures are easy to comprehend.
3. A nonparametric procedure is usually easily understood by a person who is statistically unsophisticated.

It can be shown, in a more advanced text, that the sampling distribution of runs for fixed n_1 and n_2 is approximately normally distributed when n_1 and n_2 are both greater than 10. When n_1 or n_2 is less than 10, special tables are needed. Here, we are usually dealing with a rather small sample size, and so we shall omit further discussion of such special cases.

Theorem 9-1 (no proof) For fixed sample sizes n_1 and n_2, where n_1 and n_2 are both greater than 10, the sampling distribution of runs is approximately normally distributed with

i. $\mu = \dfrac{2n_1 n_2}{n_1 + n_2} + 1$

ii. $\sigma = \sqrt{\dfrac{2n_1 n_2 [2n_1 n_2 - n_1 - n_2]}{(n_1 + n_2)^2 (n_1 + n_2 - 1)}}$

Example 9-1a Evaluate μ and σ for the previous sample sequence of F's and S's.

Solution

$$\mu = \dfrac{(2)(26)(14)}{26 + 14} + 1$$

$$\mu = 19.20$$

$$\sigma = \sqrt{\dfrac{2(26)(14)[2(26)(14) - 26 - 14]}{(26 + 14)^2 (26 + 14 - 1)}}$$

$$\sigma = \sqrt{8.03}$$

$$\sigma = 2.83$$

We restrict ourselves to the two-tail test situation—the most common case for this nonparametric test.

H_0: The sample sequence of data is random.
H_a: The sample sequence of data is not random.

Example 9-1b Test the previous sample sequence of F's and S's for a lack of randomness at the .05 level. Use the information from Example 9-1a.

Solution H_0: The sample sequence of data is random.
H_a: The sample sequence of data is not random.
There are 11 runs. Our test statistic is the z score of 11 relative to the n_1 value of 26 and the n_2 value of 14.

z-test statistic $= (11 - 19.20)/2.83 = -2.90 < -1.96$

Conclusion Reject H_0. It appears that cars traveling fast and cars traveling slowly are arriving in bunches and not randomly.

Example 9-2 The win–loss record of a high school baseball team for 60 consecutive games is as follows:

188 NONPARAMETRIC STATISTICS

WWWWWW	L	WWWWWW	L	WWWWWWW	LLL	WW	LLLL	W
6	1	6	1	7	3	2	4	1

LLLL	W	LLLL	WW	LL	WWWW	L	WWWW	L	WWWWWW
4	1	4	2	2	4	1	4	1	6

Test at the .05 level of significance to see if the string of wins and losses lacks randomness.

Solution H_0: The sample sequence of data is random.
H_a: The sample sequence of data is not random.
Let n_1 equal the number of W's: $n_1 = 39$.
Let n_2 equal the number of losses: $n_2 = 21$.
There are 19 runs.

$$z\text{-test statistic} = \frac{19 - \left[\frac{2(39)21}{39 + 21} + 1\right]}{\sqrt{\frac{2(39)(21)[2(39)(21) - 39 - 21]}{(39 + 21)^2(39 + 21 - 1)}}}$$

$$= \frac{19 - 28.3}{3.49}$$

$$= -2.67 < -1.96$$

Conclusion Reject H_0. The team is not winning and losing in a random manner. There appears to be an underlying pattern.

At this point, let's turn our attention to the case of testing the randomness of an arbitrary sample of numerical data. Our theory of runs can be applied to any sample sequence of numerical data. The trick is to turn our sample sequence into a corresponding sample sequence of exactly two letters. We will agree to use the letters a and b. Now we treat our problem just as before. How do we do this?

In cases of numerical data, we first calculate the median value. Now we go through the entire sample sequence of data and replace each value with an a if it is below the median value and with a b if it is above the median value. Our sample sequence of numerical data is now expressed entirely in terms of a's and b's, which allows us to proceed as before. Omit all values that equal the median value.

Example 9-3 The number of defective items produced by a machine in the Belcher Manufacturing Company is recorded each day for a month. The median value is 17. The results are

| 13 | ⑰ | 14 | 20 | 18 | 16 | 14 | 19 | 21 | 20 | 14 | ⑰ | 12 | 14 | 19 |
| a | | a | b | b | a | a | b | b | b | a | | a | a | b |

| 20 | ⑰ | 18 | 14 | 20 | ⑰ | 19 | ⑰ | 14 | 19 | 21 | 16 | 12 | 15 | 22 |
| b | | b | a | b | | b | | a | b | b | a | a | a | b |

Test the sample sequence of data for a lack of randomness at the .05 level.

Solution H_0: The sample sequence of data is random.
H_a: The sample sequence of data is not random.
$\mu = 13.48$ and $\sigma = 2.44$. There are 12 runs.

z-test statistic = $(12 - 13.48)/2.44 = -.61$

Conclusion There is insufficient evidence to reject H_0. The sample data supply no evidence to make us believe that the number of defective items are arriving in a nonrandom manner.

EXERCISES A

9-1. Toss a coin 50 times, record the sequence of heads and tails as you toss, and then use the runs test to check for a lack of randomness in the data.

9-2. We are given the following data:
a b c d f d g f b c a b a c d
f g d a b b g c d a b c d e f
Discuss testing such a sample for randomness.

9-3. We are given the following data:
1,1 3,3 1,1 3,3 1,1 3,3 1,1 3,3 1,1 3,3 1,1 3,3 1,1 3,3
  a. By merely inspecting the data, do you expect the runs test to show a lack of randomness in the data?
  b. Execute the runs test for this data at the .01 level of significance. (Example 9-1a, 9-1b)

9-4. Assume n_1 and n_2 are both greater than 10. If $n_1 = n_2$, write a formula for μ. Write a formula for σ.

9-5. Thirty red and blue coats are arranged on a rack as follows:
R B R B RR B R BB RRRRR B R BB R B R B R BBBB RR
Using a .05 level of significance, determine if the arrangement of coats lacks randomness. (Example 9-2)

9-6. The weather bureau recorded the rainfall for 25 consecutive days. If the total daily rainfall exceeded .10 in., an R was recorded, for rain. If the total daily rainfall was less than or equal to .10 in., D was recorded, for dry. Test the weather bureau data for a lack of randomness at the .01 level.
RRRR D R DDDD RRRRRR DDDD R DD R D

9-7. Each day on the London gold exchange the price of gold either rises or falls. Although some observers consider that the price of gold fluctuates in a random manner, others believe that the rises and falls follow a specific pattern. To check the opposing theories, an analyst records the rises (R) and falls (F) for 40 consecutive days. Check the data for a lack of randomness by executing a runs test at the .05 level.

| F | RRRRR | FF | RRRRRR | F | R | F | RRR | FFF | RRRRRR | FFF | RRRRRRRR |
| 1 | 5 | 2 | 6 | 1 | 1 | 1 | 3 | 3 | 6 | 3 | 8 |

190 NONPARAMETRIC STATISTICS

B

9-8. The family incomes of applicants for house mortgages are shown in the order in which the applications were received (by row). Test for a lack of randomness at the .01 level.
$17,800, $19,100, $24,100, $27,080, $22,063, $18,400, $23,721
$16,219, $14,400, $16,800, $27,031, $21,096, $19,500, $18,276
$22,039, $26,321, $16,246, $15,963, $27,468, $25,820, $27,306
$17,100, $16,583.

9-9. The following data, in rows, gives the number of defective bricks recorded from samples of size 100. Forty consecutive data values are recorded. Use the runs test to check for a lack of randomness at the .01 level.
11 12   8 13 16 11   8 12 12 10 14 16 21   9   8 18 10   8 14 13
 9 12 14 21   9   7 13 16 13 12 16 18 19 13 13 15 11   8   9 17

CUMULATIVE REVIEW

9-10. Following is the annual rainfall, in a certain western city, for 40 years. All the data are recorded to the nearest inch.
20 11 16   8   9 33 14 17 12 16
23 19 12 18 21 19 11   9 15 17
15 13 22 17 38 20 14 21 16 12
25 17 20 15 23 24 14 19 13 16
Use the runs test to check for a lack of randomness at the .05 level.

9-11. The following numbers were taken from a table of random digits.
54222 56179 09833 34227 43897 38517 11617 30338
Let x represent an even number and y an odd number. Use the runs test at the .05 level to see if these numbers are indeed being recorded in a random manner.

SUMMARY

When you conduct a statistical test that does not depend on a knowledge of population parameters it is said that you are conducting a *nonparametric test*. There are many nonparametric tests; indeed such tests are a statistics course. This chapter has dealt with only one nonparametric test, the *runs test*. As you travel through future chapters, you will encounter other nonparametric tests. They will appear with their "matching" parametric test.

A *run* may be a single letter or a succession of letters (symbols) which is followed and preceded by a different letter or by no letter at all. The test assumes that we are dealing with a sample sequence consisting of exactly two letters. For example, a and b. When n_1 (the number of a's) > 10, and n_2 (the number of b's) > 10, we learn the amazing fact that the distribution of runs is normally distributed.

There are advantages to nonparametric testing (three are cited in the chapter); there are disadvantages to nonparametric testing (two are cited in the chapter). The chapter ended with a discussion of how to execute a runs test on any numerical sample. The trick is to find the median value of the sample and to use this value to turn the sample sequence into a new sequence of exactly two letters (e.g., *a* and *b*).

Can You Explain the Following?
1. nonparametric test (or distribution free test)
2. a "run"
3. runs test
4. give one advantage of nonparametric testing
5. give one disadvantage of nonparametric testing.

10 THE SAMPLING DISTRIBUTION OF SUCCESS NUMBERS

10-1 THE BINOMIAL EXPERIMENT

Among the most common experiments in applied statistics is the experiment designed to have exactly two outcomes. For example, in polling a population we seek two responses—for or against. An item in a shipment is defective or it is not defective. A drug may be considered as effective or ineffective in the treatment of a disease. An exam question may be true or false. A chemical is present in a solution or it is not present. A coin lands heads or tails.

When we are dealing with an experiment that has exactly two outcomes, we usually call one of the outcomes *success* (s) and the other *failure* (f). Either outcome may be the success outcome. If p is the probability value associated with the success outcome, then $1 - p$ or q is the probability value associated with the failure outcome.

Example 10-1 State five real-life situations in which we may consider that there are exactly two outcomes.

Solution
1. A customer buys a product or doesn't buy it.
2. We are exceeding the existing speed limit or we are not exceeding the speed limit.
3. A student passes or fails a course.
4. We do or we don't have sufficient gas to take us to our destination.
5. It will rain today or it will not rain today.

Definition 10-1 A single performance of an experiment with two outcomes will be called a *trial*.

10-1 THE BINOMIAL EXPERIMENT

HISTORICAL NOTE

DANIEL BERNOULLI

Historically, we speak of "Bernoulli trials" (binomial trials) in honor of Daniel Bernoulli (1654–1705). His great treatise on probability, *Ars Conjectandi*, was published in 1713 (after his death). The Bernoulli family is a most striking case history of inherited academic ability and in particular mathematical ability. Three generations of Bernoulli's produced eight mathematicians, several of whom were outstanding, and a large number of descendants at least half of whom were academically gifted. Nowhere in the Bernoulli family history is there a record of a Bernoulli who was a failure.

Definition 10-2 A binomial experiment possesses the following properties:

i. Each trial has two outcomes. One outcome is labeled success (s) and the other failure (f).
ii. The probability value of success on a single trial is equal to p, and the value of p remains constant from trial to trial. (This implies that we are sampling with replacement when dealing with a finite population.)
iii. All trials are independent. That is, the outcome of one trial in no way influences the outcome of any other trial.
iv. Because we wish to base our work upon a sample of size n, the experiment will consist of n trials.

The experimenter is interested in the number of successes, x, in a sample of n trials: $0 \leq x \leq n$. This type of experiment serves as a reasonably good model for many real-life situations, such as one is a high school graduate or one is not a high school graduate.

We realize that in a real-life situation the binomial experiment model is not necessarily a perfect portrayal of existing conditions. For example, let's assume a binomial experiment is "take foul shots with a basketball." If we sink the ball in the hoop, it is a success; if we miss, it is a failure. We can use the binomial model here, but obviously there is a human element to be considered that may render each trial shot less than completely independent of the previous trial shot. In our applied work, we are interested in the sampling distribution of success numbers (in actual problem solving we will use only one sample); see Fig. 10-1.

Plumbers who deal with the PVC Plastic Company and purchase its population of 4-in.-diameter plastic pipe demand that the diameter measurement be accurate to the nearest one-thousandth (.001) of an inch. As we pointed out in Chapter 1, this means that each piece of pipe will fit into one of two categories: (1) the diameter measurement is accurate (success), or (2) the diameter measurement is not accurate (failure). A failure here is final for a

Figure 10-1

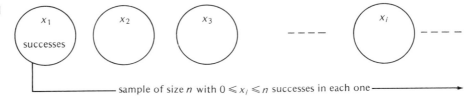

piece of pipe with a 3.7800-in. diameter. There is no market for pipe that does not fit standard measurements.

The following is a sample of 50 pieces of plastic pipe selected from the population of plastic pipe and measured accurately to the nearest ten-thousandth of an inch.

4.0006	3.9997	4.0002	3.9993	3.9999
3.9982	4.0017	4.0009	4.0009	3.9994
4.0000	3.9994	3.9997	4.0003	3.9991
4.0007	3.9993	4.0004	4.0007	4.0000
4.0000	3.9994	4.0000	4.0000	4.0001
4.0000	3.9962	3.9992	4.0008	3.9998
3.9998	4.0020	4.0000	4.0000	4.0000
4.0008	3.9988	4.0008	4.0009	4.0007
4.0007	3.9991	4.0003	4.0002	4.0000
4.0001	3.9999	3.9992	3.9998	3.9999

To obtain a better view of the success and failure results of this sample, the company constructed a frequency distribution, Table 10-1, and a histogram, Fig. 10-2, of the data.

Classes	Number	
4.0015 and up	//	2
4.0010 but less than 4.0015		0
4.0005 but less than 4.0010	𝍫𝍫 𝍫𝍫 /	11
3.9995 but less than 4.0005	𝍫𝍫 𝍫𝍫 𝍫𝍫 𝍫𝍫 𝍫𝍫	25
3.9990 but less than 3.9995	𝍫𝍫 ////	9
3.9985 but less than 3.9990	/	1
less than 3.9985	//	2

TABLE 10-1. FREQUENCY DISTRIBUTION FOR PVC PLASTIC COMPANY PIPE MEASUREMENTS

Figure 10-2

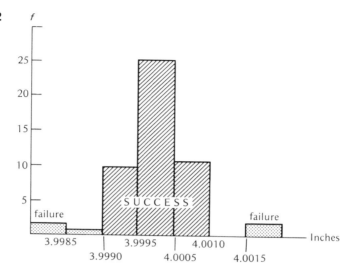

Because the derivation of the variance of the population of success numbers is lengthy, we shall state the formulas for the mean and variance without proof.

$$\mu = np$$
$$\sigma^2 = np(1 - p)$$

Example 10-2 a. The PVC Plastic Company wishes to use the data of Table 10-1 to estimate the value of p and q. b. If the company expects to manufacture this 4-in. pipe in lots of 8100, what can be estimated as the average number of usable pieces of pipe in a lot? What is the standard deviation for such lots?

Solution

a. The middle three classes of Table 10-1 correspond to pieces of pipe with an acceptable diameter measurement. The total frequency of these three classes is 45.
  Thus, $p = 45/50 = 9/10$, $q = 1/10$.
b. $\mu = np = (8100)(9/10) = 7290$
  $\sigma = \sqrt{np(1 - p)} = \sqrt{(8100)(9/10)(1/10)}$
  $= \sqrt{(81)(9)} = 27$

This is a high standard deviation. PVC should immediately go to work to reap the potential dollar savings from improving the manufacturing process.

Example 10-3 Mr. Lemos enjoys escaping the traffic congestion on the way home from work. He knows from experience that when he leaves work his chance of beating the traffic is 80%. Over the long haul, in any set of 80 days, how many days can he reasonably expect to avoid the traffic? What is the variance for collections of 80 days?

Solution $\mu = 80(4/5) = 64$ days; here $p = 4/5$

$$\sigma^2 = 80(4/5)(1/5) = 12.80 \text{ days}$$

Example 10-4 Assume, in Example 10-3, that Mr. Lemos enjoys the challenge of navigating in congested traffic. Let success be that he leaves work and encounters traffic. Compute how many days out of 80 he can reasonably expect to enjoy the challenge of traffic. Compute the variance.

Solution

$$p = 1/5, \mu = 80(1/5) = 16 \text{ days}$$
$$\sigma^2 = 80(1/5)(4/5) = 12.80 \text{ days}$$

The success outcome in Example 10-4 corresponds to the failure outcome in Example 10-3. The average number of successes in Example 10-3 plus the average number of successes in Example 10-4 total 80 days. Observe that, regardless of which outcome is labeled success, the value of the variance is unchanged. This observation is true in general.

Example 10-5 A student is given a multiple-choice exam with five possible answers for each question. The exam has 100 questions. Our student has not studied and must rely upon guessing to obtain the correct answer. What is the average number of successes that she may expect to receive? What is the standard deviation of this set of success numbers?

Solution Let success be to guess the correct answer.

$$p = 1/5 \quad \mu = 100(1/5) = 20$$
$$\sigma = \sqrt{100(1/5)(4/5)} = 4$$

It is clear, in the last example, that a student's answers to all 100 multiple-choice questions are possibly not all independent. First, it is hard to believe that the student is devoid of all knowledge; and second, some of the 100 questions may actually serve as an aid in answering some of the other questions. Thus, our binomial experiment serves as a model of a real-life situation, which is at least realistic, if not perfect.

Example 10-6 A marksman hits a target with a probability of 3/4. What is the mean number of hits that he can expect in 64 tries? What is the value of the variance?

Solution $\mu = (64)(3/4) = 48$

$\sigma^2 = (64)(3/4)(1/4) = 12$

Example 10-7 If $\mu = 50$ and $n = 250$ find p.

Solution $50 = 250p$
Therefore, $p = 1/5$

Exact binomial probabilities are computed by using the binomial theorem. A knowledge of permutations and combinations is also required. We feel that the time required to gain proficiency in calculating exact binomial probabilities is not warranted in a first course, especially when we have a method of approximating binomial probabilities that is immediately available, easy to understand, and very accurate. This approximation technique is the topic of our next section. However, the interested reader will find exact binomial probabilities is the optional section of this chapter.

EXERCISES A

10-1. Complete the following table for the case of a binomial experiment of n trials.

	μ	σ	n	p
a.	?	—	100	4/5
b.	—	?	100	4/5
c.	—	$\sqrt{18}$	?	1/10
d.	75	—	?	3/4
e.	—	$\sqrt{4}$	25	?
f.	21.6	—	27	?

Hint: Use the quadratic formula for part e.

198 THE SAMPLING DISTRIBUTION OF SUCCESS NUMBERS

10-2. A coin is tossed 36 times, and again 36 times, and again 36 times, and so forth. Should the coin land heads (H), it is considered a success. What are the mean and the standard deviation of the distribution of heads? (Example 10-2)

10-3. On a multiple-choice test with five possible answers (only one of which is correct), a student randomly selects an answer to each of 225 questions. Determine the mean and the standard deviation of this distribution.

10-4. One hundred cuts are made of a pack of 52 playing cards. What is the average number of spades? What is the standard deviation?

B 10-5. We are dealing with a binomial experiment. If one of the outcomes is labeled success and the value of μ is computed and then the other outcome is labeled success and the value of μ is again computed, prove that the sum of these two values is always n.

10-2 APPROXIMATION BY A NORMAL DISTRIBUTION

Our sampling distribution of success numbers is a discrete set of data (see Section 1-7), and our normal distribution is a continuous curve based on a continuous set of data. First we shall state (without proof) a major theorem concerning this sampling distribution of success numbers and then we will attack the problem of using a continuous probability curve to estimate a discrete distribution.

Theorem 10-1 If x represents a number of successes associated with a binomial experiment, then the limiting form ($n \to \infty$) of the distribution of $z = (x - np)/\sqrt{npq}$ (where $np > 5$ and $nq > 5$) may be approximated by a standard normal distribution.

Of course, it is then true that the actual distribution of success numbers may be fitted to a normal distribution with parameter values, $\mu = np$ and $\sigma = \sqrt{npq}$.

This theorem leaves us with a pressing question: How do we use a continuous probability curve to approximate a discrete distribution? We compute probabilities associated with a continuous probability curve such as the standard normal distribution by computing areas. The concept of using one measure of area to approximate another measure of area serves as our rationale for approximating here.

In a discrete distribution of success numbers, probabilities are recorded geometrically by plotting points directly above the values of our sampling distribution—that is, directly above the integer values 0 to n. The vertical distance to any point is the measure of probability for that number of suc-

10-2 APPROXIMATION BY A NORMAL DISTRIBUTION

cesses. We would like to turn this vertical line segment measure (without title) into a corresponding area measure (without title). The standard agreement for achieving this end is to travel one-half unit, on the axis, left and right of each success value. This process allows us to generate a set of adjacent rectangles. Each rectangle has a base dimension of one unit. Thus, the vertical line segment measure representing the probability and the area measure of the rectangle will agree exactly. That is, when $b = 1$, $A = bh = 1 \cdot h = h$. The area measure of the rectangle may now be used as the probability measure of the corresponding success value. As an illustration of this technique, consider the binomial experiment with $n = 16$ and $p = 1/2$. The rectangle associated with four successes is shown in Fig. 10-3.

Figure 10-3

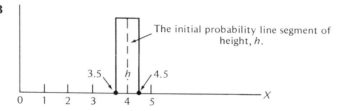

For the binomial experiment with $n = 16$ and $p = 1/2$, $\mu = 8$ and $\sigma = 2$. We fit the normal distribution with these parameter values to our discrete sampling distribution; see Fig. 10-4.

Figure 10-4

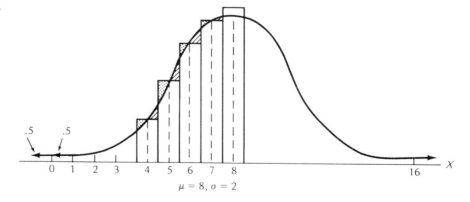

Figure 10-4 shows that there is a trade-off of areas in this approximation technique. When we use our normal curve as an estimate, we lose the shaded area but we gain the lined area. The exchange is approximately equal. Observe, in Fig. 10-4, that the normal curve that we use in our approximation doesn't necessarily pass through the points at the top of our probability line segments.

Example 10-8 For the binomial experiment with $n = 16$ and $p = 1/2$ find $P(4 \text{ successes})$. Use the normal curve approximation.

Solution (Approximately equal to is denoted by $\doteq$)
$\mu = 8$ and $\sigma = 2$
P(of the binomial variable equalling 4) $\doteq P$(that the normal variable lies between 3.5 and 4.5) $= P(-2.25 < z < -1.75)$

where
$$z_1 = \frac{3.5 - 8}{2} = -2.25,$$

and
$$z_2 = \frac{4.5 - 8}{2} = -1.75$$

From the standard normal table:

$$\begin{array}{r} .4878 \\ - .4599 \\ \hline \text{answer} \quad .0279 \end{array}$$

Thus,
$$P(-2.25 < z < -1.75) = .0279$$

Observe, in Fig. 10-5 that the area below our normal distribution is being used to approximate the area of the rectangle centered at 4.

Figure 10-5

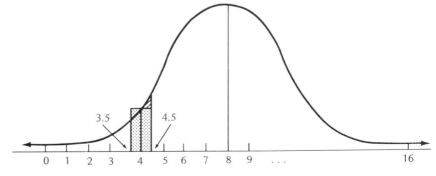

Example 10-9 For the binomial experiment with $n = 16$ and $p = 1/2$:
a. Find P(of from 4 to 6 successes inclusive) or $P(4 \leq \text{binomial } x \leq 6)$.
b. Find P(of obtaining at least 6 successes) or $P(\text{binomial } x \geq 6)$.

Solution $\mu = 8$, $\sigma = 2$
a. $P(4 \leq \text{binomial } x \leq 6) \doteq$
$P(3.5 < \text{normal } x < 6.5) =$
$P(-2.25 < z < -.75) = .2144$

where

$$z_1 = \frac{3.50 - 8}{2} = -2.25$$

$$z_2 = \frac{6.50 - 8}{2} = -.75$$

$$\begin{array}{r} .4878 \\ - .2734 \\ \hline \text{answer} \quad .2144 \end{array}$$

Figure 10-6

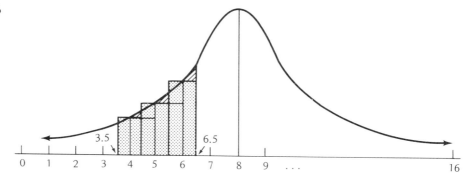

b. $P(\text{binomial } x \geq 6) \doteq$
$P(\text{normal } x > 5.5) =$
$P(z > -1.25) = .8944$

Example 10-10 A manufacturing process produces defective items at a .10 rate. If 100 items are randomly selected from the process, what is the probability that the number of defectives exceeds 13?

Solution We will use our normal curve approximation:

$$\mu = (100)(.10) = 10$$

$$\sigma = \sqrt{(100)(.10)(.90)} = 3.00$$

$$P(\text{binomial } x > 13) \doteq P(\text{normal } x > 13.5) = P(z > 1.17) = .1210$$

where

$$z = \frac{13.50 - 10}{3} = 1.17$$

```
         .5000
       - .3790
Answer   .1210
```

Facts to Think About

1. The base of the approximating rectangle associated with zero successes in Fig. 10-4 is $(-.5, .5)$.
2. Either of the two outcomes in a binomial experiment may be called success.
3. If one outcome is called success and the value of μ is computed and then the other outcome is called success and the value of μ again computed, the two values of μ are equal only when $p = .5$. However, it is always true that the sum of the two values will total n. It will be left as an exercise to show that this is true. Regardless of which outcome is labeled success, the value of the variance is the same. Do you see why?
4. We obtain our best normal approximation to the discrete sampling distribution of success numbers when n is large and $p = .5$. However, it has been determined that the approximation is adequate if both $np > 5$ and $nq > 5$. That is, the normal approximation is adequate if the value of μ exceeds 5 for whichever outcome is called success. The next fact supplies the motivation for this fact.
5. When we say that we are using a normal curve to approximate our discrete sampling distribution of success numbers, we mean a normal curve with the tails cut off at $x = 0$ and $x = n$. Thus, we need $np > 5$ to allow our sampling distribution of success numbers an opportunity to generate left rectangle areas that will approximate the area below the left tail of our normal distribution ($nq > 5$ to take care of the right side).

6. The normal curve that we use as an approximation is one with $\mu = np$ and $\sigma = \sqrt{npq}$. By specifying these two parameters we determine a unique normal distribution.
7. When we are evaluating probabilities, in an effort to adjust to the continuity of the normal curve, all normal variable values will end in ".5."

Example 10-11 There are 200 questions in a multiple-choice exam with four parts to each question. Robert isn't prepared to answer 80 of these questions. What is the probability that by sheer guesswork he can correctly answer 25 to 30 of these 80 questions?

Solution

$$n = 80 \quad p = 1/4 \quad \mu = 20 > 5$$
$$\sigma = \sqrt{(80)(1/4)(3/4)} = 3.87$$

$P(25 \le \text{binomial } x \le 30) \doteq P(24.5 < \text{normal } x < 30.5) = P(1.16 < z < 2.71) = .1196$

where

$$z_1 = \frac{24.5 - 20}{3.87} = 1.16$$

$$z_2 = \frac{30.5 - 20}{3.87} = 2.71$$

Answer $\quad \begin{array}{r} .4966 \\ -.3770 \\ \hline .1196 \end{array}$

Example 10-12 An established drug cures a blood disease 80% of the time. What is the probability that we can cure at least 76 of the next 100 patients who come to us with this blood disease?

Solution

$$n = 100 \quad p = .80 \quad \mu = 80 > 5$$
$$\sigma = \sqrt{(100)(.80)(.20)} = 4$$

$P(\text{binomial } x \ge 76) \doteq P(\text{normal } x > 75.5) = P(z > -1.13) = .8708$

                        .5000
                      + .3708
               Answer   .8708

EXERCISES A

10-6. In Fig. 10-4 what is the base of the approximating rectangle associated with 16 successes?

10-7. A coin is tossed 400 times. Use the normal curve approximation to find the probability of obtaining:
  a. Between 185 and 210 heads inclusive. (Example 10-9c)
  b. Exactly 205 heads (Example 10-8)
  c. Less than 176 or more than 227 heads.

16-8. The probability that a patient recovers from a delicate heart operation is .9. What is the probability that between 84 and 95 (note, this is not inclusive) of the next 100 patients who have this operation survive? Use the normal curve approximation. (Example 10-11)

10-9. A survey of residents of the United States showed that 20% of these people preferred a white telephone over any other color available. What is the probability that between 170 and 185 inclusive of the next 1000 telephones installed in the United States will be white?

10-10. Moore Pharmaceutical Inc. knows from research done in its laboratory that .05 of a certain type of pill will have a basic ingredient that is below full strength. Because such pills are worthless, what is the probability that fewer than 10 pills in a sample of 200 pills will be unacceptable?

10-3 HYPOTHESIS TESTING

In this section we deal with experiments that yield a whole number of successes such as two successes, or two to five successes. We are not concerned with fractional successes, such as 2.5 successes.

We are going to test the null hypothesis, H_0: $p = p_0$. A sample of n binomial trials is recorded of which x are labeled "success" ($0 \leq x \leq n$) and $n - x$ are labeled "failure." A level of significance, α, is selected and, on the basis of our null hypothesis ($p = p_0$) we calculate the values of both μ and σ for the sampling distribution of success numbers for samples of size n. *The object now is to compare the actual number of successes obtained in a sample with the average number of successes expected (μ), assuming H_0 is true.* If the number of successes actually obtained in our sample differs markedly from the expected number of successes (μ), we reject H_0. This process of testing a hypothesis about the value of p is very similar to our previous hypothesis testing of μ.

10-3 HYPOTHESIS TESTING

Note that we have made no restriction as to our sample size. You should not associate the fact that our test statistic is a z score with the necessity for having a large sample size ($n \geq 30$).

Example 10-13 An advertising executive maintains that exactly 50% of all the people who saw a certain spectacular show on television will remember the name of the product it advertised 24 hours later. In order to try to substantiate the executive's claim, 200 people who watched the show were interviewed 24 hours later. Of these 200 people, 112 remembered the name of the advertised product. Test the executive's claim at the .05 level.

Solution Because we believe the executive's claim to be one of simple equality, we will execute a two-tail test.

$$H_0: p = .5 \qquad H_a: p \neq .5$$
$$\mu = np = (200)(.5) = 100, \quad \sigma = \sqrt{(200)(.5)(.5)} = 7.07$$

z-test statistic = $(112 - 100)/7.07 = 1.70 < 1.96$

Conclusion There is insufficient evidence to reject H_0. The sample supports the executive's claim at the .05 level.

Example 10-14 Assume a well-known medicine causes improvement in 60% of all cases in which it is used. Researchers believe that they have discovered a new medicine that is superior to the old medicine. To test their belief at the .01 level, they obtain a random sample of 200 people who agree to try the new medicine. Of these people, 150 show improvement. Are the researchers correct in their belief?

Solution The researchers will place their belief in the alternative hypothesis in the hope it will be accepted.

$$H_0: p = .60 \qquad H_a: p > .60 \text{ (researchers)}$$
$$\mu = (200)(.60) = 120, \quad \sigma = \sqrt{(200)(.6)(.4)} = 6.93$$

z-test statistic = $(150 - 120)/6.93 = 4.33 > 2.33$

Conclusion Reject H_0. There is a significant difference at the .01 level. Accept the researcher's claim at this level.

With the previous example as a guide, we can make a few comments. First, if the number of successes in our sample had been 120 or less, the sample data would support H_0. That is, our sample would have provided no evidence for rejecting H_0. We would have stopped work immediately. Second, the rule of thumb for locating a one-tail region of rejection still holds. See which way the greater or less-than sign is pointing in H_a. Finally, because our test statistic is two full units beyond our critical point, we feel that there is not a great chance of making a type I error.

As always, H_0 is a statement of strict equality. $H_0: p = p_0$. H_a is expressed in one of the three following ways, depending on the problem at hand: H_a is $p \neq p_0$ or $p > p_0$ or $p < p_0$.

Example 10-15 A museum purchased a million dollar painting in the expectation that 75% of the visitors to the museum would be attracted to the painting. In an attempt to justify the expenditure for the painting, the museum trustees decided to test the wisdom of their purchase at the .01 level. Thirty-six of the next 48 people who entered the museum stopped to look at the painting. Were the trustees correct?

Solution $H_0: p = .75$ $H_a: p > .75$ (trustees)
$n = 48$, the sample has 36 successes

$$\mu = (48)(.75) = 36$$
$$\sigma = \sqrt{(48)(.75)(.25)} = 3.00$$

Conclusion There is insufficient evidence to reject H_0. Stop the testing process; the number of successes in the sample support H_0.
The sample provides no evidence to indicate that H_0 may be untrue.

Example 10-16 Hugo believes that he has ESP. In order to test this null hypothesis he is given a deck of 50 cards that are red or blue. The deck is well shuffled and placed face down on the table. Hugo is to state the color of each card before he turns it over. In testing Hugo's claim at the .05 level, we find that Hugo can correctly predict 32 of the 50 cards. Is the test result significant?

Solution $H_0: p = .5$ $H_a: p > .5$ (Hugo's claim)

$$\mu = (50)(.5) = 25$$
$$\sigma = \sqrt{50(.5)(.5)} = 3.54$$

z-test statistic $= (32 - 25)/3.54 = 1.98 > 1.65$

Conclusion Reject H_0. Accept Hugo's claim. On the basis of this one sample he did significantly better than can be attributed to pure chance.

Example 10-17 A real estate agent is attempting to sell a parcel of land as a potential site for a new restaurant. He claims that at least 10% of the people in the town where the land is located will patronize this new restaurant. Because our investment in land and future construction is substantial we would like to check the real estate agent's claim at the .01 level.

A random sample yields the fact that 12 of 100 townspeople sampled are interested in patronizing the restaurant. Test the real estate agent's claim.

Solution $H_0: p = .10$ $H_a: p > .10$ (agent's claim)

$$\mu = (100)(.10) = 10$$
$$\sigma = \sqrt{100(.10)(.90)} = 3$$

z-test statistic = $(12 - 10)/3 = .67 < 2.33$

Conclusion There is insufficient evidence to reject H_0. The sample doesn't back up the agent's claim in H_a. In fact, it isn't even close. We must seriously examine our resources versus the risk involved in this business venture before we do any additional funding of the project.

EXERCISES A

10-11. A company claims that less than 10% of its radish seeds fail to germinate. For testing purposes, 100 seeds were planted and 20 failed to germinate. Does this imply that the company claim is correct at the .05 level?

10-12. We believe that a coin is not true (not perfectly balanced) and we wish to run a binomial test to show that this is so. For testing purposes, the coin is flipped 100 times and lands heads up 60 times. Test at the .05 level. (Example 10-13)

10-13. The percentage of A's given by a physics professor at a major university has averaged .10 over the years. Because 40 of her students received A's on this term's exam, the professor believes that it must have been easier. Test the professor's belief at the .01 level. $n = 200$. (Example 10-14)

10-14. A precision machine is considered in proper working order if no more than 1 in 1000 items produced is defective. A shipment of 10,000 items is found to have 15 defective items. Based on this sample, is the machine in proper working order? Test at the .01 level. Test that it is not in proper working order.

10-15. A town manager feels that the probability that people in the town favor a new incinerator is at least .80 because of the taxes involved. Of 100 people randomly selected and interviewed, 72 favored the new incinerator. Test the town manager's belief at the .05 level.

10-16. Squirp Corporation claims that its pregnancy pills are at least 99% effective. The company would like to establish this claim. All testing is to be done at

the 5% level. Try to find support for the company claim based upon sample data from 500 women who used the pill,
a. If 498 women found the pill effective.
b. If 499 women found the pill effective.
c. Compare the results of (a) and (b).

10-17. We believe that a particular job is held by the same number of men as women. To test this belief we select a sample of 200 people in this job category and find that 110 are men and 90 are women. Will the equality-of-sex hypothesis hold up at the .01 level?

10-18. A personnel director maintains that 60% of all single women employed in clerical jobs get married and quit work within two years after they are hired. The result is a loss of money invested in additional training for these women. The director randomly selected 600 women employed in clerical positions during the past five years, and found that 390 women did indeed get married and quit work within two years of being hired. Test the director's claim at the .05 level.

10-19. An educator believes that less than 30% of incoming freshman students at his university drop out of college by the end of their freshman year. He finds that 110 out of a randomly selected sample of 400 beginning college students dropped out by the end of the freshman year. Test the educator's belief at the .01 level.

10-4 THE SIGN TEST

The *sign test* is a nonparametric test that is related to the binomial test. For this test, we make no assumptions about population parameters.

The sign test deals with a comparison of data from two different populations, A and B. One sample is selected from population A and another from population B. The data from the two samples must possess a one-to-one correspondence. We wish to subtract matching pairs of data values and concentrate all of our attention upon the algebraic sign (plus or minus) of this difference; hence the name sign test.

Our rationale is that should we have two equal populations, A and B, with elements from each population matched one-to-one in a random manner; half of the differences between pairs should be plus and half of the differences between pairs should be minus. For convenience, we consistently start with a piece of data from the A population and subtract the matching piece of data from the B population. For example, should we have the pair a in A where $a = 4$ and b in B where $b = 5$, then $a - b = -1$. The fact that the algebraic sign of the answer is negative is the only thing that is of interest to us. If a equals 4.5 and b equals 3.7, then we are interested only in the fact that $a - b = +$. In the event that we encounter a pair of data where a and b are equal $(a - b = 0)$, we omit this pair.

Example 10-18 Sixteen women are asked to rate two hand creams, A and B, on their ability to prevent dryness. The women are to rate each hand cream on a numerical scale from 1 (very poor) to 15 (very good) inclusive.

a. Conduct a two-tail test, at the .05 level, to test the conjecture that the two hand creams perform equally well. Test H_0: $p = 1/2$ versus H_a: $p \neq 1/2$.
b. Conduct a one-tail test, at the .05 level, based on the conjecture that brand A is superior to brand B. (Test the belief that the magnitude of values in column A exceeds the matching values in column B.) H_0: $p = 1/2$ H_a: $p > 1/2$.

Pair	A	B	Sign of A − B
1	10	9	+
2	8	9	−
3	7	6.2	+
4	9.7	6	+
5	12	11	+
6	13	10	+
7	5.1	6.4	−
8	10	9.68	+
9	8	7	+
10	10	7.5	+
11	11	12	−
12	12	5	+
13	11	10	+
14	11.1	12	−
15	14	10	+
16	14	13	+

Solution

a. If our null hypothesis is true, the probability that a data value from the A population exceeds the matching data value from the B population is exactly 1/2. This is true for every matched pair.

H_0: $p = 1/2$ H_a: $p \neq 1/2$

We are going to conduct a two-tail test.
We have 12 successes (12 pluses). We let plus be success.
$\mu = (16)(1/2) = 8$,
$\sigma = \sqrt{(16)(1/2)(1/2)} = 2$

z-test statistic $= (12 - 8)/2 = 2 > 1.96$

Conclusion Reject H_0.

b. Because we are now testing the belief that the data in the A population are larger than the data in the B population, we hope to have the pluses dominate.

H_0: $p = 1/2$ H_a: $p > 1/2$

We are going to conduct a one-tail test.

z-test statistic $= (12 - 8)/2 = 2 > 1.64$

Conclusion Reject H_0. Indeed, the magnitude of the data from the A population does appear to be significantly larger than the magnitude of the data from the B population.

Since we do not use the exact value of measurements in the sign test, we do lose information. A sign test based on small samples may not detect small differences between two populations; however, this doesn't detract from the value of the test as a rapid method of detecting a difference between two populations. On the positive side, our old criterion for obtaining adequate results is that $\mu = np > 5$ (and $nq > 5$). Because p always equals $1/2$, this criterion can be met by having $n > 10$.

Let's see if we can't turn the execution of the sign test into a rather straightforward process. Our test has two outcomes, plus and minus. For convenience, we always call plus (+) the success outcome. The following outlines in more detail a setup for hypothesis testing for the sign test.

Let Plus (+) Represent Success: A-B

1. If we wish to test the belief that there is no difference between populations A and B, then we execute a two-tail test with H_a: $p \neq 1/2$.
2. If we favor A (the pluses), then we execute a one-tail test with H_a: $p > 1/2$.
3. If we favor B (the minuses), then we execute a one-tail test with H_a: $p < 1/2$.

Example 10-19 A health study is done to determine whether or not vitamin C reduces the incidence of colds in humans. Sixteen people agree to take large doses of vitamin C at regular intervals for a year. The number of colds contracted during this experimental year will be compared with the number of colds contracted by the same people during the previous year. The data are given in the following. Run a sign test, at the .05 level, to see if there is a significant difference.

Person	A: Number of Colds without Vitamin C	B: Number of Colds with Vitamin C	Sign of A minus B
1	6	1	+
2	5	1	+
3	2	0	+
4	1	0	+
5	3	4	−
6	4	2	+
7	2	3	−
8	6	2	+
9	5	3	+
10	4	2	+
11	4	3	+
12	4	2	+
13	2	1	+
14	2	3	−
15	3	1	+
16	2	0	+

Solution $H_0: p = .5$ $H_a: p \neq .5$ (We do not favor any outcome.)
$\mu = 16(.5) = 8$
$\sigma = \sqrt{(16)(.5)(.5)} = 2$
We have 13 successes (+'s).

z-test statistic = $(13 - 8)/2 = 2.50 > 1.96$

Conclusion Reject H_0. We would accept H_a. The data seems to indicate that large doses of vitamin C reduce the incidence of colds.

Example 10-20 Twenty-one women were asked to compare two facial creams, A and B. The women then graded the two creams on a scale of 0 to 5. Use the sign test to determine whether there is sufficient evidence to indicate that cream A is preferred over cream B. Test at the .01 level. The data are listed in the following table.

Pair	A	B	A − B
1	3	2	+
2	4	2	+
3	3	5	−
4	5	4	+
5	4	3	+
6	3	2	+
7	3	4	−
8	4	3	+
9	3	2	+
10	4	2	+
11	5	4	+
12	3	4	−
13	2	1	+
14	3	2	+
15	5	3	+
16	5	4	+
17	5	3	+
18	2	3	−
19	4	2	+
20	4	3	+
21	3	3	tie (omit)

Solution $H_0: p = 1/2$ $H_a: p > 1/2$
$\mu = (20)(.5) = 10$ $\sigma = \sqrt{(20)(.5)(.5)} = 2.24$
We have agreed, as a convention, to let plus be success. We have 16 successes (+'s).

z-test statistic = $(16 - 10)/2.24 = 2.68 > 2.33$

Conclusion Reject H_0. The women do feel that there is a difference between the creams and that A is the superior cream.

Example 10-21 The Health Education Department of a large university is doing research on a diet to improve a person's daily energy level.

After a great deal of screening, 24 candidates were selected for testing and formed into 12 matched pairs. The matching process was quite extensive because it involved matching sex, body weight, and average daily activity.

The 12 candidates in group A used the special university breakfast diet and the 12 in group B ate their usual breakfasts. At the end of three months, the same energy test was given to all candidates, with the following results.

Conduct a sign test at the .05 level to see if we can substantiate the Health Education Department's belief in the new breakfast diet.

Matched Pair	Group A	Group B	A − B
1	53	47	+
2	41	35	+
3	40	43	−
4	38	36	+
5	61	52	+
6	45	39	+
7	58	38	+
8	43	51	−
9	50	48	+
10	65	37	+
11	63	58	+
12	51	54	−

Solution $H_0: p = .5 \quad H_a: p > .5$ (the university)
$\mu = 12(.5) = 6 \quad \sigma = \sqrt{(12)(.5)(.5)} = 1.732$
We have nine successes (pluses).

$z\text{-test statistic} = (9 - 6)/1.732 = 1.732 > 1.64$

Conclusion Reject H_0. Accept H_a. However, our test statistic is close to the critical point.

EXERCISES A

10-20. The television habits of 14 children randomly selected from a large elementary school were investigated to see if there was a difference in viewing between days when school is in session and weekends when school isn't in session. Run a sign test at the .05 level to see if there is a significant difference in viewing time. (Example 10-19)

Child	Hours School in Session	Hours School Not in Session	Child	Hours School in Session	Hours School Not in Session
1	1.0	.5	8	11.00	8.20
2	4.5	4.0	9	2.20	1.00
3	9.5	8.3	10	15.50	16.50
4	6.0	7.1	11	19.50	4.50
5	8.0	7.2	12	3.50	2.40
6	5.8	12.5	13	10.00	6.50
7	28.0	22.5	14	18.40	13.50

10-21. Twelve overweight female patients at a large clinic undertake a diet plan. Run a sign test, at the .05 level, to see if there is a significant drop in weight because of this diet. Their before and after weights, in pounds, are as follows: (Example 10-20)

Before	After
108	92
123	105
137	135
152	153
128	126
112	114
127	128
110	105
108	111
146	122
155	129
102	107

10-22. A professor gives one of his classes what he feels is a strongly motivating talk on improving arithmetic accuracy in executing basic statistical formulas. He gives the class a pretest on handling arithmetic concepts, followed by his lecture, and then by a post-test on the same arithmetic concepts. He would like to use the results of the two tests to determine whether the lecture actually improved their scores. Conduct a sign test at the .05 level for this purpose. The test results are
Pretest: 65, 80, 50, 92, 85, 75, 45, 89, 60, 65, 71, 69
Post-test: 75, 79, 90, 90, 80, 85, 68, 89, 82, 75, 68, 77

 10-23. An experiment in cloud-seeding to increase rainfall is conducted using the towns of Ashland and Axton. Data were recorded, as shown, for both towns for a year. The clouds over Ashland were not seeded whereas those over Axton were seeded at regular intervals.
  a. Execute a sign test to show that there was no difference in rainfall.
  b. Execute a sign test to show that Axton received more rainfall as a result of the regular seeding.

| | *Inches of Precipitation* | | |
Month	Ashland	Axton	
January	1.49	1.38	+
February	1.40	1.40	omit
March	2.18	2.63	−
April	2.61	2.52	+
May	3.90	4.80	−
June	4.50	4.30	+
July	3.58	3.89	−
August	3.40	3.50	−
September	4.00	3.90	+
October	2.51	2.62	−
November	1.91	1.73	+
December	1.50	1.40	+

 10-24. A taxi company is trying to decide whether the use of radial tires improves fuel economy. Twelve cars were equipped with radial tires and driven over a prescribed test course. Without changing drivers, the same cars were then equipped with the regular belted tires and driven over the test course. Let $\alpha = .05$. The gasoline consumption, in kilometers per liter, was as follows:

Car	Radial Tires	Belted Tires
1	4.2	4.1
2	4.7	4.9
3	6.6	6.2
4	7.0	6.9
5	6.7	6.8
6	4.5	4.4
7	5.7	5.7
8	6.0	5.8
9	7.4	6.9
10	4.9	4.7
11	6.1	6.0
12	5.2	4.9

10-25. The 15 members of a citizens' committee for tax relief were polled to determine which form of tax relief they favored
  Plan A—increased deductions for taxpayers and dependents
  Plan B—decreased tax rates for all income levels
The results of the poll are given here. Test the null hypothesis that the members of the committee show no preference for either plan at the .05 level of significance. Here "1" is the superior rating.

Member	1	2	3	4	5	6	7	8	9	10	11	12	13	14	15
Plan A	1	1	2	1	2	2	1	2	1	1	1	2	1	1	1
Plan B	2	2	1	2	1	1	2	1	2	2	2	1	2	2	2

10-5 EXACT BINOMIAL PROBABILITIES (optional)

Consider the experiment of rolling a die with success being a roll of 3, 4, 5, or 6. $P(\text{success}) = p = 2/3$; $P(\text{failure}) = q = 1/3$. Let's analyze the cases of 0, 1, 2, and 3 successes for 3 ($n = 3$) binomial trials.

10-5 EXACT BINOMIAL PROBABILITIES (optional)

Possible Outcomes	Probabilities
F F F	P(F F F) = (1/3)(1/3)(1/3) = 1/27
F F S	P(F F S) = (1/3)(1/3)(2/3) = 2/27
F S F	P(F S F) = (1/3)(2/3)(1/3) = 2/27
F S S	P(F S S) = (1/3)(2/3)(2/3) = 4/27
S F F	P(S F F) = (2/3)(1/3)(1/3) = 2/27
S F S	P(S F S) = (2/3)(1/3)(2/3) = 4/27
S S F	P(S S F) = (2/3)(2/3)(1/3) = 4/27
S S S	P(S S S) = (2/3)(2/3)(2/3) = 8/27

Observe that there is exactly one outcome (SSS) with a probability of 8/27; three outcomes, each with a probability of 4/27; three outcomes, each with a probability of 2/27; and exactly one outcome (FFF) with a probability of 1/27. Now consider the binomial expansion of $(F + S)^3$:

$$(F + S)^3 = \binom{3}{0} F^3 + \binom{3}{1} F^2S + \binom{3}{2} FS^2 + \binom{3}{3} S^3$$
$$= F^3 + 3F^2S + 3FS^2 + S^3$$

Coefficients of terms in the expansion of $(F + S)^3$ yield the number of ways of obtaining 0, 1, 2, and 3 successes in three independent trials (left to right). If we replace F by 1/3 and S by 2/3 in this expansion, the terms represent the probabilities of obtaining 0, 1, 2, and 3 successes in three independent trials. If $F = 1/3$ and $S = 2/3$:

$$(F + S)^3 = (1/3)^3 + (3)(1/3)^2(2/3) + (3)(1/3)(2/3)^2 + (2/3)^3$$
$$= 1/27 + (3)(2/27) + (3)(4/27) + 8/27$$

When we generalize this thinking, we find that exact binomial probabilities may be obtained by calculating specific terms in the expansion of $(q + p)^n$ where q denotes the probability of failure, p denotes the probability of success, and n denotes the number of independent trials in our sample.

$$(q + p)^n = \binom{n}{x} p^x q^{n-x}; \text{ where } x = 0, 1, 2, \ldots, n$$

If $p = .7$ ($q = .3$), the exact binomial probability of 85 successes in 100 independent trials is $\binom{100}{85} (.7)^{85}(.3)^{15}$. To complete this calculation, even

218 THE SAMPLING DISTRIBUTION OF SUCCESS NUMBERS

with the aid of a hand-held calculator, is an overwhelming chore. Indeed, this illustration and the next examples show the power of the normal curve approximation to the binomial distribution.

Example 10-22 (a) Calculate the (exact) binomial probability of four successes in 16 trials with $p = 1/2$. (b) Compare the result with the approximation of the same probability obtained in example 10-8.

Solution

a. $\binom{16}{4} (1/2)^4 (1/2)^{12} = .0278$

b. The difference in answers is .0001.
The previously obtained approximation is remarkably accurate.

Example 10-23 (a) Calculate the (exact) binomial probability of obtaining from four to six successes inclusive in 16 trials with $p = 1/2$. (b) Compare the result with the approximation previously obtained in Example 10-9.

Solution

a. $\binom{16}{4} (1/2)^4 (1/2)^{12} + \binom{16}{5} (1/2)^5 (1/2)^{11} + \binom{16}{6} (1/2)^6 (1/2)^{10}$
$= .0278 + .0667 + .1222$
$= .2165$

b. The difference in answers is $(.2165 - .2144) = .0021$. (This is still a very accurate approximation.)

EXERCISES A

Evaluate the following binomial probabilities.

10-26. What is the probability of obtaining two successes in five trials if $p = .5$? (Example 10-22)

10-27. What is the probability of obtaining two or three successes in four trials if $p = 4/5$?

10-28. What is the probability of obtaining less than four successes in four trials if $p = 4/5$? [Hint: Do a subtraction.]

10-29. What is the probability of obtaining five successes in six trials if $p = 1/2$?

10-30. What is the probability of obtaining exactly two heads when six pennies are tossed?

10-31. A box contains 10 balls that are alike in every respect but color. Seven are white and three are black. A blindfolded person selects a ball at random and the color is noted. Then the ball is replaced and the balls thoroughly mixed. Then a second ball is drawn, and so forth. In five such drawings, what is the probability of obtaining three white and two black balls?

SUMMARY

This chapter introduces the *binomial experiment*. This is an experiment that is designed to have exactly two outcomes, *success* and *failure*. The P(success) $= p$ and P(failure) $= q$ or $1 - p$. Each performance of a binomial experiment is called a *trial,* and a sample consists of n independent trials, where it is assumed that the value of p remains constant for each trial.

By repeatedly recording the number of successes for samples of size n, it is possible to obtain a sampling distribution of *success numbers* that is normally distributed with $\mu = np$ and $\sigma = \sqrt{npq}$. The restrictions are that $\mu = np > 5$ and $nq > 5$.

It is a lot of work to calculate exact binomial probabilities. Because of this, the calculation of exact binomial probabilities has been left to an optional section. A normal distribution with parameter values $\mu = np$ and $\sigma = \sqrt{npq}$ allows us to quickly obtain an approximation of a binomial probability that is very accurate. Hence, there is a question as to how much time, if any, should be spent on the laborous chore of calculating exact binomial probabilities.

In regard to the parameter p, this chapter will test statistical hypotheses about p. Estimating p is covered in the next chapter.

The chapter ended with another nonparametric test, the *sign test*. This is a specialized binomial test in which the two outcomes are plus (+) and minus (−).

Can You Explain the Following?

1. binomial experiment
2. trial
3. the two outcomes of a binomial experiment
4. the formula for μ for the sampling distribution of success numbers
5. the formula for σ for the sampling distribution of success numbers
6. the two restrictions for using the normal curve approximation to the binomial
7. the sign test

CUMULATIVE REVIEW

MISCELLANEOUS EXERCISES

10-32. Give two examples of a binomial experiment.
10-33. If $n = 81$ and $p = 1/9$, find σ^2 and σ.
10-34. If $n = 63$, and $p = 2/7$, find σ^2 and σ.

In Exercises 10-35 through 10-37 use the normal curve approximation to the binomial distribution.

10-35. If $n = 16$ and $p = 1/2$, find $P(6 < \text{binomial } x < 9)$.

10-36. A fair coin is tossed 900 times. Find the probability that heads will turn up between 400 and 450 times inclusive.

10-37. Mendell discovered that when the seeds of tall F_1 peas were planted, 75% of the resulting plants were tall and 25% were dwarf. If 100 F_1 peas are planted, what is the probability that at least 28 plants are dwarf?

10-38. For each H_a given write H_0.
    (a) $H_a: p < .4$.   (b) $H_a: p \neq 2/3$.

10-39. The American Medical Association decided to test a Department of Health, Education and Welfare claim that 10% of all persons over 65 years of age are covered by adequate private health insurance. A random sample of 900 elderly persons showed 99 possessed adequate private health insurance. Test the HEW claim at the 5% level.

10-40. The Consumer Protection Union wishes to test a television manufacturer's claim that less than 10% of all new color television sets need any repair during their first two years of operation. To test this claim a random sample of 100 sets finds that 15 sets require some repair within the first two years. Conduct the test at the 1% level.

10-41. It is claimed that 20% of all families in San Francisco are Mexican-American. What do you conclude about this claim if a random sample of 400 families shows that 90 are Mexican-American?

10-42. The production department of a large corporation conducted a test to determine whether or not background music played during working hours made a difference in production. The test data were gathered over a period of 10 consecutive working days. Use the sign test to test the hypothesis that the distribution output is identical under both conditions. Let $\alpha = .05$.

Day	Output with Music	Output without Music
1	780	790
2	940	910
3	920	910
4	850	880
5	860	850
6	840	820
7	830	825
8	840	850
9	910	890
10	900	888

10-43 A stock market analyst wished to compare the return on investment of two prominent stocks. In the absence of any hard data relative to either company, she decided to use the sign test at the .01 level to compare the two companies. Test the belief that company A is superior to company B.

	% Return on Investment	
Year	Company A	Company B
1969	7.9	8.1
1970	7.5	7.3
1971	7.3	7.2
1972	7.3	7.3
1973	6.8	6.5
1974	5.6	5.8
1975	5.5	4.7
1976	4.3	3.8
1977	3.7	2.9
1978	3.4	2.5
1979	3.8	3.1

11 THE SAMPLING DISTRIBUTION OF SUCCESS RATIOS

11-1 INTRODUCTION

Maria and Marsha were discussing how severe their respective statistics instructors had been in grading the course final. Maria said that two students in her class had failed whereupon Marsha replied that four students had failed in her statistics class. Four failures are more than two failures, but the discussion did not end there. Maria went on to say that there had been 20 students in her class, and Marsha revealed that there had been 40 students in her class. Thus, in Maria's class 2/20, or 1/10, failed the final exam and in Marsha's class 4/40, or 1/10, failed the final exam. A comparison of the failure ratios (the proportion of each class that failed) showed the instructors to be equally severe in their grading.

To say that 24,557 of 26,434 doctors surveyed oppose National Health Insurance is a strong statement, but an even stronger statement might be to say that 24,557/26,434, or 92.9%, of all doctors surveyed oppose it. This was the result of a survey by *Private Practice Magazine*. Further, the proportion of doctors who stated that they would refuse to practice under National Health Insurance was 27.2%.

If we return to the PVC Plastic Company and its sample data in Table 10-1, we see 45 pieces of pipe in the sample with an accurate diameter measurement. There are 45 successes in a sample of 50 pieces of pipe. It is sometimes more revealing to look at the sample success ratio. Here, 45/50, or 9/10, of the sample is usable and the remaining 1/10 is junk.

In this chapter we encounter another sampling distribution, which is closely related to the sampling distribution of success numbers of the previous chapter. The binomial experiment discussed in Chapter 10 is the foundation upon which this sampling distribution is built.

For an arbitrary sample of size n, we create a ratio whose numerator is the number of successes, x, in the sample and whose denominator is the sample size. The ith such ratio would then be x_i/n, where $n \neq 0$. Because $0 \leq x_i \leq n$, we have $0 \leq x_i/n \leq 1$ for all i. That is, every success ratio will

Figure 11-1

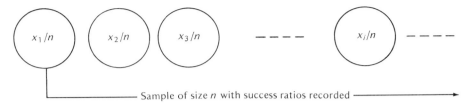

have a value between 0 and 1 inclusive. In this chapter we are interested in the sampling distribution of such ratios. See Fig. 11-1.

It is common practice to refer to each ratio as a *sample proportion*. However, we will continue to use the word ratio because we feel that "proportion" implies the two equal ratios of traditional high school usage, and it is not our intention to convey this meaning.

In dealing with any population parameter, two concepts always arise. (1) estimating the parameter value, and (2) conducting tests of hypotheses related to the parameter. Although both this and the last chapter are concerned with the binomial experiment parameter, p, this chapter will deal with the two types of parameter estimation (point and interval) whereas Chapter 10 dealt with testing hypotheses related to p.

Definition 11-1 For an arbitrary sample of size n, the ratio of x/n (number of successes/sample size) will be denoted $\hat{p}$. The tilde, over the p is sometimes read as "p-hat."

Any such ratio is both a statistic and a point estimate of p. Further, we will show in the optional section that $\hat{p}$ is an unbiased estimator of p.

Example 11-1 A guidance counselor would like to estimate the proportion of the senior class who have PSAT scores in English and mathematics above 550. Ninety students are selected from the class and 15 are found to be in this category. Compute the estimate for the guidance counselor.

Solution $\hat{p} = 15/90 = 1/6$

Example 11-2 We are working on our new superhigh-velocity tennis serve, and would like to estimate our probability of using this serve and not faulting. We execute our high-velocity serve 100 times, and find that we are successful in 83 of these serves. Compute the point estimate of p.

Solution $\hat{p} = .83$

Theorem 11-1 a. The mean of the sampling distribution of success ratios, $\mu_{\text{ratios}} = p$.
b. The variance of the sampling distribution of success ratios, $\sigma^2_{\text{ratios}} = [p(1-p)]/n$.

Proof The proof of Theorem 11-1 may be found in the optional section.

Example 11-3 Our binomial experiment is to toss a coin in the air and observe whether it lands heads (H) or tails (T). Let success be heads. Find the value of μ and σ^2 for the sampling distribution of success ratios when dealing with a sample size of 60.

Solution $\mu = p = .5$, $\sigma^2 = [(.5)(.5)]/60 = .004$

Example 11-4 A marksman hits a target with a probability of .75. For samples of size 64, find the value of μ and σ^2 for the sampling distribution of success ratios.

Solution $\mu = .75$ $\sigma^2 = [(.75)(.25)]/64 = .003$

Now we are in a position to state an important theorem dealing with the sampling distribution of success ratios (proportions). Actually, this theorem may be considered a corollary of Theorem 10-1.

Theorem 11-2 The sampling distribution of success ratios for large sample sizes is approximately normally distributed (where $np > 5$ and $nq > 5$).

First, this implies that the sampling distribution of the statistic $(\hat{p} - p)/\sqrt{pq/n}$ is standard normally distributed. Second, the theorem does say the sample size should be large; however, we are often willing to settle for adequate results. It has been determined that adequate results will be achieved when $\mu = np > 5$ (and $nq > 5$). We recognize this as the same criterion that we encountered previously.

This last theorem is sometimes referred to as the *central limit theorem of sample proportions*.

EXERCISES A

11-1. The following grades were obtained in a 10-point quiz: 10, 9, 8, 8, 8, 7, 7, 7, 7, 7, 6, 6, 6, 5, 5, 4.
   a. Find the ratio of 7's to the total sample. (Example 11-1)
   b. Find the ratio of 7 or larger to the total sample.

11-2. In April 1975 there were 84,086,000 employed workers and 8,176,000 unemployed workers. (a) Assuming that we have all the potential workforce recorded, what is the ratio of employed workers to the workforce? (b) What is the ratio of unemployed workers to the workforce?

11-3. A large soap company claims that the probability that a person selected at random will be using its product is .25. For samples of size 64, find the value of μ and σ^2 for the sampling distribution of success ratios. (Example 11-3)

11-4. It is an established fact that a cold remedy provides relief in 80% of all cases. For samples of size 100, find the value of μ and σ for the sampling distribution of success ratios.

11-5. In a large city, the probability of an alarm being false is .40. Compute the value of μ and σ^2 for the population of success ratios. Let $n = 36$.

11-6. It is known that the probability of passing a difficult physics course is .90. For classes of 50 students taking this course, compute μ and σ for the sampling distribution of success ratios.

B

11-7. If we are dealing with the sampling distribution of success ratios where $p = 1/3$ and $\sigma^2_{\text{ratios}} = 1/450$, what is the value of n?

We shall now develop the interval estimate of p, just as we did the interval estimate of μ. A formula for computing the end-points of the confidence interval for p is given in Theorem 11-3. In Section 11-3 we utilize this concept to provide yet another partial answer to the question of sample size—relative to p.

11-2 ESTIMATING p

We have already discussed $\hat{p}$, our point estimate of p, it only remains to discuss the interval estimate of p.

Theorem 11-3 The end-points of a $100(1 - \alpha)\%$ confidence interval of p are $\hat{p} \pm z \sqrt{\hat{p}(1 - \hat{p})/n}$, where $\hat{p}$ is determined by a sample and z is a positive z score, determined by specifying α.

Proof We begin with a symmetric interval about zero on the Z-axis. The actual size of the interval is determined by our selection of a level of significance, α.

1. $-z < (\hat{p} - p)/\sqrt{p(1-p)/n} < z$
The expression in the middle of line 1 is the z score of $\hat{p}$.
The logic is to solve this inequality for the value of p in the numerator. However, this will mean that we will be solving for p in terms of p. To avoid this untenable position, we replace p by its point estimate $\hat{p}$ in $\sqrt{p(1-p)/n}$. Thus (after a little algebra),

2. $\hat{p} - z\sqrt{[\hat{p}(1-\hat{p})]/n} < p < \hat{p} + z\sqrt{[\hat{p}(1-\hat{p})]/n}$
This last line yields the end-points stated in Theorem 11-3.

Example 11-5 A school superintendent is trying to obtain a 95% confidence interval to estimate the probability of a student, having an IQ less than or equal to 110. For this purpose, the names of 100 students are randomly selected. It is found that 61 have IQ's of 110 or less. Evaluate this interval for the superintendent.

Solution $\hat{p}$ is 61/100 or .61. Our 95% confidence interval for p is then:

$$.61 \pm (1.96)\sqrt{\frac{(.61)(.39)}{100}} = .61 \pm .10$$

or

$$.51 < p < .71.$$

Example 11-6a A psychologist believes that punishment slows down the learning process. To test this belief, students are randomly selected to work on a vocabulary list until they have mastered it. A student who falters in any way loses recess time as a punishment. The time needed to master the vocabulary list has been determined from previous testing. Four hundred students receive punishment and 320 students still fail to complete their assigned task in the allotted time. On the basis of this sample, compute the 95% confidence interval for the probability of failing to complete the assigned task in spite of receiving punishment.

Solution $\hat{p} = 320/400 = .80$. The end-points of the interval are:

$$.80 \pm (1.96)\sqrt{\frac{(.80)(.20)}{400}} = .80 \pm .04$$

11-3 ANOTHER PARTIAL SOLUTION TO THE QUESTION OF SAMPLE SIZE

or

$$.76 < p < .84.$$

Example 11-6b Return to Example 11-6a and compute the 99% confidence interval for p. Compare the lengths of the two intervals.

Solution Our end-points are:

$$.80 \pm (2.58)\sqrt{\frac{(.80)(.20)}{400}} = .80 \pm .05$$

or

$$.75 < p < .85.$$

The 99% confidence interval is longer than the 95% confidence interval found in Example 11-6a. This is always true.

EXERCISES A

11-8. A sample of 700 voters showed 390 in favor of repeal of a town ordinance. Find the 95% confidence interval for the true ratio (proportion) of voters who favor repeal. (Example 11-5)

11-9. An anthropologist found that in a sample of 100 people chosen at random from a certain remote region 65% were males. Find the 95% confidence interval for the population proportion, p, of males.

11-10. A random sample of 2500 eligible workers was taken, and the proportion ($\hat{p}$) of those voting for Mr. Jones was 44%. Find the 95% confidence interval for the true value of p.

11-11. Lenco Gear and Motor Company would like to obtain a 99% confidence interval to estimate the probability that its new aluminum motor will not fail in a one-year period. Compute this interval based on a trial sample of 100 motors in which 98 performed perfectly. (Example 11-6b)

11-12. A study showed that a sample of 80 Asians carefully selected to represent all income levels was 53% literate. Compute the 95% confidence interval for the population proportion, p, of literate Asians.

11-13. A recent report shows that approximately 30% of people in Third World countries now live in urban areas as compared with 16% in 1950. If this report is based on a sample of size 36, compute a 90% confidence interval to estimate the true proportion of Third World people now living in urban areas.

11-3 ANOTHER PARTIAL SOLUTION TO THE QUESTION OF SAMPLE SIZE

Just as in the estimation of the sample size, n, relative to μ, we need the previous development of a confidence interval for the parameter, p. Here, we use the confidence interval just derived for p. Our solution, as before, has a give and a take aspect. The give is another set of restrictions and the take is the determination of an appropriate sample size based upon these restrictions. We begin the determination of this sample size by starting with line 1 of Theorem 11-3.

1. $-z < \dfrac{\hat{p} - p}{\sqrt{p(1-p)/n}} < z$

2. $|\hat{p} - p| < z\sqrt{\dfrac{p(1-p)}{n}}$

This last line tells us that the distance between $\hat{p}$ and p is less than $z\sqrt{p(1-p)/n}$. Now we set this last expression less than any tolerance distance, d, we desire. Thus,

3. $|\hat{p} - p| < z\sqrt{\dfrac{p(1-p)}{n}} < d,$

which implies

4. $n > \dfrac{z^2(p)(1-p)}{d^2}$

At this point, we must replace p, in line 4, by its point estimate $\hat{p}$. The value z is a positive z score determined by our choice of a level of significance. We pick the tolerance distance, d. With all of these conditions stated, we have our formula for computing the sample size.

5. $\boxed{n > \dfrac{z^2(\hat{p})(1-\hat{p})}{d^2}}$

Example 11-7 Suppose we want to estimate the proportion of drivers who are exceeding the 55-mph speed limit on a stretch of highway in California. A sample of 100 drivers yields 82 drivers exceeding the 55-mph speed limit. We wish $\hat{p}$ to be within a .10 tolerance of p. How large a sample is needed to be 95% sure of achieving the desired tolerance?

Solution $n > [(1.96)^2(.82)(.18)]/(.1)^2 = 56.70$
Use $n = 57$ (or larger).

Example 11-8 A designer of shirts for men believes that approximately 40% of his shirts are purchased by women. It is important to him that he accurately estimate this probability as it will affect his choice of color and design. How large a sample will be needed to be able to assert with a probability of .99 that his estimate will not be off by more than .02?

Solution $d = .02$, $z = 2.58$, $\hat{p} = .40$
$n > [(2.58)^2(.4)(.6)]/(.02)^2 = 3,993.84$
Answer: $n = 3,994$ (or larger).

The last two examples dealt with computing a sufficiently large sample size, n, that will guarantee us that the distance between $\hat{p}$ and p will be within a stated tolerance. We need a point estimate, $\hat{p}$, in order to carry out our calculation. It may be helpful to think of $\hat{p}$ as a good starting point. See Fig. 11-2.

Figure 11-2

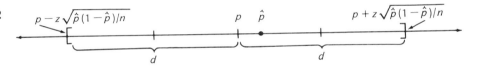

Example 11-9 Beckerfield, Horn, and Wimple, Inc., an investment company, wants to estimate the percentage of its customers who own more than one home. Previous surveys and a careful assessment of additional data indicate that the percentage is about .10. However, since a large and expensive promotional campaign for these customers is being planned, the investment company desires to estimate the percentage within .01 of the true value of p with .95 confidence.
How large a sample must be drawn?

Solution Our z score is 1.96, $\hat{p} = .10$, and we have decided upon a d value of .01.

$n > [(1.96)^2(.10)(.90)]/(.01)^2 = 3,457.44$
Answer $n = 3,458$ (or larger).

(At this point, we should remind you that when we speak of a success ratio the general practice is to call it a proportion.)

230 THE SAMPLING DISTRIBUTION OF SUCCESS RATIOS

EXERCISES A **11-14.** What would have been an appropriate sample size for the Lenco Gear and Motor Company to select if it wished to be 99% confident that its tolerance distance for $\hat{p}$ would not exceed .01? Use the data of Exercise 11-11 for your point estimate of p. (Example 11-8)

 11-15. A national union estimates that .60 of its members favor a new contract. How large a sample must be obtained to be 95% sure that the difference between p and $\hat{p}$ is not in excess of .01? (Example 11-7)

 11-16. A well-known pollster maintains that his estimate of the proportion of voters favoring a certain presidential candidate is not in error by more than .02. In an upcoming presidential race, how large a sample is needed to be 95% certain of this claim. Assume $\hat{p} = .55$.

 11-17. A tax collector for a southern city estimates that 30% of her collections are being sent in late, thus costing her time and the city interest dollars. She would like to estimate the true proportion of late taxpayers with an error estimate not to exceed .05. She feels she can use 30% as an estimate of p. What sample size should she elect to be 99% confident of obtaining her desired tolerance distance?

B **11-18.** Start with the end-points of the confidence interval for p. For fixed $\hat{p}$ and fixed z, discuss what happens to these end-points as $n \to \infty$.

11-19. What would be an appropriate definition of a 100% confidence interval for p?

11-20. Derive the results of Theorem 11-3 directly from the sampling distribution of success numbers of the previous chapter.

CUMULATIVE REVIEW

 11-21. A bowler would like a 95% confidence interval estimate of his probability of throwing a strike in an arbitrary box. Toward this end, he records six strikes in the next 100 boxes of bowling. Compute the interval for him.

11-4 SOME THEORY (optional)

In this opional section we are going to accomplish two things: (1) prove Theorem 11-1, and (2) derive an alternate formula for computing the sample size.

A Proof of Theorem 11-1

Our proof rests on the following two facts that the reader may wish to prove:

1. $\mu_{ax+b} = a\mu + b$
2. $\sigma^2_{ax+b} = a^2 \sigma^2$

For the case of Theorem 11-1, we have $a = 1/n$ and $b = 0$. Thus,

a. $\mu_{\hat{p}} = \mu_{(1/n)x} = (1/n)\mu = (1/n)np = p$

b. $\sigma_{\hat{p}}^2 = \sigma_{(1/n)x}^2 = (1/n)^2\sigma^2 = (1/n)^2(np)(1-p) = \dfrac{p(1-p)}{n}$

It is important to note that the proof of (a) establishes the fact that $\hat{p}$ is an unbiased estimator of p.

Another formula for computing n

In the previous section (line 5), we derived the formula

$$n > \frac{z^2(\hat{p})(1-\hat{p})}{d^2}$$

We will now derive another formula for computing n. This formula is independent of our point estimate, $\hat{p}$. The trick is to observe that, for constant z and d,

$$\frac{z^2(p)(1-p)}{d^2}$$

is maximal when $p = 1/2$.

Proof $p(1-p) = p - p^2 = 1/4 - (1/2 - p)^2$
Our expression takes on its maximum value when the square term is zero. This occurs when $p = 1/2$. With this result, we can offer a more conservative (meaning larger but very safe) estimate of our sample size, n, which is independent of $\hat{p}$, as follows:

1.
$$\boxed{n > \frac{z^2}{4d^2} \geq \frac{z^2(\hat{p})(1-\hat{p})}{d^2}}\text{, our previous result}$$

Example 11-10 Redo Example 11-7 for the case that $p = 1/2$.

Solution $n > (1.96)^2/4(.1)^2 = 96.04$
Note: This answer is much more conservative (larger) than the answer obtained in Example 11-7.

Example 11-11 Suppose that we want to estimate what proportion of heavy smokers will get lung cancer, and we want to be 95% sure of our work with a .04 estimate of error. What is a sufficiently large sample size?

232 THE SAMPLING DISTRIBUTION OF SUCCESS RATIOS

Solution

$$n > \frac{(1.96)^2}{4(.04)^2} = 600.25$$

EXERCISES A

11-22. What does the formula $n > z^2/4d^2$ become when $d = 1$?
11-23. Use line 1 of this section to redo Exercise 11-16. (Example 11-10)
11-24. Use line 1 of this section to redo Exercise 11-17.

SUMMARY

This chapter dealt with another binomial sampling distribution, the sampling distribution of *success ratios,* (or, as many texts prefer, "proportions"). The sampling distribution of success ratios is normally distributed with $\mu = p$ and $\sigma = \sqrt{pq/n}$. It is still necessary to satisfy the restrictions that $\mu = np > 5$ and $nq > 5$.

A success ratio is formed by dividing the number of successes, (call it x), obtained in a sample of n binomial trials, by the sample size. Thus, a success ratio is of the form x/n where $0 \le x/n \le 1$. Our success ratio is the point estimate of p. It is customary to denote the point estimate of p, $\hat{p}$, read "p-hat." Hence $\hat{p} = x/n$.

Armed with $\hat{p}$ and the fact that our sampling distribution of success ratios is normally distributed, it is possible to calculate a confidence interval estimate of p. The two most common confidence intervals are 95% and 99%.

In section 11-3 a formula was derived for determining the appropriate sample size, n. The problem and its solution is exactly analogous to the problem of determining the sample size for μ in Chapter 7. To determine a sample size for p it is necessary to first specify: (1) the desired distance, d. That is, the distance of $\hat{p}$ to p. (2) the desired probability, and (3) provide a point estimate of p, $\hat{p}$. Thus, as before, the question of sample size is a "give and take situation." We must have information, (or restrictions), before we can take away an answer.

The optional section shows that p is an unbiased estimator of p. Another formula was also derived for determining the sample size. This formula is more conservative than our previous formula.

Can You Explain the Following?

1. $\hat{p}$
2. the formula for μ for the sampling distribution of success ratios

3. the formula for σ for the sampling distribution of success ratios
4. the formula for determining n

CUMULATIVE REVIEW

MISCELLANEOUS EXERCISES

11-25. A credit card agency wants to estimate the proportion of creditors who miss a monthly payment. A random sample of 500 creditors shows that 32 failed to make their last monthly payment. Find $\hat{p}$.

11-26. If $n = 81$ and $p = 1/9$, find μ_{ratios} and σ^2_{ratios}.

11-27. If $n = 63$ and $p = 2/7$, find μ_{ratios} and σ^2_{ratios}.

11-28. Given $n = 400$ and $\hat{p} = 1/4$, compute the 95% confidence interval for the true value of p.

11-29. The health department of a large city wants to estimate the proportion of adults who smoke at least one package of cigarettes every day. A sample of 200 adults is randomly selected and it is found that 25 of these adults smoke at least one package of cigarettes every day. Find the 99% confidence interval for p.

11-30. The principal of a 5000-student high school wishes to determine the proportion of students absent on any given day with 95% confidence. A sample of 50 students shows two absent. Calculate a confidence interval estimate for the principal.

11-31. The manager of Convenient Motels Inc. would like to determine a sample size, with an error estimate not to exceed .08, to be used to determine the number of reservation "no-shows" in his motel chain. He would like to be 99% confident in using this sample size. The manager of a similar-size motel chain tells him that the proportion of no-shows is approximately 3%. Find the sample size.

11-32. How large a sample is required to estimate the proportion of families in a town owning color television sets if we wish to be 95% confident in our answer and to be off by no more than .02? A preliminary sample yields an estimate $\hat{p} = .32$.

12 THE SAMPLING DISTRIBUTION OF THE DIFFERENCE BETWEEN MEANS

12-1 INTRODUCING THE DISTRIBUTION

The simplest and perhaps most common of all experiments is a comparison of two essentially similar groups of data—such as the weights of two groups of men on different cholesterol-control programs. Another example would be the final test results of two groups of fifth graders where one group is taking a traditional mathematics course and the other is taking an experimental mathematics program. The experimenter makes a change in exactly one variable in one of the two groups. This variable is often referred to as the *independent variable*. The experimenter attempts to compare the two groups by measuring the difference between the means of the groups with regard to this variable. This difference between sample means may be referred to as the *dependent variable*.

Since the two groups were similar before we manipulated the independent variable in one group, any subsequent significant difference between sample means will be attributed to our different treatment of the independent variable. The key word is "significant," for we would expect mere chance alone to account for a small difference between sample means.

Our researcher might be seeking the answer to such questions as: Do Germans differ from Americans with respect to obedience to authority? Do Protestants or Catholics have a higher suicide rate? Is a name-brand suntan lotion superior to its nearest competitor? Are political conservatives greater disciplinarians than political liberals? Does the same dosage of vaccine produce different effects in two different age groups? Do children learn to read faster in an open classroom or in a self-contained classroom?

The ability to investigate such questions rests upon a knowledge of the sampling distribution of the difference between sample means. Assume that we have two populations that are normally distributed. Let's call the populations 1 and 2. The variance of population 1 must equal the variance of population 2: $(\sigma_1^2 = \sigma_2^2)$.

We select a random sample of size n_1 from 1 and another sample of size n_2

12-1 INTRODUCING THE DISTRIBUTION

from 2. The sample means will be denoted $\bar{x}_1$ and $\bar{x}_2$, respectively. The difference between these means, $\bar{x}_2 - \bar{x}_1$, represents a single piece of data in the sampling distribution of the difference between means. From each of the two populations we draw another sample, using the sample sizes previously selected. Compute each sample mean, compute the difference between sample means, and then record this difference value for it represents another piece of data in our sampling distribution. By continuing this process, we generate a sampling distribution of the difference values between sample means.

Because we must be consistent, let's agree to record the difference: $\bar{x}_2 - \bar{x}_1$. At this point, let's take an intuitive look at the geometry of this distribution. We have randomly drawn independent samples from two populations. Clearly, the difference between sample means could be zero, a positive value, or a negative value. Most frequently though, we would expect to obtain a value close to zero. If the two populations have the same mean, a large positive (negative) difference between sample means is unlikely to occur. In other words, we expect the bulk of our data to cluster about zero. Further, it is reasonable to expect the frequency of occurrence to drop rapidly as our difference value becomes large in a positive (negative) sense. Figure 12-1 is an intuitive sketch of the geometry of this distribution.

Figure 12-1

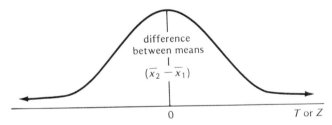

The distribution of the difference between means

Let μ_1 and μ_2 denote the means and σ_1 and σ_2 the standard deviations of populations 1 and 2, respectively. It can be shown that the following two statements are true.

$$\begin{aligned}
&\text{i.} \quad \mu_{(\bar{x}_2-\bar{x}_1)} = \mu_{\bar{x}_2} - \mu_{\bar{x}_1} = \mu_2 - \mu_1 \\
&\text{ii.} \quad \sigma^2_{(\bar{x}_2-\bar{x}_1)} = \sigma^2_{\bar{x}_2} + \sigma^2_{\bar{x}_1} = \sigma^2_2/n_2 + \sigma^2_1/n_1
\end{aligned}$$

It is interesting to note that the algebraic sign between the variances of the two populations is "plus." This is an intuitive surprise.

Mathematical investigation of this sampling distribution has shown that there are two cases to be considered. Case I carries with it a more demanding set of restrictions than does case II. These restrictions, which are discussed in the next section, are

The restrictions of Case I
1. At least one sample size is small.
2. Both populations must be normally distributed.
3. The variances of the populations must be equal ($\sigma_1^2 = \sigma_2^2$).

12-2 (CASE I) AT LEAST ONE SAMPLE SIZE IS SMALL

In this case, our sampling distribution of the difference between sample means is a t distribution with $\nu = n_1 + n_2 - 2$, where n_1 is the sample size drawn from population 1 and n_2 is the sample size drawn from population 2. The rationale for ν is that the combined sample total must be reduced by two because we know the value of $\bar{x}_1$ and $\bar{x}_2$.

Our *t* test statistic is given in terms of a quantity called the *pooled standard deviation* of the two samples. Under the assumption that we are not in a position to know σ_1 and σ_2, we have the next definition.

Definition 12-1 The *pooled standard deviation* of two combined samples is denoted S_p and defined as follows:

$$S_p = \sqrt{\frac{(n_2 - 1)s_2^2 + (n_1 - 1)s_1^2}{n_1 + n_2 - 2}}$$

Our test statistic now becomes the following *t* score:

1. $t = \dfrac{(\bar{x}_2 - \bar{x}_1) - \mu_{\bar{x}_2 - \bar{x}_1}}{\sigma_{\bar{x}_2 - \bar{x}_1}} = \dfrac{(\bar{x}_2 - \bar{x}_1) - (\mu_2 - \mu_1)}{S_p \sqrt{1/n_1 + 1/n_2}}$

Let us take a closer look at $S_p\sqrt{1/n_1 + 1/n_2}$ (our approximation for $\sigma_{\bar{x}_2-\bar{x}_1}$). Students always ask why we don't use an estimation other than S_p. Actually, we know of no reason why we must always use this estimate. Indeed, s_1 and s_2 are both point estimates of σ. However, we are dealing with the case that at least one sample size is small. Further, the mean of the point estimates s_1 and s_2, $(s_1 + s_2)/2$, would seem to be a reasonable estimate of σ. It may be

Hypothesis testing It should come as no surprise to discover that the hypothesis to be tested is $H_0: \mu_1 = \mu_2$, which is equivalent to $H_0: \mu_2 - \mu_1 = 0$. This means that the second term in the numerator of our t-test statistic, line 1, is zero.

> 1. If we believe $\mu_1 \neq \mu_2$, we conduct a two-tail test of hypothesis with $H_a: \mu_1 \neq \mu_2$.
> 2. If we believe $\mu_2 > \mu_1$, we conduct a one-tail test of hypothesis with $H_a: \mu_2 > \mu_1$.
> 3. If we believe $\mu_2 < \mu_1$, we conduct a one-tail test of hypothesis with $H_a: \mu_2 < \mu_1$.

Example 12-1 We wish to run a t-test based on the sampling distribution of the difference between means. (Both populations are normal. $\sigma_1^2 = \sigma_2^2$, and the samples are drawn independently.) If $\bar{x}_2 = 50$ and $\bar{x}_1 = 50$, test $H_a: \mu_1 \neq \mu_2$ at the .05 level.

Solution Stop the testing procedure. Our two sample means match perfectly. Our sample data provide no evidence to make us doubt the truth of the conjecture in H_0. Observe that we did not even bother to give the two sample sizes.

Example 12-2 Test $H_a: \mu_1 \neq \mu_2$ based upon the samples $X_1 = \{1, 2, 4, 5\}$ and $X_2 = \{5, 10, 15\}$. Conduct the test at the .05 level.

Solution

$$H_0: \mu_1 = \mu_2 \qquad H_a: \mu_1 \neq \mu_2$$

$$\nu = n_1 + n_2 - 2 = 4 + 3 - 2 = 5 \qquad \bar{x}_2 = 10, \bar{x}_1 = 3$$

$$s_1^2 = 10/3 \qquad s_2^2 = 25 \qquad S_p = \sqrt{\frac{(3)(10/3) + (2)(25)}{5}} = \sqrt{12}$$

$$t\text{-test statistic} = \frac{7}{\sqrt{12}\sqrt{7/12}} = \sqrt{7} = 2.646 > t_{.025,5} = 2.571$$

Conclusion Reject H_0. The difference between means is significant. See the illustration at the top of page 238.

238 THE SAMPLING DISTRIBUTION OF THE DIFFERENCE BETWEEN MEANS

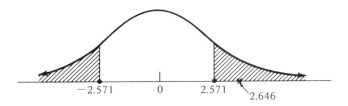

Example 12-3 Test H_a: $\mu_2 < \mu_1$ based on the samples $X_1 = \{4, 5, 6, 7, 8\}$ and $X_2 = \{2, 4, 6\}$. Conduct the test at the .05 level.

Solution H_0: $\mu_1 = \mu_2$ H_a: $\mu_2 < \mu_1$
$n_1 = 5$ $\bar{x}_1 = 6$ $n_2 = 3$ $\bar{x}_2 = 4$ $s_1^2 = 10/4$ $s_2^2 = 4$
$\nu = 6$ $\bar{x}_2 - \bar{x}_1 = -2$

$$S_p = \sqrt{\frac{4(10/4) + 2(4)}{6}} = \sqrt{3} = 1.73$$

$$t\text{-test statistic} = \frac{-2}{\sqrt{3}\sqrt{1/5 + 1/3}}$$
$$= -1.58 \not< -1.943$$

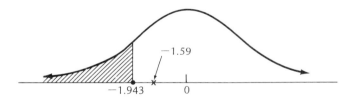

Conclusion There is insufficient evidence to reject H_0.

Example 12-4 The population of 6-year-olds suffering from malnutrition is to be investigated to determine the comparative effects of two medical diets for correcting the condition. Ten children follow diet plan I and eight children follow diet plan II. At the 1% level, is there a significant difference in the weight gain resulting from the use of the two diet plans?
Weight gain (pounds) diet I: 5, 6, 8, 4, 7, 1, 3, 9, 2, 5
Weight gain (pounds) diet II: 6, 7, 6, 10, 8, 10, 8, 9

Solution Because no preference is indicated, we are going to conduct a two-tail test.

H_0: $\mu_1 = \mu_2$ H_a: $\mu_1 \neq \mu_2$
$n_1 = 10$ $\bar{x}_1 = 5$ $s_1^2 = 60/9$ $n_2 = 8$ $\bar{x}_2 = 8$
$s_2^2 = 18/7$ $\nu = 16$

12-2 (CASE I) AT LEAST ONE SAMPLE SIZE IS SMALL

$$t\text{-test statistic} = \frac{3}{\sqrt{\frac{[(9)(60/9) + (7)(18/7)]}{[10 + 8 - 2]}} \sqrt{1/10 + 1/8}}$$

$$= 2.86 \not> t_{.005, 16} = 2.921$$

Conclusion There is insufficient evidence to reject H_0. We have a difference in the mean weights but it is not a significant difference.

Note in the solution to Example 12-4 how the 9's and 7's cancel in the calculation of S_p. This cancellation always takes place. It leaves as the numerator of S_p the addition of two sums of differences squared from their respective means.

Example 12-5 A researcher believes that a diet that is higher in fat will cause the weight of pigs to increase substantially. This will, in turn, feed more people. To test this hypothesis at the .05 level, two small control groups of pigs were carefully monitored.

	I High-Fat Diet	II Regular Diet
Number of guinea pigs	15	8
Mean weight gain, grams	150	98
Standard deviation	23.20	17.50

Solution $H_0: \mu_1 = \mu_2$ $H_a: \mu_2 < \mu_1$ or $\mu_2 - \mu_1 < 0$ (researcher)

$$\nu = 21 \qquad S_p = 21.47 \qquad \bar{x}_2 - \bar{x}_1 = -52$$

t-test statistic $= -5.53 < -t_{.05, 21} = -1.721$

Conclusion Reject H_0. There is a significant difference in weight gain. On the basis of these data, the high-fat diet is superior. The only remaining question is the cost of the high-fat diet.

In this past example, the researcher hopes for the rejection of H_0. This is a common situation. Can you imagine a researcher developing a new method and hoping that it won't yield a significant difference?

Example 12-6 Following are the data on the length of time necessary to complete an assembly line procedure based upon two different methods of training. Test at the .05 level to see if there is sufficient evidence to indicate a difference in mean times for the two methods. We will execute a two-tail test.

Standard Procedure (I)	New Procedure (II)
$n_1 = 9$	$n_2 = 9$
$\bar{x}_1 = 35.22$ seconds	$\bar{x}_2 = 31.56$ seconds
$s_1^2 = 24.44$	$s_2^2 = 20.03$

Solution $H_0: \mu_1 = \mu_2 \qquad H_a: \mu_1 \neq \mu_2$

$$\nu = 16 \qquad S_p = 4.71 \qquad \bar{x}_2 - \bar{x}_1 = -3.66$$

t-test statistic $= -1.65 \not< -t_{.025,16} = -2.120$

Conclusion There is insufficient evidence to reject H_0. Note, if we played the role of the company in this example (perhaps the more realistic situation), we would hope that our new training method reduced assembly time. In this case, we would want $H_a: \mu_2 < \mu_1$.

We leave it to the reader to check that the conclusion would still be the same for the given data.

Here is an *innovative thought*. A guarantee for rejecting $H_0: \mu_1 = \mu_2$ would be to show that the confidence intervals for μ_1 and μ_2 don't intersect. However, although this condition is guaranteed to cause rejection, it is not a necessary condition. This is shown by the data of Example 12.2 where the 95% confidence interval for μ_1 is completely contained within the 95% confidence interval for μ_2 and the conclusion is to reject H_0. We encourage you to check this statement.

Estimation The statistic $(\bar{x}_2 - \bar{x}_1)$ is the point estimate of $(\mu_2 - \mu_1)$. The end-points of the confidence interval estimate of $\mu_2 - \mu_1$ are obtained by solving the inequality:

1. $\quad -t_{\alpha,\nu} < \dfrac{(\bar{x}_2 - \bar{x}_1) - (\mu_2 - \mu_1)}{S_p \sqrt{1/n_1 + 1/n_2}} < t_{\alpha,\nu}$

12-2 (CASE I) AT LEAST ONE SAMPLE SIZE IS SMALL

where $t_{\alpha,\nu}$ is the positive value in the t distribution table such that there is a probability of alpha to its right. The expression in the middle of line 1 is the t score of $(\bar{x}_2 - \bar{x}_1)$.

> The end-points of the $100(1 - \alpha)\%$ confidence interval for $\mu_2 - \mu_1$ are
>
> $$(\bar{x}_2 - \bar{x}_1) \pm t_{\alpha,\nu} S_p \sqrt{1/n_1 + 1/n_2}$$

Example 12-7 For the two samples given in Example 12-2 compute the 95% confidence interval for $\mu_2 - \mu_1$.

Solution $S_p \sqrt{1/n_1 + 1/n_2} = \sqrt{7} = 2.65 \qquad \nu = 5$

$$7 \pm (2.571)(2.65) = 7 \pm 6.81$$

or $.19 < \mu_2 - \mu_1 < 13.81$

Example 12-8 Six first graders in each of two economically different school districts were tested with regard to IQ. The data are as follows:
District 1: 92, 85, 104, 91, 80, 109; $n_1 = 6$
District 2: 96, 112, 103, 118, 92, 111; $n_2 = 6$
Compute a 99% confidence interval estimate of the mean difference between the two school districts.

Solution $\bar{x}_1 = 93.50 \qquad s_1^2 = 122.70 \qquad \bar{x}_2 = 105.33 \qquad s_2^2 = 101.47$

$$S_p\sqrt{1/n_1 + 1/n_2} = 6.11 \qquad t_{.005,10} = 3.169$$

$$11.83 \pm (3.169)(6.08) = 11.83 \pm 19.36$$

or $-7.53 < \mu_2 - \mu_1 < 31.9$

Our work is based on the assumption that $\sigma_1^2 = \sigma_2^2$. However, we must remember that s^2 is an estimate of σ^2 (it is the point estimate.). As such it is not surprising that $s_1^2 \neq s_2^2$ in the previous example.

Example 12-9 A sociology experimenter wishes to answer the question: Does the kind of motivation employed, reward or punishment, affect the learning performance? For this purpose, two samples of five and seven, respectively, are compared on the basis of their scores on a motor learning task. In the first group a subject is rewarded for each correct move made, and in the second group each incorrect move is punished.

$$\bar{x}_1 = 18 \quad s_1^2 = 4.80 \quad \bar{x}_2 = 20 \quad s_2^2 = 5.00$$

Compute the 99% confidence interval for the difference between performances.

Solution

$$\sigma_{\bar{x}_2-\bar{x}_1} = S_p\sqrt{1/n_1 + 1/n_2} = 1.30 \quad t_{.005,10} = 3.169$$

$$2 \pm (3.169)(1.30) = 2 \pm 4.12$$

$$\text{or } -2.12 < \mu_2 - \mu_1 < 6.12$$

Observe that the interval obtained would lend credence to the null hypothesis, $\mu_2 - \mu_1 = 0$.

EXERCISES A

12-1.
a. If $n_1 = 16$ and $n_2 = 12$, then $\nu = ?$
b. If $n_1 = 14$ and $\nu = 30$, then $n_2 = ?$
c. If $n_1 = n_2 = n$ and $\nu = 12$, then $n = ?$

12-2. We are executing a t test for the difference between sample means. We learn $\bar{x}_1 = \bar{x}_2$. Assuming we don't stop the testing procedure, what will the value of the t-test statistic be?

12-3. Execute a t test for the difference between sample means at the .05 level based on the following two groups of data: I = {1, 2, 3}   II = {2, 4, 6} (Example 12-2)

12-4. Execute a t test for the difference between sample means at the .01 level based upon the following two groups of data: Group I = {2, 3, 4, 5, 6, 7, 8}   Group II = {1, 1, 2, 2, 6, 6, 7, 7}

12-5. Show $S_p = \sqrt{\dfrac{\sum_{i=1}^{n_1}(x_i - \bar{x}_1)^2 + \sum_{j=1}^{n_2}(x_j - \bar{x}_2)^2}{n_1 + n_2 - 2}}$

12-6. Redo Exercise 12-3 using the formula for S_p given in Exercise 12-5.

12-7. Redo Exercise 12-4 using the formula for S_p given in Exercise 12-5.

12-8. Use the data of Exercise 12-3 to calculate a 95% confidence interval for $(\mu_2 - \mu_1)$.

12-9. Use the data of Exercise 12-4 to calculate a 99% confidence interval for $(\mu_2 - \mu_1)$.

12-10. What do we call $\sigma_{\bar{x}_2-\bar{x}_1}$? [Hint: See Definition 7-2.]

12-11. We wish to determine whether or not a special training program will increase college board scores. In order to determine if there is any significant difference between the specially trained students and the untrained students, a sample is obtained from each group. Twelve of the specially trained students received a mean score of 532.50 with a variance of 3.20. Ten of the untrained students received a mean score of 530.00 with a variance of 2.80. Execute a t-test for the difference between means at the .05 level. (Example 12-6)

 12-12. Use the data of Exercise 12-11 to calculate a 95% confidence interval for $(\mu_2 - \mu_1)$. (Example 12-7)

 12-13. CPAs Bent and Sutherland believe that companies that wholesale their product do better than companies that retail their product. They obtain a sample of each category from accounts in their files. The data are given in the following. Test to see if there is a significant difference in mean values at the .01 level. (Example 12-5)

	Wholesalers	Retailers
Number of accounts	30	20
Average weekly profits	$2100.00	$1750.00
Sample variance	$ 250.00	$ 200.00

 12-14. Use the data of Exercise 12-13 to compute the 99% confidence interval for $(\mu_2 - \mu_1)$. (Example 12-9)

 12-15. Twelve families in city A showed an average weekly food expenditure of $98 with a standard deviation of $5. Fifteen families in city B showed an average weekly expenditure of $92 with a standard deviation of $2. Test at the .05 level to see if there is a significant difference in the average food budget between the two cities. It is assumed that all families in the study were of the same size and had approximately the same gross family income.

 12-16. Use the data of Exercise 12-15 to compute the 95% confidence interval for $(\mu_2 - \mu_1)$.

 12-17. Two salesmen have an argument. Mr. A. claims that he is superior to Mr. B in industrial sales. To settle the dispute comparable industrial accounts for the two men are compared. Is there any real difference between the average sales amount of the two men? Test at the .05 level.

	Mr. A	Mr. B
Number of sales	18	20
Average sales amount	$210.00	$175.00
Standard deviation	$ 25.00	$ 20.00

 12-18. The federal government is informed of a serious infestation of houseflies in a particular region of the country. The government decides to test two methods of housefly control—an organic method and a chemical method. The organic method involves introducing nonpoisonous spiders into a selected community. Data collected in six houses consist of a numerical count of houseflies in a specific period of time. The mean value was 30 and $s_1^2 = 91.60$. The chemical method consists of a combination of poisonous sprays and bait. Data are collected in nine houses. The mean value was 20 and $s_2^2 = 101.00$.

At the .01 level, do the test results indicate that the chemical method is more effective than the organic method in reducing the number of houseflies?

CUMULATIVE REVIEW

B 12-19. The following data represent lifetimes, in years, of two major brands of auto batteries. The manufacturer of brand A claims that, on the average, his brand is superior to brand B. Execute a t test at the .05 level to establish whether or not this claim seems reasonable.

Brand A: 2.4 1.9 2.0 2.1 1.8 2.3 2.1 1.7 2.0 1.8 3.0
          2.6 1.5 2.7 1.9 2.4

Brand B: 2.0 2.1 1.8 2.4 1.6 2.1 1.5 2.0 1.4 1.1 1.0
          1.6 2.2 3.1 1.5 2.8 1.4 1.5

12-20. It is known that peanuts are an excellent source of protein and so may serve as a good substitute for meat and fish when prices are high. To find out whether roasted peanuts have less protein than raw peanuts, a nutritionist selected a random sample of 20 bags of peanuts for testing. She then selected eight of the 20 bags at random, roasted them, and measured the protein content of all 20 bags. The data are in grams.

Raw peanuts: 10 8 6 10 8 10 10 8 6 10 4 6
Roasted peanuts: 8 6 5 10 7 9 6 5

Test at the .01 level to see if the roasting process reduces the protein content of the peanuts.

12-3 (CASE II) BOTH SAMPLE SIZES ARE LARGE

We mentioned three restrictions in case I. What happens to each of these restrictions in case II. Restriction (1) is changed by the fact that we are demanding that both sample sizes be large. Restriction (2) may now be weakened to include populations that aren't normally distributed. Restriction

(3) may now be completely dropped. Finally, our formula for $\sigma_{\bar{x}_2 - \bar{x}_1}$ is considerably simpler.

$$\sigma_{\bar{x}_2 - \bar{x}_1} = \sqrt{\frac{s_1^2}{n_1} + \frac{s_2^2}{n_2}}$$

The test statistic is now a z score. Thus,

$$z\text{-test statistic} = \frac{\bar{x}_2 - \bar{x}_1}{\sigma_{\bar{x}_2 - \bar{x}_1}} = \frac{\bar{x}_2 - \bar{x}_1}{\sqrt{s_1^2/n_1 + s_2^2/n_2}}$$

Example 12-10 Use the following data to execute a test for the difference between the sample means. Test at the .05 level. Let $H_a: \mu_2 > \mu_1$ or $(\mu_2 - \mu_1) > 0$.

$n_1 = n_2 = 100$ $\bar{x}_1 = 1.90$ $s_1 = 3$ $\bar{x}_2 = 3.50$ $s_2 = 4$

Solution

$$H_0: \mu_2 = \mu_1 \qquad H_a: \mu_2 > \mu_1$$

This is a one-tail test.

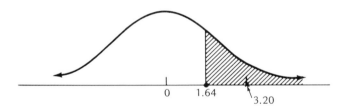

$$z\text{-test statistic} = \frac{1.60}{\sqrt{\frac{16}{100} + \frac{9}{100}}}$$
$$= 3.20 > 1.64$$

Conclusion Reject H_0.

Example 12-11 In testing the difference between the nicotine contents of two major brands of cigarettes, 50 cigarettes selected from brand 1 have a mean of 23.8 mg of nicotine, $s_1^2 = 1.44$. For 50 cigarettes from brand 2, the mean is 24.10 mg of nicotine with $s_2^2 = 1.96$.

Test at the .05 level to see if there is a significant difference between the means of the two brands.

Solution $H_0: \mu_1 = \mu_2$ $H_a: \mu_1 \neq \mu_2$

z-test statistic $= (24.10 - 23.80)/\sqrt{1.44/50 + 1.96/50} = .30/.26 =$
$$1.15 < 2.57$$

Conclusion There is insufficient evidence to reject H_0.

Example 12-12 A machine for filling boxes with dry cereal has a large hopper directly below its mouth, and gates at the mouth so that boxes moving past on a conveyor belt may be filled. Many things, such as the timing device controlling the gates, the bulk of the cereal, and the speed of the belt, may alter the amount of cereal contained in each box.

Based upon months of observation it is ascertained that the variance of the population of all boxes of cereal is .04. Each box of cereal is supposed to contain 16 oz. As part of the company quality control program, 40 boxes are selected each week for inspection. Of course, the company hopes that the weekly sample means are equal to 16 oz. Following are the data for two successive weeks. Test $H_a: \mu_1 \neq \mu_2$ at the .05 level.
Week no. 1: $n_1 = 40$, $\bar{x}_1 = 15.99$ oz
Week no. 2: $n_2 = 40$, $\bar{x}_2 = 16.10$ oz

Solution $H_0: \mu_1 = \mu_2$ $H_a: \mu_1 \neq \mu_2$

$$z\text{-test statistic} = \frac{16.10 - 15.99}{\sqrt{.04/40 + .04/40}} = \frac{.11}{.045} = 2.44$$

Conclusion Reject H_0. The fluctuation between the mean weekly weights is too great to attribute to mere chance. The process warrants an inspection.

Now let's turn our attention to computing the $100(1 - \alpha)\%$ confidence interval for $\mu_2 - \mu_1$.

> Our confidence interval for $(\mu_2 - \mu_1)$ is determined by the end-points:
>
> $$(\bar{x}_2 - \bar{x}_1) \pm z_\alpha \sqrt{s_1^2/n_1 + s_2^2/n_2},$$
>
> where z_α is a z score with α probability to its right.

Example 12-13 Use the data of Example 12-11 and evaluate a 99% confidence interval for $(\mu_2 - \mu_1)$.

Solution

$$.30 \pm (1.96)(.26) = .30 \pm .67$$

$$\text{or } -.37 < \mu_2 - \mu_1 < .97$$

Example 12-14 Use the data of Example 12-12 (the cereal example) to evaluate a 99% confidence interval for $(\mu_2 - \mu_1)$.

Solution

$$.11 \pm (2.57)(.045) = .11 \pm .12$$

$$\text{or } -.01 < \mu_2 - \mu_1 < .23$$

Now because the t distribution does approach the normal distribution, in a limit sense, a question arises. Do we really need the simpler formula in case II for $\sigma_{\bar{x}_2 - \bar{x}_1}$? For large sample sizes the formulas derived as estimates of $\sigma_{\bar{x}_2 - \bar{x}_1}$ in cases I and II tend to yield nearly equal results. However, given a choice, the formula of case II is certainly easier to use. The Next set of exercises illustrates these ideas.

We leave it for you to show, as an exercise, that

$$S_p\sqrt{1/n_1 + 1/n_2} \doteq \sqrt{s_1^2/n_1 + s_2^2/n_2}$$

(recall $\doteq$ means approximately equal to) when n_1 and n_2 are both greater than 30.

EXERCISES A

12-21. Conduct a test of the difference between means at the .05 level. Let H_a: $\mu_2 > \mu_1$. Use the following data: $n_1 = n_2 = 49$, $s_1 = 3$, $s_2 = 4$, $\bar{x}_2 = 4.96$, $\bar{x}_1 = 3.06$.

12-22. Conduct a test of the difference between means at the .01 level. Let H_a: $\mu_1 \neq \mu_2$. Use the following data: $n_1 = n_2 = 100$, $s_1 = 5$, $s_2 = 12$, $\bar{x}_2 = 55.91$, $\bar{x}_1 = 53.86$.

12-23. A sociological study was conducted to determine if persons in suburban district I have different mean incomes from persons living in suburban district II. In each district a sample of 50 homeowners were interviewed. The data are recorded in thousands of dollars. Conduct a test to see if there is a significant difference in the mean incomes of the two groups of people. Use $\alpha = .01$.

	District I	District II
Mean	$14.50	$15.25
Variance	.50	1.00

 12-24. Two sections of 54 and 67 students each took the same examination in statistics. The first section had a mean score of 73.10, with $s = 11.20$. The second section had a mean score of 76.60, with $s = 13.00$. Can the difference in mean scores be attributed to mere chance? Let $\alpha = .05$.

 12-25. A sample study was made of the number of business lunches that were claimed as deductible by executives in the insurance industry versus executives in the banking industry. It was found that 50 executives in the insurance industry averaged 9.70 such deductions, with a variance of 2.50, whereas 50 executives in the banking industry averaged 8.30 such deductions, with a variance of 3.50. Test at the .01 level of significance to see if there is a significant difference between the means.

 12-26. Two methods of teaching remedial reading to 10-year-old youngsters, the Boston Method and the Detroit Method, were applied to several randomly selected slow readers who were divided into two groups. On a reading test given to both groups at the conclusion of the teaching program, the 60 pupils exposed to the Boston Method scored a mean of 20, with a standard deviation of 7, and the 50 students using the Detroit Method had an average score of 22, with a standard deviation of 8. At the .05 level, can we say that the Boston Method is superior?

 12-27. Use the data of Exercise 12-23 to compute the 95% confidence interval for $(\mu_2 - \mu_1)$.

 12-28. Using the data of Exercise 12-23, and formulas of case I compute the t-test statistic. Compare the result with z-test result previously obtained in Exercise 12-23.

 12-29. Use the data of Exercise 12-25 to compute the 99% confidence interval for $(\mu_2 - \mu_1)$.

CUMULATIVE REVIEW

B 12-30. Using the data of Exercise 12-24 and the formulas of case I, compute the t-test statistic. Compare the result with the z-test result previously obtained in Exercise 12-4.

 12-31. Show that when both n_1 and n_2 are greater than 30:

$$S_p\sqrt{1/n_1 + 1/n_2} \doteq \sqrt{s_1^2/n_1 + s_2^2/n_2}$$ [Hint: Let $n_1 = n_2 = n$.]

12-4 MANN-WHITNEY U-TEST

H. B. Mann

Whitney

In order to compare the means of small samples we have to assume that these samples were obtained from normal populations with equal standard deviations. To handle situations in which these assumptions cannot be met, statisticians have developed an alternative test, the Mann-Whitney U-test. In this section we are still dealing with a comparison of two groups of data. We shall restrict our discussion to the most common situation, a two-tail test.

The *Mann-Whitney U-test* is the nonparametric "match-up" of the test for the difference between sample means. It is also our first example of a family of tests called *rank order tests*. We must be careful to call this test the match-up and not the equivalent of the test for the difference between sample means because of the null hypothesis in the U-test.

H_0: The two samples were drawn from identical populations.
H_a: The two samples aren't from the same population.

This test is executed by combining the two selected samples, of sizes n_1 and n_2, into one large sample of size $n_1 + n_2$. All of the $n_1 + n_2$ pieces of data are placed in increasing order according to their magnitudes. The smallest piece of data is assigned the number 1. The second smallest piece of data is assigned the number 2, and so on. The largest piece of data is assigned the number $n_1 + n_2$. Ties are handled by assigning each of the tied data values the mean rank position of the positions they occupy. For example, a tie in the third and fourth positions would result in assigning each piece of data the rank position 3.50. A three-way tie in the seventeenth, eighteenth, and nineteenth positions would result in assigning all three pieces of data the rank value of 18. A tie in the last positions to be ranked means that the last rank value will not be $n_1 + n_2$. For example, if $n_1 = n_2 = 10$ and there is a tie in the last two positions, each position will be assigned the rank value 19.5. Either of the two groups being compared may be labeled group one; the other group is labeled group two. Frequently, if one of the groups is smaller, it is selected as group one.

Our actual research work is based upon the sampling distribution of a statistic called U, which can be defined as follows:

Definition 12-2 $U = n_1 n_2 + [n_1(n_1 + 1)]/2 - R_1$, where R_1 is the sum of the ranks in group one.

As amazing as it may seem, when n_1 and n_2 are both greater than 8, the sampling distribution of the statistic U is approximately normally distributed. Formulas for calculating the mean and the variance for the sampling distribution of the U statistic are

$$\text{Mean of } U = \frac{n_1 n_2}{2}$$

$$\text{Variance of } U = \frac{n_1 n_2 (n_1 + n_2 + 1)}{12}$$

Example 12-15 An educator wishes to test to see if there is any difference in the academic performance of students sitting in even-numbered seats (group one) versus students sitting in odd-numbered seats (group two). On the basis of what the educator felt was a good exam, 10 students were randomly selected from each group. Their scores are listed in the following. Use the Mann-Whitney U-test, at the .05 level, to test the null hypothesis that the scores come from identical populations.

Group 1: 52, 78, 56, 90, 65, 86, 64, 90, 49, 78
Group 2: 72, 62, 91, 88, 90, 74, 98, 80, 81, 71

Solution

H_0: The two samples were drawn from identical populations.

H_a: The two samples aren't from the same population.

When the data are combined into a single set of data of size 20 and placed in increasing order of magnitude we obtain:

Data: 49   52   56   62   64   65   71   72   74   78   78
Rank: 1   2   3   4   5   6   7   8   9   10.5   10.5
Data: 80   81   86   88   90   90   90   91   98
Rank: 12   13   14   15   17   17   17   19   20

$$R_1 = 1 + 2 + 3 + 5 + 6 + 10.50 + 10.50 + 14 + 17 + 17 = 86$$

$$U = (10)(10) + \frac{(10)(10 + 1)}{2} - 86 = 69$$

$$\text{Mean } U = \frac{(10)(10)}{2} = 50 \qquad \text{Variance of } U = \frac{(10)(10)(10 + 10 + 1)}{12} = 175$$

$$z\text{-test statistic} = \frac{69 - 50}{\sqrt{175}} = \frac{19}{13.23} = 1.44 < 1.96$$

Conclusion There is insufficient evidence to reject H_0. From these two samples, all the data appear to be part of the same population.

If there is a sizable difference between the means of the two populations, most of the lower ranks will be occupied by the values of one sample whereas the higher ranks will tend to be occupied by data from the other sample. This is the *rationale* of the test.

Example 12-16 Assume that we wish to compare two brands of gasoline. Mileages, based on one gallon of gasoline are listed here. Execute the Mann-Whitney U test at the .01 level to see if the distributions of gasoline mileages are identical.

Gasoline A:	17.0	17.8	15.2	16.8	18.4	16.2	18.3	18.1	17.3	
Rank:	5	10	1	4	14	3	13	12	7	
Gasoline B:	18.6	18.8	17.1	19.5	17.6	19.0	15.7	19.8	17.5	18.0
Rank:	15	16	6	18	9	17	2	19	8	11

Solution H_0: The two samples are drawn from identical populations.

H_a: The two samples aren't from the same population.

Let group 1 be gasoline A. The data in this group hold the ranks: 1, 3, 4, 5, 7, 10, 12, 13, and 14.
$R_1 = 69$ $\quad U = 66 \quad$ Mean of $U = 45 \quad$ Variance of $U = 150$
z-test statistic $= (66 - 45)/\sqrt{150} = 1.71 \ngtr 2.58$

Conclusion There is insufficient evidence to reject H_0. As far as statistics is concerned, the two samples of data could have come from either brand of gasoline.

We realize that the next example will violate the condition that n_1 and n_2 must both exceed 8 in order to use the Mann-Whitney U-test. However, we included it because it is a simple arithmetic example that is very illuminating.

It illustrates the ability of the nonparametric test to approximate the result of the related parametric test, and it shows how the parametric test usually gives a slightly more refined answer. Here the refinement makes a difference.

Example 12-17   Redo Example 12-2 using the Mann-Whitney U-test.

Solution H_0:   The two samples are drawn from identical populations.
  H_a: The two samples aren't from the same population.
Let $X_1 = \{1, 2, 4, 5\}$ be group one.
Let $X_2 = \{5, 10, 15\}$ be group two.

Data:	1	2	4	5	5	10	15
Rank:	1	2	3	4.5	4.5	6	7
Group:	1	1	1	1	2	2	2

Observe the disjointness of the numbers 1 and 2 in the group row. From the test rationale, already stated, it looks intuitively like we are about to have H_0 rejected. Or does it?

Recall in Example 12-2 that the null hypothesis was rejected. However, the difference between the test statistic and our critical point value was less than .10. Our decision to reject was a case of rejection with caution. Now we will follow the nonparametric solution for the same data.

$R_1 = 1 + 2 + 3 + 4.50 = 10.50$ Mean of $U = 6$,
$U = 11.5$ Variance of $U = 8$.

z-test statistic $= (11.50 - 6.00)/\sqrt{8} = 1.94 < 1.96$

Conclusion   There is insufficient evidence to reject H_0, which is just the opposite of the conclusion in Example 12-2.

The two types of testing do not reach the same conclusion. Or do they? Perhaps it is best to say, because both test statistics are so close to the critical value, that fresh data should be obtained and the testing redone.

EXERCISES A

Exercises 12-32 and 12-33 are designed to provide computational practice. We know they violate the condition that both n_1 and n_2 exceed 8. (Example 12-17)

12-32. Execute the Mann-Whitney U-test for the data of Exercise 12-3.

12-33. Execute the Mann-Whitney U-test for the data of Exercise 12-4.

12-34. It is desired to compare the life span of two kinds of tires on the basis of two random samples of size 10. The following data below are in units of 1000

miles. Use the Mann-Whitney U-test at the .05 level to test the null hypothesis that the two samples come from identical populations. (Example 12-15)
Type I: 14, 4, 7, 17, 12, 6, 9, 11, 10, 13
Type II: 15, 18, 8, 1, 3, 19, 20, 2, 16, 5

 12-35. The following are the scores obtained in a personality test administered by a psychological testing team.

Unmarried: 21 5 13 18 2 16 14 8 9 11 22 24 3
           19 23 25 1 26
Married: 17 16 5 12 11 2 7 27 21

Use the Mann-Whitney U-test at the .05 level to test the null hypothesis that the two samples are from identical populations.

12-36. Redo Exercise 12-20. Conduct the Mann-Whitney U-test at the .05 level to see if it is reasonable to believe that the two types of peanuts are from identical populations.

12-5 A NOTE ON EXPERIMENTAL DESIGN (optional)

Experimental design is an extremely important concept in statistical work. *Designing the experiment should precede the gathering of any data,* because it determines what data need to be obtained. Once obtained, the data are routinely inserted into the indicated (by the design) statistical procedures. However, our experience is that all too frequently the opposite takes place. A ton of data, much of which are useless, is gathered and only then does the would-be-researcher begin to wonder how to analyze the data.

An introductory statistics text such as this must have priorities. We can't cover everything of importance in statistics. Any real treatment of experimental design is beyond the scope of this text, but we hope a true story will serve to illustrate the importance of experimental design in statistics.

Three of my former students who were physical education majors came to me with data gathered from an experiment in which they were involved. It seems that the experiment had gone on for some time. Its object was to develop increased muscle stamina through bicycle riding. The experiment involved three groups of female physical education majors. The women in group 1 rode bicycles for one-half hour each day. The women in group 2 pedalled bicycles rigged with additional weights for one-half hour each day. Group 3 consisted of women who just sat and rested for one-half hour each day.

Because the women had not yet had analysis of variance, they had been running a series of *t* tests based upon the difference between sample means.

They had tested all possible pairs derived from the three groups. Each t test showed little or no difference between the mean values. In fact, the three sample means were virtually identical. To make matters worse, this seemed to indicate that the group who just sat and rested for the half hour each day were as strong as those who exercised faithfully.

The student researchers were quite chagrined. Their problem was one of experimental design. Female physical education majors were used in all three groups because they were willing to cooperate. However, this presented two major problems: (1) All the participants were in great physical condition to begin with. (2) The group of women who were resting did so faithfully, but when their rest time was up they sprang into immediate physical action.

Now, although the young researchers did not gain any useful statistical data, they did reap considerable wisdom in experimental design.

SUMMARY

One of the most common statistical tests deals with a comparison of two population means. H_0: $\mu_1 = \mu_2$ or $\mu_2 - \mu_1 = 0$. That is, testing that the difference between population mean is zero. To understand this test it is necessary to investigate the *sampling distribution of the difference between sample means*. We have population 1 from which we draw sample X_1 (with mean $\bar{x}_1$), and population 2 from which we draw sample X_2 (with mean $\bar{x}_2$). The sampling distribution of the difference between sample means is normally distributed with $\mu_{(\bar{x}_2-\bar{x}_1)} = \mu_2 - \mu_1$ and $\sigma^2_{(\bar{x}_2-\bar{x}_1)} = \sigma_2^2/n_2 + \sigma_1^2/n_1$.

There are two cases to be considered. In case I, (1) at least one sample size is small, (2) both populations must be normally distributed, and (3) the variances of the two populations must be equal ($\sigma_1^2 = \sigma_2^2$). When working with case I, we find that our estimate of $\sigma_{(\bar{x}_2-\bar{x}_1)}$ involves the calculation of a *pooled standard deviation*, denoted S_p. In case I, our test statistic is a t score. In case II, we are dealing with two large samples. Here we may have two populations that are not necessarily normally distributed; here it is not necessary to calculate a pooled standard deviation. In case II, the test statistic is a z score. The null hypothesis stated above was tested for both cases and for both cases we discussed the point and interval estimates of $\mu_2 - \mu_1$.

Section 12-4 dealt with the *Mann-Whitney U-Test*. This is the nonparametric "match" to the parametric test for the difference between means. This nonparametric test is especially interesting for it is our first encounter with a *rank-order test*. For this test our two samples were combined into one large sample, and all of the data were ranked in increasing order of magnitude. After the ranking was completed a statistic, U, was calculated. Once again, for sample sizes in excess of eight, the amazing discovery was made that the sampling distribution of the statistic U was approximately normally distributed.

Can You Explain the Following?

1. independent variable
2. dependent variable
3. Case I
4. Case II
5. pooled standard deviation
6. point estimate of $\mu_2 - \mu_1$
7. Mann-Whitney U-test
8. ranking data

CUMULATIVE REVIEW

MISCELLANEOUS EXERCISES

12-37. Write $\sigma^2_{\bar{x}_2 - \bar{x}_1}$ for the case $n_1 = n_2 = n$ ($n > 30$).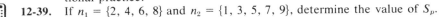
12-38. Determine ν if $n_1 = 18$ and $n_2 = 16$.

Exercises 12-39 through 12-41 are designed strictly for additional computational practice.

12-39. If $n_1 = \{2, 4, 6, 8\}$ and $n_2 = \{1, 3, 5, 7, 9\}$, determine the value of S_p.
12-40. Use the data of Exercise 12-39 to execute a t test for the difference between means at the 5% level. Test H_a: $\mu_1 \neq \mu_2$.
12-41. Use the data of Exercise 12-39 to compute a 95% confidence interval for $(\mu_2 - \mu_1)$.
12-42. Two samples of menhaden are drawn, one from Buzzard Bay and one from Pawtucket Bay. The 37 fish from Buzzard Bay had a mean weight of 11.60 oz, with a standard deviation of 1.30 oz. The 52 fish from Pawtucket Bay had a mean weight of 12.10 oz, with a standard deviation of 2.10 oz. Test the hypothesis that there is no weight difference between Buzzard Bay and Pawtucket Bay menhaden. Use the .05 level.
12-43. Use the data of Exercise 12-43 to compute a 95% confidence interval for $(\mu_2 - \mu_1)$.
12-44. A reading test is given to elementary school classes of 12 Anglo-American children and 10 Mexican-American children. Is the difference between means significant at the 5% level? The data are as follows:

Anglo-American	Mexican-American
$\bar{x}_2 = 74$	$\bar{x}_1 = 70$
$s_2 = 8$	$s_1 = 10$
$n_2 = 12$	$n_1 = 10$

12-45. Use the data of Exercise 12-44 to calculate a 99% confidence interval for $(\mu_2 - \mu_1)$.

12-46. State a formula for the Mann-Whitney U-test statistic for the case $n_1 = n_2 = n$.

12-47. A sociologist has devised a test to measure a person's "social grace" skills. The sociologist decides to give the test a trial run on two groups of people from different backgrounds. To maintain fairness, the person who is administering the test is not informed of the background of the person being tested. Our sociologist has no particular leaning toward either group of people.

Score Background A	Score Background B
7	8
11	9
9	13
4	14
8	11
6	10
12	12
11	14
9	13
10	9
11	10
	8

a. Conduct the Mann-Whitney U-test at the .05 level by ranking the data lowest to highest. That is, the lowest value receives 1, the second lowest 2, and so forth.
b. Conduct the Mann-Whitney U-test at the .05 level by ranking the data highest to lowest. That is, the highest value receives 1, the second highest 2, and so forth.
c. Compare the answers to (a) and (b).

13 THE LINE OF REGRESSION

'13-1 INTRODUCTION

Is there a relationship between the supply and the demand of crude oil? How is the amount of rainfall related to the per-acre yield of wheat? How, if at all, does IQ influence English grades? What is the relationship between dosage of medication and its effect? Does daily temperature influence retail sales? A desire to analyze and predict seems to be embedded in human nature.

A stock analyst seeks to find a relationship between the number of customer calls that a broker makes in a week's time and the number of sales the broker makes in the same week's time. The analyst reasons that the number of sales should depend to some degree on the number of calls made. Or does it? When a relationship does exist, is there a simple way of portraying it? We could begin by labeling the number of calls the *independent variable* (x), and the number of sales the *dependent variable* (y).

A psychologist is interested in a person's ability to adapt knowledge. The independent variable may be the number of trials that it takes to learn a task. The dependent variable may be the ability of the learner to turn a given task into a variation of itself. The variable that the experimenter controls is the independent variable.

In this chapter we deal with only two variables: an independent one and a dependent one. We are interested in the answers to the following two questions:

1. Is there a simple relationship between the variables?
2. If there is a relationship, just how strong is it?

The first question is the subject of this chapter. The second question deals with a concept called *correlation,* which is discussed in Chapter 14. Here, we attempt to find a *linear relationship only* (a straight-line relationship) between two variables.

If it is reasonable to represent our data by a linear equation, the equation will serve as an instrument in allowing us to make predictions about the dependent variable. For example, if x is the independent variable and y is the dependent variable and the linear relationship is $y = 2x + 1$, when the independent variable is 3, the dependent variable is 7.

HISTORICAL NOTE

Bettmann Archive, Inc.
SIR FRANCIS GALTON

The history of linear predictions dates back to the investigations of Sir Francis Galton in England. Galton was concerned with the general question of whether people of the same family were more alike than were people from different families. Specifically, he wanted to know whether such characteristics as genius, height, athletic skill, and musical talent tended to run in families. His investigation was based on his personal feeling on the topic. What was this personal feeling? Let's simply say that the famous Charles Darwin was his first cousin and that he had an illustrious grandfather, Erasmus Darwin.

How Galton gained some of his data is an interesting story. He wished to investigate the question of whether fathers and their adult sons were more alike than men of different families. To carry out this investigation, he needed to measure fathers and their sons as well as unrelated men in regard to weight, height, strength of grip, etc. He set up a booth at a local fair, and people paid him 3 pence to participate in his research. His subjects left with little self-knowledge, but Galton left with a wealth of data and pockets full of coin. Today, subjects are paid to participate in experiments.

13-2 THE LINE OF REGRESSION

We will deal with a pair of samples that are in a one-to-one correspondence (1–1); for example, the number of years of education of the father versus the number of years of education of the son. Each student's college board score is compared with the student's freshman-year quality point average. A salesperson's number of years of experience is compared with total yearly sales. The stress on a body organ, such as the liver, is compared with the amount of physical damage done to the organ.

A medical researcher is trying to relate the amount of dosage of the drug ephedrine (x), the independent variable, to the number of heart beats per minute in mice (y). Mouse no. 1 has his dosage labeled x_1 and his number of heart beats per minute labeled y_1. We have generated the ordered pair (x_1, y_1). The last or nth mouse in the sample will have the ordered pair (x_n, y_n) associated with him. This one-to-one correspondence seems to arise quite naturally in research work.

The letter n can be correctly interpreted as the number of pieces of data in

Definition 13-1 The graph of ordered pairs of data is called a *scatter diagram*.

Example 13-1 Plot the scatter diagram for the two matched samples X and Y in the following table.

Y	2	4	1	5
X	0	1	2	3

Solution

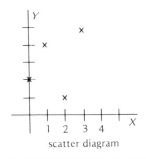

scatter diagram

Example 13-2 A medical researcher has obtained data on dosages of the drug ephedrine as it relates to the number of heart beats per minute in humans. Plot the scatter diagram for the part of the data given in the table.

Total Daily Dosage of Ephedrine, in grains (x)	Number of Heart Beats per Minute (y)
3	70
2	60
1	50
3	80
5	100
4	90

Solution

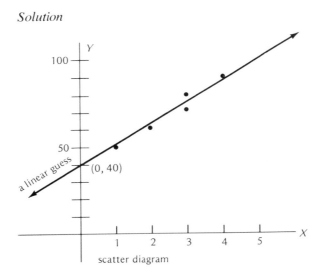

scatter diagram

Note that in this diagram, by chance, the x value 3 occurred twice. Further inspection seems to indicate that a linear equation representation for this collection of scatter points is reasonable. The slope of this line would be positive. Finally, mere inspection indicates that the y-intercept of this line is approximately (0, 40).

In Galton's study of heights of fathers versus their sons, he observed a "regression" or turning back of the data toward the mean. This observation motivated the following definition.

Definition 13-2 The straight line that is used to represent the set of points in a scatter diagram is called a *line of regression*.

It is highly unlikely that all the points of a scatter diagram lie on a straight line. Should this situation occur, the straight line would be a perfect representative of these ordered pairs. More realistically, what we hope to obtain is a straight line that will pass through the scatter points in a set sense. Some of these points will lie above the line and some of the points will lie below it. Intuitively, we would not expect a line of regression to be a good representative of our set of points if all of the points were below the line or if all of the points were above the line. The line may pass through some points of the scatter diagram or it may not pass through any points of the diagram.

Plotting a scatter diagram is a simple chore. Routine inspection of the diagram may serve to show a trend between the two variables x and y. Simple inspection may serve to tell us: (1) whether a linear equation relationship seems possible, and (2) if the straight line is found, the algebraic sign (positive or negative) of its slope.

Our next example illustrates *a set of ordered pairs that doesn't possess a good linear representation.* Observe in the accompanying diagram that there is a relationship, other than a linear relationship, for all the scatter points lie on the boundary of a square.

Example 13-3 Plot the scatter diagram for the following set of ordered pairs.

Y	2	−2	1	2	−1	−2	2	−2	1	2	−1	−2
X	1	1	2	2	2	2	−1	−1	−2	−2	−2	−2

Solution

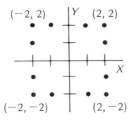

Assume that the scatter diagram of our data is plotted, and that we have plotted our line of regression on this same diagram. Our line of regression is not parallel to the Y-axis. Through every scatter point construct a line parallel to the Y-axis. Each member of this family of parallel lines will now intersect our regression line at exactly one point. Upon each parallel line we have created a one-to-one correspondence between the point of intersection with our line and the scatter point. See Fig. 13-1.

Exactly what makes this line a "good fit" to the set of scatter points? The established criterion that we use for fitting a straight line to a set of points is: *Select the straight line that makes the sum of vertical distances*

Figure 13-1

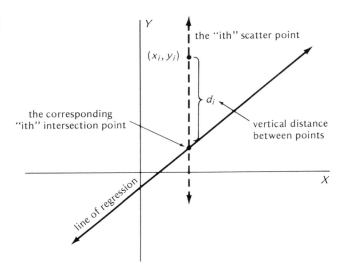

squared, between corresponding points, minimal. See Fig. 13-1. This means $\sum_{i=1}^{n} d_i^2$ is minimal when the line being superimposed upon the scatter diagram is the line of regression. Should all of the scatter diagram points happen to lie upon the line of regression, this sum of vertical distances squared would equal zero.

The slope-intercept form of a straight line, $y = mx + b$, is a common form used in mathematics for expressing the equation of a straight line.

Figure 13-2

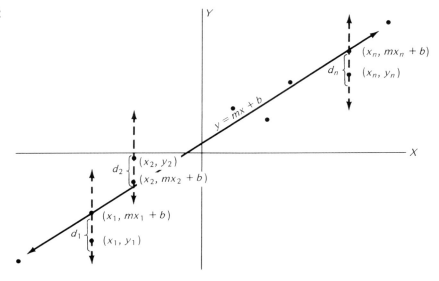

If our line of regression is expressed in this form, our criterion for a straight line to be a good fit is that

$$\sum_{i=1}^{n} d_i^2 = \sum_{i=1}^{n} (y_i - mx_i - b)^2$$

be minimal. See Fig. 13-2.

HISTORICAL NOTE

Our criterion for fitting a straight line to a set of points is often called the *method of least squares*. This method of fitting any curve (a straight line may be thought of as a special curve) to a set of points was first suggested early in the 19th century by the French mathematician Adrien Legendre (1752–1833).

Culver Pictures Inc.
ADRIEN MARIE LEGENDRE

Definition 13-3 The formula for computing the slope of the line of regression will be denoted $\hat{m}$ (m-hat) and be given by the formula:

$$\hat{m} = \frac{\sum_{i=1}^{n} x_i y_i - n\bar{x}\bar{y}}{\sum_{i=1}^{n} (x_i - \bar{x})^2}$$

Definition 13-4 The form of the equation of the line of regression is $y - \bar{y} = \hat{m}(x - \bar{x})$.

The point $(\bar{x}, \bar{y})$ will always satisfy the equation of the line of regression. To see that this is true we need only to replace (x,y) in the equation of the line of regression (Definition 13-4) by $(\bar{x}, \bar{y})$. If we set $x = 0$ in the equation and solve for y, we easily obtain a second point on this line. Thus, the line itself may be quickly graphed by simply plotting these two points.

Example 13-4 For the following data:
  a. plot the scatter diagram.
  b. calculate the equation of the line of regression.
  c. graph this equation upon the scatter diagram of (a)

Y	1	4	4
X	1	2	3

Solution We solve our problem in steps. Calculate the value of the slope, $\hat{m}$, and use this value to find the equation of the line of regression.

$$\hat{m} = \frac{\sum_{i=1}^{3} x_i y_i - (3)(2)(3)}{\sum_{i=1}^{3} (x_i - 2)^2} = \frac{1 + 8 + 12 - 8}{1 + 1} = \frac{3}{2} \quad \bar{x} = 2, \bar{y} = 3$$

$$y - 3 = (3/2)(x - 2)$$

$$y = (3/2)x, \text{ line of regression}$$

Figure 13-3

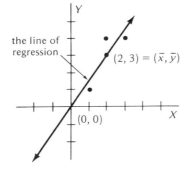

Example 13-5 Calculate the equation of the line of regression for these two matched sets of data.

Y	−1	1	2	2	1
X	−2	−1	1	2	3

Solution

$\bar{x} = 3/5$, $\bar{y} = 1$; Again, solve the problem in steps.

$$\hat{m} = \frac{2 - 1 + 2 + 4 + 3 - (5)(3/5)(1)}{(-2 - 3/5)^2 + (-1 - 3/5)^2 + (1 - 3/5)^2 + (2 - 3/5)^2 + (3 - 3/5)^2}$$

$$= 35/86$$

When the equation is simplified it becomes $y = (35/86)x + 65/86$.

Example 13-6 The farmers' association is trying to find a linear equation to show the relationship between the amount of water that a farmer uses per square inch (x) and the yield of alfalfa in tons per acre (y). The following data are collected from seven test farms. Find the equation of the line of regression. Plot the scatter diagram and upon it sketch the line of regression.

y tons	5.27	5.68	6.25	7.21	8.02	8.71	8.42
x inches	12	18	24	30	36	42	48

Figure 13-4

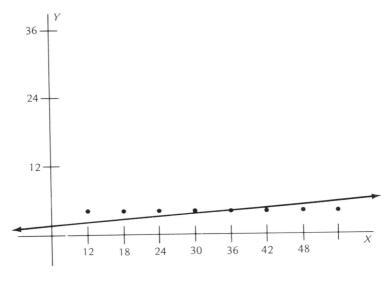

Solution

$$\bar{x} = 30 \text{ in.} \qquad \bar{y} = 7.08 \text{ tons} \qquad n = 7$$

$$\hat{m} = \frac{\sum_{i=1}^{7} x_i y_i - (7)(30)(7.08)}{\sum_{i=1}^{7} (x_i - 30)^2} = .10 \text{ ton/in.}$$

We note that $\hat{m}$ has a title. However, the title has no real bearing upon the graph. We omit titles where possible.

The equation of the line of regression is: $y - 7.08 = .10(x - 30)$
$$y = .10x + 4.08$$

EXERCISES A

13-1. **a.** Example 13-3 shows us a set of data for which we can't find a *good* linear relationship. However, our formulas for the line of regression will still yield an answer, although it may be a poor one. Find the equation of this line.
 b. By mere inspection of the scatter diagram in Example 13-3 state the equations of two other lines that appear to be as representative of the data as our answer to (a). (Example 13-2)

13-2. Plot the scatter diagram and find the equation of the line of regression for these two matched samples below. (Example 13-4)

Y	−1	0	1
X	−1	2	5

13-3. Observe, in Exercise 13-2, that the three scatter points all lie in a straight line. What is the equation of this straight line?

13-4. **a.** Plot the scatter diagram of the following set of data.
 b. Find the equation of the line of regression.
 c. Plot the line of regression upon the scatter diagram of (a).

Y	−1	0	6	3
X	2	4	6	8

13-5. Calculate the equation of the line of regression for these data.

Y	−1	1	3	2	5
X	−3	−2	0	2	3

13-6. Ten people were asked to guess the weight of a package to the nearest pound. Three weeks later they were again asked to calculate the weight of the same package to the nearest pound. The second time they guessed they were told that the value of the package was $2000. The results are recorded below. Compute the line of regression. (Example 13-6)

Y, second guess	9	9	8	9	10	9	9	8	9	10
X, first guess	7	8	7	8	9	5	7	5	7	7

13-7. Quotas imposed upon the catches of tuna fish have encouraged A and B Food Company to raise the price x, of a can of tuna fish versus the number of cans of tuna fish sold y, for five successive months. The data are as indicated:

Price in cents, x	Number of Cans Sold, y
68	82
73	73
78	66
82	58
89	46

Compute the line of regression.

B 13-8. An engineer was working on stress versus strain for a certain alloy. His experiments led him to the following results. X is the stress variable in 100,000 psi and Y is the strain variable in thousands of an inch. Find the equation of the line of regression.

Y	1	4	6	8	10	11	17
X	.10	.30	.50	.80	1.00	1.20	1.50

 13-9. M.S. University is trying to predict a student's final exam grade from the mid-term exam. In order to do this they want to obtain the line of regression for the data below.

Y, final exam	71	89	62	74	17	65	78	35	59	97
X, mid-term exam	65	92	54	80	21	62	72	43	57	91

 13-10. A biologist is interested in measuring the rate of growth of a certain type of bacteria culture. The same type of culture was checked at five incubation periods: $x = 1, 3, 5, 7,$ and 9 hours. Find the equation of the line of regression for these data.

Y, growth rate	10.0	10.3	12.2	12.6	13.9
X, time in days	1	3	5	7	9

 13-11. A large grocery chain conducted a study to determine the relationship between the amount of money spent on advertising, x, and the weekly volume of sales, y. Six different levels of advertising expenditure were tried in a random order over a six-week period. The following data were recorded in units of $100.
  a. Plot the scatter diagram of the data.
  b. Compute the line of regression.
  c. Plot the line of regression upon the scatter diagram of (a).

Weekly sales volume, y	10.2	11.5	16.1	20.3	25.6	28
Amount spent on advertising, x	1.0	1.25	1.5	2.0	2.5	3.0

CUMULATIVE REVIEW

 13-12. We are given two matched samples X and Y. The standard deviation of the X sample is 0 and that of the Y sample is greater than 0.
  a. Describe this scatter diagram.
  b. What is the equation of the line of regression?
  c. Explain why Definition 13-3 is undefined for such a case.

All of the preceding exercises dealt with computing the line of regression. Our next step will be to use the line of regression as a linear model predictor. That is, assuming our data can be effectively portrayed by a straight line, we will predict the y part of a potential scatter point, given the x part of the point.

Example 13-7 The management of a heavy equipment firm wishes to predict a salesperson's first-year sales total on the basis of a specially designed aptitude test. The following data are obtained. The test scores are labeled x and the total first-year sales are labeled y. Compute the equation of the line of regression.

x	y	x	y
48	312	50	288
32	164	26	146
40	280	50	361
34	196	22	149
30	200	43	252

Solution

$$\bar{x} = 37.50 \qquad \bar{y} = 234.8 \qquad \sum_{i=1}^{10} x_i y_i = 94{,}448$$

$$\sum_{i=1}^{10} (x_i - 37.50)^2 = 930.50, \quad \hat{m} = 6.88$$

Therefore,

$$y = 6.88x - 23.20$$

Example 13-8 Use the regression line of Example 13-7 to predict a potential salesperson's first-year total sales on the basis of a test score of 45.

Solution When $x = 45$ is substituted into the equation,
$$y = (6.88)(45) - 23.20$$
$$y = 286.40$$

Example 13-9 The Swift Bicycle Company would like a linear equation to portray its annual bicycle production versus the production year for the years 1976 to 1980 inclusive. a. Find the equation of the line of regression. b. Make a graph showing the line of regression as a dotted line and the raw data as a series of straight-line segments. c. Use the line of regression to predict the bicycle production for 1981.

Year	Annual Production
1976	10
1977	12
1978	13
1979	11
1980	14

**SWIFT BICYCLE COMPANY
ANNUAL PRODUCTION
(THOUSANDS OF BICYCLES)**

Solution

a. In order to simplify the computation for the line of regression, replace 1976 by 0, 1977 by 1 - - -, 1980 by 4.

$$\hat{m} = \frac{\sum_{i=0}^{4} x_i y_i - 5(2)(12)}{\sum_{i=0}^{4} (x_i - 2)^2} \qquad \bar{x} = 2, \bar{y} = 12$$

$$\hat{m} = \frac{127 - 5(2)(12)}{10}$$

$$\hat{m} = .70$$

Therefore,

$$y = .70x + 10.60$$

b.

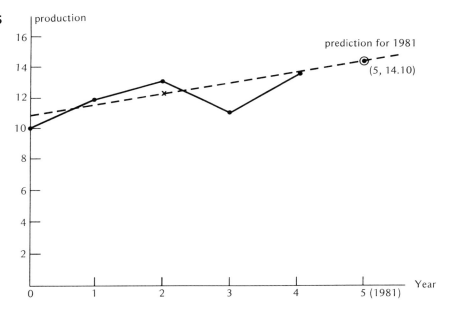

Figure 13-5

c. The year 1981 will correspond to the x value of 5. Therefore, our prediction for the production of bicycles in 1981 is

$$y = .70(5) + 10.60$$

$y = 14.10$ thousand; see circled point in Fig. 13-5.

Example 13-10 In Example 13-6, if each square inch of farm land is to receive 20 in. of water, what would be a reasonable estimate of the yield of alfalfa?

Solution We will substitute $x = 20$ into the equation of the line of regression $y = .10x + 4.08$. The result is $y = 6.08$ tons.

We now observe that we must apply common sense in using our line of regression as a linear predictor. For example, when $x = 0$, $y = 4.08$. This implies that if we don't water at all, we still harvest 4 tons. Further, if this equation were used without common sense, we would be led to believe that a large increase in water supply would result in harvesting a tremendous tonnage. This equation does allow us to extrapolate with care. (To extrapolate means that we wish to extend our x-value predictions above or below the existing values of x, the independent variable.) In the last example, there must be a limit in our water variable at which point harvesting additional tonnage stops.

Example 13-11 As another illustration of a straight line being fitted to a set of data, consider the equation $y = -1.50x + 58.00$. In this equation, y represents the yearly infant mortality rate (number of deaths per thousand) in New York City and $x = 0, 1, 2, \ldots$ correspond to the years 1930, 1931, 1932, ..., respectively. Assume that the trend indicated by the line of regression is accurate through the end of 1940.
  a. Compute the value of y for 1935 ($x = 5$).
  b. Does the equation indicate an increase or a decrease in infant deaths per 1000? Explain briefly.
  c. Find the year in which the rate of infant mortality was 43.00.

Solution
  a. $y = (-1.50)(5) + 58.00 = 50.50$ (per 1000).
  b. Because the coefficient of x is negative, the slope is negative; the infant mortality rate for these years is decreasing at the rate of 1.50 deaths per 1000 per year.
  c. If we set $y = 43.00$, then $x = 10$, which corresponds to the year 1940.

This last example is especially good because it illustrates a little-used interpretation, the slope of the line of regression. The slope of the line of regression represents the *constant rate of change* of y (the dependent variable) with respect to x (the independent variable).

Example 13-12 Return to Example 13-6 and interpret the slope of the line of regression as a constant rate of change of y versus x.

Solution The slope is .10. Because the slope is positive we have an increasing rate of change. It tells us that for each unit increase in the x variable there is a .10 unit increase in the y variable.

A positive slope means that we have an increasing rate of change between our two variables and a negative slope means that we have a decreasing rate of change. When the slope is expressed in decimal form, this value represents the amount of change in the y variable per unit of change in the x variable.

EXERCISES A

In Exercises 13-13 through 13-16 we are given that our line of regression is a good linear model of our data. Use the given line of regression to predict (to estimate) the y value of a potential scatter point whose x value is specified. (Example 13-8)

13-13. $y = x + 5$ for: (a) $x = 2$, (b) $x = 5$.
13-14. $y = -3x + 11$ for: (a) $x = 3$, (b) $x = 7/3$.
13-15. $y = .27x + 7.14$ for: (a) $x = 4$ (b) $x = 3.97$.
13-16. $y = .63x - 5.45$ for: (a) $x = 5$, (b) $x = 2.06$.
13-17. Use the line of regression derived in Exercise 13-4 to predict the y value for $x = 1, 3, 7.5$.
13-18. Use the line of regression of Exercise 13-9 to predict the final exam grade if the mid-term is 70. If it is 85.

CUMULATIVE REVIEW

B **13-19.** A statistics professor did a study of the number of hours students spent preparing for his exam versus their exam grade. Fifteen of the professor's students were selected at random and the data are as indicated:

Hours, x	.50	.75	1.00	1.25	1.50	1.75	2.00	2.25	2.50	2.75	3.00	3.25
Grade, y	57	64	59	68	74	76	79	83	85	86	88	89

Hours, x	3.50	3.75	4.00
Grade, y	90	94	96

a. Find the equation of the line of regression. (Example 13-7)
b. If a student studied .25 hour, what is the student's predicted grade? (Example 13-8)

 13-20. Based on the line of regression of exercise 13-10 predict the growth rate after six days. Use this linear model to extrapolate the potential growth rate after 12 days.

 13-21. Return to Exercise 13-7 and interpret the slope of the line of regression as a constant rate of change of y versus x.

 13-22. Return to Exercise 13-8 and interpret the slope of the line of regression as a constant rate of change of y versus x.

 13-23. Return to Exercise 13-9 and interpret the slope of the line of regression as a constant rate of change of y versus x.

It is true that a scatter diagram often cannot be effectively portrayed by a straight line. An equation such as $y = e^x$, $y = \sin(x)$, or $y = \log(x)$ may indeed be more realistic. However, all of our effort in a first course is concentrated on linear (straight-line) regression. Figure 13-6 illustrates how a polynomial may be fitted to a scatter diagram.

Figure 13-6

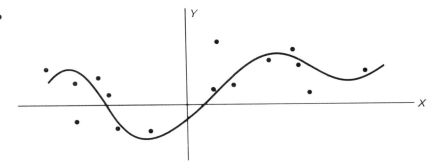

Multiple regression occurs when we are dealing with three or more values that are matched one-to-one. Here our scatter points are n-tuples of the form $(x_1, x_2, x_3, \ldots, x_n)$, where $n > 2$. For the three-dimensional case we are dealing with data triples and three deminsions X_1, X_2, and X_3 are needed to plot our scatter points.

For three-dimensional multiple regression, what would be the simplest geometric figure to fit to our scatter diagram? The answer: a *regression plane*. We still use the method of least squares. By moving along a line parallel to one of the three axes we compute the distance from the scatter point to the corresponding point on the regression plane. Clearly, our choice of parallel movement will yield three possible regression planes. Pick a direction of parallel movement. With this movement settled, our plane will be that plane which makes the sum of the distances squared minimal. This concept is illustrated in Fig. 13-7.

Figure 13-7

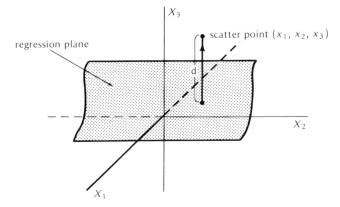

SUMMARY

This chapter deals with n pairs of data of the form (x_i, y_i) where $1 \leq i \leq n$. When these n pairs (points) of data are plotted, we have a graph called a *scatter diagram*. The object is to represent the n points of our scatter diagram by a single equation (if at all possible). It is true that we could begin to think about logarithmic equations, trigonometric equations, polynomial equations, and so on; however, we are only interested in the simplest equation of all. We are interested in the equation of a straight line. Hence, we speak of the work in this chapter (and the next) as "linear" (from linear equation). The word "regression" is taken from the work of Sir Francis Galton who was keenly interested in finding relationships between pairs of data. Thus, we are interested in finding the equation of a *line of regression* to represent the points of our scatter diagram. The method of fitting a line to our scatter diagram is called *the method of least squares* because we select the straight line that makes the sum of vertical distances squared between corresponding points (points of the scatter diagram and points of the line) minimal.

You may always find the line of regression in two steps. First, calculate the slope of the line, $\hat{m}$. Second, use this value of $\hat{m}$, $\bar{x}$ and $\bar{y}$ and plug them into the formula for the line of regression $[y - \bar{y} = \hat{m}(x - \bar{x})]$.

What do we do if we cannot effectively portray the points of our scatter diagram by a straight line? For us the job is done; we stop. If you would like a polynomial equation (a common choice) to portray the data, you bring your data to your mathematics department.

Can You Explain the Following?

1. Sir Francis Galton
2. scatter diagram
3. line of regression
4. method of least squares
5. the formula for the slope ($\hat{m}$) of the line of regression
6. the formula for the line of regression
7. multiple regression

CUMULATIVE REVIEW

MISCELLANEOUS EXERCISES

13-24. State (in your own words) the criterion used to fit a line of regression to a scatter diagram.

13-25. a. Plot the scatter diagram for these data

y	0	1	0	−1
x	1	0	−1	0

b. Explain why it seems unreasonable to try to fit a line to these data. By inspecting the scatter diagram, what figure might be decided on as a good fit?

13-26. In the event $\hat{m} = 0$, $\bar{y} \neq 0$, what is the equation of the line of regression?

13-27. a. Plot the scatter diagram for the data below.
b. From an inspection of the line of regression, make a guess at the line of regression.
c. Compute the line of regression.

y	2	3	4	5	6
x	4	5	6	7	8

13-28. Ten students studied for a quiz in statistics. Following are their study times (x) in hours and their quiz grades (y). Compute the line of regression for these data.

Grade (y)	85	95	90	78	47	63	90	80	60	82
Time (x)	2	4	3	2	0	1	3	2	1	2

13-29. Return to Exercise 13-28 and use the line of regression to:
  a. Predict the quiz grade for a student who studies 2.50 hours.
  b. Predict the quiz grade for a student who studies 5.00 hours.

 13-30. Suppose a chemical company wishes to study the effect of various factors on the efficiency of an extraction operation. In the overall study there are several processing conditions (time, moisture, amount of raw material used, type of solvent, etc.). However, we will study only the extraction time (x), in minutes, versus a measure of extraction efficiency (y), in percent form. Compute the line of regression.

y	55	60	44	78	75	50	62	65	55	70
x	14	34	18	40	48	18	44	30	26	38

14 LINEAR CORRELATION

14-1 INTRODUCTION

HISTORICAL NOTE

Bioretrika Pencil portrait by Miss F. A. de Biden Footner, 1924

KARL PEARSON

In this chapter we deal with the problem of measuring the strength of linear regression for a set of scatter points. In his research Sir Francis Galton needed a method to evaluate the degree to which his data fitted the equation being used to represent the data. The method became known as correlation (co-relation). He used this method in relating the heights of fathers and sons. Galton's student, Karl Pearson, with Galton's aid, later developed a formula that yielded a statistic known as a *correlation coefficient*. Because we are applying it to a straight line, it is often called linear correlation. In honor of Karl Pearson (1857–1936) it is also called the *Pearson r* or the *Pearson product-moment correlation coefficient*. The latter name indicates the physics orientation of this coefficient.

14-2 A LOOK AT THE CORRELATION COEFFICIENT

In the equation of the regression line, $y - \bar{y} = \hat{m}(x - \bar{x})$, we have agreed to let x be the independent variable (this is the one the experimenter controls) and y the dependent variable. How do we tell the degree to which the y values of points on the regression line represent corresponding y values of scatter points? What should we use as a criterion for determining the degree of linear relationship? For a fixed value of the independent variable, x, we agree to consider the difference between the y value on the regression line and $\bar{y}$ as the amount of *variation in the dependent variable that is explained* by the line of regression. The difference between the y value of our scatter point and the corresponding y value on the line of regression is considered the *variation in the dependent variable that cannot be explained* by the line of regression. See Fig. 14.1.

278 LINEAR CORRELATION

We must remember that we are restricting ourselves to only two variables, although, in practice, several variables may be operating on our dependent variable. Consider the grocery chain that is attempting to measure the amount of advertising (x) against the weekly sales volume (y).

Weather, inflation, worker strikes, etc., all have an effect on the weekly sales volume.

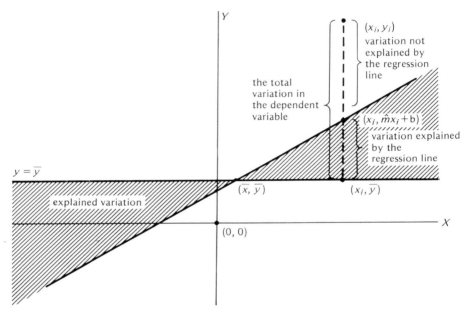

Figure 14-1

Definition 14-1 The ratio

$$\frac{\sum_{i=1}^{n} (\hat{m}x_i + b - \bar{y})^2}{\sum_{i=1}^{n} (y_i - \bar{y})^2}$$

is denoted r^2. This ratio is called the *coefficient of determination*.

It is the coefficient of determination that we use to measure the linear strength of our set of scatter points. However, it does seem unreasonable to have to compute the line of regression in order to evaluate r^2, which, in turn,

14-2 A LOOK AT THE CORRELATION COEFFICIENT

tells us whether or not we should bother with the line of regression in the first place. Fortunately, we have an easier way of evaluating r^2. It is based on a formula for evaluating r directly from the two matched samples of data.

Definition 14-2 The *correlation coefficient* is denoted by r and

$$r = \frac{1/n \sum_{i=1}^{n} (x_i - \bar{x})(y_i - \bar{y})}{\sqrt{1/n \sum_{i=1}^{n} (x_i - \bar{x})^2} \sqrt{1/n \sum_{i=1}^{n} (y_i - \bar{y})^2}}$$

It can be shown that the correlation coefficient squared yields the coefficient of determination, but we have decided to omit this computation because the algebra involved is lengthy.

Example 14-1 For the two matched sets of data (Example 13-4)

y	1	4	4
x	1	2	3

$\bar{x} = 2, \bar{y} = 3$

a. Compute the correlation coefficient,
b. Compute the coefficient of determination.

Solution

a. $$r = \frac{1/3[(-1)(-2) + (1)(1)]}{\sqrt{1/3[2]} \sqrt{1/3[6]}} = \frac{\sqrt{3}}{2} = .87$$

b. $r^2 = .76$ This means that the regression line explains 76% of the variation in the dependent variable. It fails to explain 24% of the variation.

Example 14-2 For the following sets of data compute r. Compute r^2.

Y	0	1	2	3	4
X	-2	-1	0	1	2

$\bar{x} = 0, \bar{y} = 2$

280 LINEAR CORRELATION

Solution

a. $$r = \frac{1/5[(-2)(-2) + (-1)(-1) + (1)(1) + (2)(2)]}{\sqrt{1/5[10]}\sqrt{1/5[10]}} = 1$$

b. $$r^2 = 1.00$$

Note, in the previous example, that $r^2 = 1$ or the line of regression explains 100% of the variation in the dependent variable. When the scatter diagram of the previous example is plotted the result is not surprising, as all the points of Example 14-2 lie on the line $y = x + 2$. See Fig. 14-2.

Figure 14-2

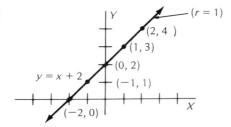

Return to Definition 14-2 and consider the two square-root factors in the denominator of r. Each of these factors is "nearly" our old standard deviation, and they would actually be standard deviations if only n were replaced by $n - 1$.

Let's name these two factors:

$$B_x = \sqrt{\frac{1}{n}\sum_{i=1}^{n}(x_i - \bar{x})^2} \qquad B_y = \sqrt{\frac{1}{n}\sum_{i=1}^{n}(y_i - \bar{y})^2}$$

$$\boxed{r = \frac{\frac{1}{n}\sum_{i=1}^{n}(x_i - \bar{x})(y_i - \bar{y})}{B_x B_y}} \quad \text{or} \quad \boxed{r = \frac{\frac{1}{n}\sum_{i=1}^{n}x_i y_i - (\bar{x})(\bar{y})}{B_x B_y}}$$

It is frequently convenient, for purposes of calculation, to compute r in steps. That is, calculate the following four quantities separately, and with these results compute r.

1. $(1/n)\sum_{i=1}^{n} x_i y_i$

2. $\bar{x}\bar{y}$
3. B_x
4. B_y

Because the numerator of r^2 is always less than or equal to the denominator, it follows that $r^2 \leq 1$. Equality will occur only when all of the scatter points lie on the line of regression. If $r^2 \leq 1$, it immediately follows that $-1 \leq r \leq 1$. For the case that $r = 1$, we say that we have *perfect positive correlation;* for the case that $r = -1$, we say that we have *perfect negative correlation.*

Example 14-3 Calculate the value of r and r^2 for the grocery chain of Exercise 13-11 in Chapter 13.

Solution $\quad \bar{x} = 1.88 \quad \bar{y} = 18.62$

$$\text{Numerator} = 1/6 \sum_{i=1}^{6} x_i y_i - (1.88)(18.62) = 4.54$$

$$B_x = .70 \quad B_y = 6.68$$

$$r = .97 \text{ and } r^2 = .94$$

The regression line is 94% effective in explaining the relationship between the independent variable, advertising, and the dependent variable, sales volume.

Example 14-4 Example 13-2 in Chapter 13 dealt with a medical researcher and his attempt to relate dosages of the drug ephedrine to the number of heart beats per minute in humans. We plotted the scatter diagram for this data. Inspection of this diagram would seem to indicate that a straight-line representation is reasonable. Is a straight line a good fit? Evaluate r^2 and answer this question.

Solution

$$r = \frac{(1/6)(1,480) - (3)(75)}{(1.29)(17.08)} = .98$$

Thus

$$r^2 = .96.$$

The line of regression does an excellent job of representing the collection of scatter points.

282 LINEAR CORRELATION

Example 14-5 It is desired to relate the independent variable, speed of a car (x), to the dependent variable, gasoline efficiency in miles per gallon (y). Will the line of regression serve as a good equation relating these two variables? Calculate r^2. Five matched pairs of data follow.

Y	20	18	17	14	11
X	30	40	50	60	70

Solution

$$r = \frac{(1/5)(3{,}780) - (50)(16)}{(14.14)(3.16)} = -.98$$

Thus

$$r^2 = .96.$$

The line of regression does an excellent job of representing the collection of scatter points.

We observe, at this time, that r is a pure number. It never has a title. But before going further, we must mention what r does not do. Even when a regression line serves as an excellent representative of a scatter diagram, we must be careful not to read in any cause-and-effect relationships. Does the x variable cause the y variable to move in a linear fashion? This question demands further research. It is not answered by evaluating r.

As an illustration of this, for a period of years, the correlation between teachers' salaries and the consumption of liquor was observed to be .90. This is a high positive correlation. When the situation was examined closely it was discovered, for the years in question, that the country was going through an era of prosperity. This prosperity forced both wages and buying power steadily upward. This increased buying power was reflected in the increased purchase of liquor. Had we recorded the salaries of plumbers, lawyers, clerks, etc., instead of teachers, the value of r would have remained about .90 in all cases.

Our next example illustrates two sets of data that do not possess a linear relationship.

Example 14-6 For the following sample data:
a. Find the value of r,
b. Find the value of r^2,
c. Plot the scatter diagram.
d. Part (b) shows that there is no linear relationship; can we think of a nonlinear relationship?

14-2 A LOOK AT THE CORRELATION COEFFICIENT

Y	1	0	0	−1
X	0	−1	1	0

Solution

a. $\bar{x} = \bar{y} = 0 \quad \sum_{i=1}^{4} x_i y_i = 0 \quad \text{thus } r = 0$

b. If $r = 0, r^2 = 0$

c.

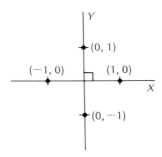

d. Although a linear relationship is nonexistent, there does exist a nonlinear relationship, namely: $x^2 + y^2 = 1$. All the scatter points lie on the circumference of a circle whose center is (0, 0) and whose radius is 1.

Example 14-7 A psychologist wishes to calculate the correlation between the midparent IQ (this is the average of the parents' IQ's) and the IQ of the offspring. The data are given in the following. Compute the correlation for the psychologist.

Midparent IQ (X)	Offspring IQ (Y)
125	110
120	105
110	95
105	125
105	120
95	105
95	75
90	95
80	90
75	80

284 LINEAR CORRELATION

Solution $\bar{x} = 100$ $\bar{y} = 100$ $B_x = 15.33$ $B_y = 15.33$ $r = .59$

Conclusion There is some correlation, however, it is not strong. See the scatter diagram in Fig. 14-3.

Figure 14-3

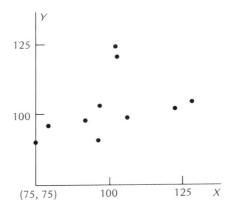

Example 14-8 The data in the following table indicate the number of banks in the United States that closed permanently or temporarily because of financial difficulties and the amount of money on deposit at the time of failure. Compute the correlation coefficient for these data.

Period	Number of Banks Closed x	Amount of Money on Deposit (in millions), y
1951–1955	17	$58
1956–1960	19	41
1961–1965	28	99
1966–1970	11	50
1971	3	5
1972	2	57
1973	3	21

Source: *Federal Reserve Bulletin,* Board of Governors of the Federal Reserve System, 1974.

Solution $r = .76$.

14-2 A LOOK AT THE CORRELATION COEFFICIENT

A version of the formula for r that is applicable to computer programming is given in the following. To use this formula with a hand calculator, line the work up in columns. For example, one column would be headed x_i's and another $x_i y_i$'s, and so forth.

$$r = \frac{n \sum_{i=1}^{n} x_i y_i - \sum_{i=1}^{n} x_i \sum_{i=1}^{n} y_i}{\sqrt{\left[n \left(\sum_{i=1}^{n} x_i^2\right) - \left(\sum_{i=1}^{n} x_i\right)^2\right] \left[n \left(\sum_{i=1}^{n} y_i^2\right) - \left(\sum_{i=1}^{n} y_i\right)^2\right]}}$$

Theorem 14-1 $\hat{m} = r \cdot a$, where $a = B_y/B_x \geq 0$.

Proof

$$\hat{m} = \frac{\sum_{i=1}^{n}(x_i - \bar{x})(y_i - \bar{y})}{\sum_{i=1}^{n}(x_i - \bar{x})^2} = r \left[\frac{B_y}{B_x}\right], \text{ see Exercise 14-16}$$

This theorem tells us that when $r > 0$, $\hat{m} > 0$, and when $r < 0$, $\hat{m} < 0$.

Example 14-9   Find the equation of the line of regression for the data of Example 14-5. Use the solution of Example 14-5 and Theorem 14-1 to help obtain this equation.

Solution $\bar{x} = 50$ $\bar{y} = 16$

$$\hat{m} = r \left[\frac{B_y}{B_x}\right] = (-.98) \left(\frac{3.16}{14.14}\right) = -.22$$

$$y - 16 = (-.22)(x - 50)$$

Therefore,

$$y = -.22x + 27$$

Our last theorem shows the close relationship between the linear correlation coefficient and the slope of the line of regression. In view of this theorem, let's take another look at the line of regression. Assume r^2 is large, indicating that our linear regression line is a good fit. Further assume $r > 0$. Under these conditions, let's look at the points on the line of regression. When $r > 0$, $\hat{m} > 0$. Thus, if the x values of points on this line are such that $x_1 < x_2 < x_3 < \cdots < x_{n-1} < x_n$, it follows that $y_1 < y_2 < y_3 < \cdots < y_{n-1} < y_n$.

For those points on the line of regression, as x increases y increases. Because our line is assumed to be a good fit to the scatter diagram, although this trend may not hold exactly for our set of scatter points, it should hold in a general sense. What does this all mean? A large positive value of r indicates *direct variation* between our two matched sample sets of data. Conversely, negative correlation indicates *inverse variation* between our two matched sample sets of data. Realistic examples include grade-point average versus alcohol consumption for college students and time versus the electric charge of an automobile battery. Figure 14-4 illustrates the case when $r > 0$.

Figure 14-4

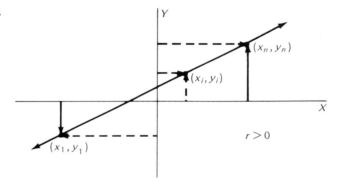

Figure 14-5 presents several smooth curve sketches of sets of data and their respective correlation coefficients. The overall shape is the key idea. Actual distributions of plotted points within these curves will not be as smooth.

Figure 14-5

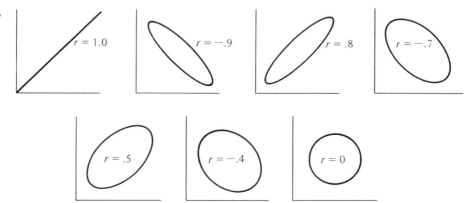

14-2 A LOOK AT THE CORRELATION COEFFICIENT

EXERCISES A

14-1. Can the coefficient of determination ever be negative?

14-2. State theorem 14-1 for the case that $B_x = B_y$.

14-3. Find the correlation coefficient for:

y	−1	−2	5	2
x	−1	0	1	4

(Example 14-1)

14-4. Find the correlation coefficient for:

y	4	5	6
x	2	3	4

14-5. Find the correlation coefficient for:

y	2.1	4.6	2.9
x	3.4	1.8	.8

14-6. Find the correlation coefficient for:

y	1	6	0	−1	4
x	2	5	7	3	3

14-7. Verify that the sample data of Example 13-3 in Chapter 13 has a value of $r = 0$.

14-8. Explain how, with a very small adjustment, Definition 14-1 may become the ratio of two variances.

14-9. The blood pressure and the corresponding ages for nine women are shown in the accompanying table. Evaluate the Pearson r for this data. (Example 14-5)

Systolic blood pressure, y	118	128	145	125	115	153	146	140	149
Age, x	36	47	49	42	38	68	56	42	63

14-10. W is the weight in grams of a certain chemical that will dissolve in 1000 grams of water at temperature T°C. Calculate the value of the correlation coefficient.

W	480	552	650	725	840	970	1035	1099
T	0	10	20	40	60	80	100	110

14-11. Suppose a study is being made of the relationship between daily family income, x, and weekly food expenditures, y. A random sample of 10 families is given in the table below. Compute r.

$, Y	47	55	83	92	74	64	105	87	98	78
$, X	49	46	60	71	53	47	70	65	53	58

14-12. The following table shows the number of minutes the caution light is on during a complete cycle of traffic lights at an intersection and the number of accidents at the intersection during a month. Compute r.

Intersection	1	2	3	4	5	6	7	8	9	10
accidents/month, y	11	12	12	13	15	12	14	15	16	16
Yellow light minutes/cycle, x	1.0	1.0	1.2	1.2	1.2	1.4	1.4	1.8	1.7	1.8

14-13. a. Compute the coefficient of determination for Exercises 14-3 through 14-6.
  b. Discuss each answer in terms of whether or not a regression line appears to be a good fit.

14-14. The fire chief in a large New Hampshire town believes that as the temperature decreases, the number of fires in his town increases. To support this claim, he gathered the following data.

Number of fires, y:	28	32	35	39	51	55	56	66
Degrees, x:	40	35	30	25	20	15	10	5

  a. Draw a scatter diagram for the data.
  b. Find the value of r.
  c. Discuss the variation of the two samples.

14-15. Data are recorded on eight people. Use these data below to see if there is any linear correlation between cigarette smoking in packs per day, y, and number of days absent per year, x.

Smoking, y	.5	1.5	2.0	.5	0.0	1.0	3.5	0.0
Days absent, x	4	8	15	0	3	10	20	0

14-16. Show $(1/n) \sum_{i=1}^{n} (x_i - \bar{x})(y_i - \bar{y}) = (1/n) \sum_{i=1}^{n} x_i y_i - \bar{x}\bar{y}$

14-3 THE NONPARAMETRIC VERSION OF THE CORRELATION COEFFICIENT

The nonparametric analogue of the ordinary correlation coefficient is the *rank correlation coefficient*. As its name implies, it is merely the correlation coefficient calculated for the ranks of the variables rather than for numerical values of the variables. This rank correlation coefficient is especially useful when variables are difficult to measure. For example, this is sometimes the case in psychology.

Ten major brand wines may be ranked for robustness by two professional wine tasters. Two different supervisors for a large firm may rank the same set of employees on overall job performance. A sample of students may be ranked for their ability in mathematics versus their ability in English. A psychologist may wish to compare the ranks of a sample of people with regard to motivation versus achievement. The technique is to assign the numbers 1, 2, 3, . . . , n to our sample using one criterion and then to repeat the process using the same numbers 1, 2, 3, . . . , n for a second criterion. Should ties occur within a single criterion, each piece of datum is assigned the mean of the ranks that they occupy. The same technique was used in breaking ties in the Mann-Whitney U test.

Clark University Archives
Psychologies of 1930 edited by
Carl Murchison, published by
Clark University Press
CHARLES SPEARMAN

This special correlation coefficient is often called the *Spearman rank correlation coefficient*, in honor of the well-known statistician—or simply Spearman r. The formula for the Spearman r, denoted r_s, is stated without proof.

$$r_s = 1 - \frac{6 \sum_{i=1}^{n} (y_i - x_i)^2}{n(n^2 - 1)}$$

Example 14-10 Five people are lined up and ranked 1, 2, 3, 4, 5 on the subject of race prejudice. These same five people, lined up as before, are ranked 4, 3, 2, 1, and 5 on liberalism of thinking. See the following table. Calculate the Spearman r, r_s, for these two matched sample sets of data.

Person number	1	2	3	4	5
Liberal thinking, y	4	3	2	1	5
Race prejudice, x	1	2	3	4	5

Solution

$$r_s = 1 - \frac{6[9 + 1 + 1 + 9]}{5(24)} = 0$$

This implies $r_s^2 = 0$. There is no linear relationship between these two variables.

Example 14-11 Twelve management trainees in a large firm are ranked according to their degree of motivation and then according to their management-trainee final exam. The results are given here. Compute r_s.

Trainee	Motivation	Performance on Final Exam
1	3	7
2	7	2
3	1	4
4	11	9
5	2	1
6	9	12
7	6	5
8	4	6
9	5	3
10	10	8
11	12	11
12	8	10

Solution $r_s = 1 - [6(82)]/[12(143)] = .71$

Because r_s is a special correlation coefficient (a special r) it immediately follows that $-1 \leq r_s \leq +1$. If we wish, we may evaluate r_s by using the formula for r. This formula is a more tedious way of computing r_s, but it always works.

Example 14-12 For the matched sample sets of data

y	2	3	1
x	1	2	3

a. Compute r_s.   b. Compute r.

Solution

a. $$r_s = 1 - \frac{6[6]}{3[8]} = -1/2$$

b. $$r = \frac{[1/3][2 + 6 + 3] - (2)(2)}{\sqrt{2/3}\sqrt{2/3}} = -1/2$$

Observe that the answers are identical.

The next example illustrates that you may not, in general, use the formula established for r_s to evaluate r.

Example 14-13 Use the data of example 14-1 "blindly" in the formula for r_s.

y	1	4	4
x	1	2	3

Solution Our wrong use of $r_s = 1 - [6(4 + 1)]/[3(9 - 1)] = -.25$. When this answer is compared with the previously obtained solution of Example 14-1, we see that we have a different (and wrong) answer.

EXERCISES A

14-17. Five aspiring artists are ranked by two professional art judges on their creative potential—1 is low and 5 is very high. Compute r_s for the two sets of ratings. (Example 14-10)

Artist	I	II	III	IV	V
Expert no. 1	1	4	2	5	3
Expert no. 2	2	4	1	5	3

14-18. Compute the Spearman r for the following two sets of data. (Example 14-12)

Y	1	2	3	4
X	1	3	2	4

14-19. By inspection only, state the value of r_s for the following data. [Hint: Plot the scatter diagram.]

y	5	4	3	2	1
x	1	2	3	4	5

14-20. A sociologist ranks eight people chosen at random by their socioeconomic status, x, and by the time spent watching television, y, with 1 the lowest ranking and 8 the highest. Find r_s.

TV time, y	2	1	3	5	4	8	6	7
Status, x	1	2	3	4	5	6	7	8

14-21. A psychologist is trying to determine if there is any relationship between IQ's and class standing for eighth graders. The data are given below.
a. Compute r_s. b. Discuss the variation of the data.

IQ score, y	6	10	8	7	9	5	3	4	1	2
Class standing, x	10	9	8	7	6	5	4	3	2	1

14-4 THE SAMPLING DISTRIBUTION OF r'S (optional)

Pick a sample size n. Repeatedly draw samples of paired data and compute r. You are constructing a sampling distribution of correlation coefficients. From research, we know that this sampling distribution is unpredictable. R. A. Fisher discovered a remarkable equation that can be used to convert the erratic sampling distribution of r's into a normal distribution. The equation is one-to-one. That is, for each value of r there can exist one, and only one, value of z. This equation is called the *Fisher z-transformation*; it is presented in terms of the natural log function and is stated here without proof.

14-4 THE SAMPLING DISTRIBUTION OF r'S (OPTIONAL)

$$z = 1/2 \ln \left[\frac{1+r}{1-r}\right]$$

where ln denotes the natural logarithm.

By using this equation we are able to transform the sampling distribution of r's into a normal (not standard normal) distribution Z. We have used the letter Z because it is the classic way of writing this transformation. For this Z distribution we have

$$\mu = 1/2 \ln \frac{1+\rho}{1-\rho} \qquad \sigma = 1/\sqrt{n-3}$$

Now the z score (not to be confused with the Fisher z) for an arbitrary value of r is

$$z = \frac{1/2 \ln \left[\frac{1+r}{1-r}\right] - 1/2 \ln \left[\frac{1+\rho}{1-\rho}\right]}{1/\sqrt{n-3}}$$

where ρ denotes the population correlation coefficient.

Figure 14-6

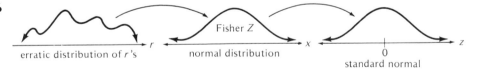

erratic distribution of r's   normal distribution   standard normal

We use this theory and Fisher's equation to test the null hypothesis of no correlation. Namely, there is no linear correlation between two variables. $H_0: \rho = 0$. For this application, line 1 is our z-test statistic. Note how the null hypothesis makes the second term in the numerator zero. (Recall that ln 1 = 0.) Our discussion will be confined to the two-tail case—$H_a: \rho \neq 0$.

Example 14-14 A sociologist wishes to know if there is any linear correlation between the face index and the cephalic index of a certain race. To resolve this question, 50 members of the race are randomly selected and the correlation coefficient = .20. Test at the .05 level.

294 LINEAR CORRELATION

Solution $H_0: \rho = 0 \quad H_a: \rho \neq 0$

$$z\text{-test statistic} = \frac{1/2 \ln\left[\frac{1.2}{.8}\right]}{1/\sqrt{47}} = \frac{1/2[.40547]}{.15}$$
$$= 1.35 < 1.96$$

Conclusion There is insufficient evidence to reject H_0. We cannot establish that a linear correlation exists at this level on the basis of our present data. A natural log table is needed to complete these calculations.

EXERCISES B

14-22. We wish to test $H_a: \rho \neq 0$ using the data in Example 14-1; however, such an endeavor will not be possible. Explain why.

14-23. Test $H_a: \rho \neq 0$ using the data of Example 14-8. Test at the .05 level.

14-24. A researcher would like to see if there is any linear correlation between the heights and weights of male college students. His data are given here in table form. Test $H_a: \rho \neq 0$ at the .01 level.

x (in.)	y (lb)	x (in.)	y (lb)
72	191	66	147
70	172	68	162
63	125	72	191
74	210	70	164
69	154	73	175
72	186	71	163

SUMMARY

How do you tell whether or not a straight line is a good representative for portraying the points of your scatter diagram? This is determined by calculating the *linear correlation* (denoted r). Frequently the linear correlation is called the *Pearson r* in honor of Karl Pearson. If r is close to zero, there is no linear correlation. You should not calculate the equation of the line of regression. If r is close to -1 or if r is close to $+1$, you have respectively a strong negative or a strong positive linear correlation. Given a strong linear correlation you should proceed to compute the equation of the line of regression.

I have chosen to present the line of regression before a discussion of the topic of linear correlation. The topics are actually used in the reverse of the order that I have presented them. That is, you calculate the linear correlation and then (if called for) you calculate the equation of the line of regression. Why has this been done? I believe it makes better teaching sense. If you had learned correlation before regression you could not fully appreciate the job of the linear correlation coefficient (r). The author does realize that most statistics texts present correlation before regression.

The *Spearman rank correlation coefficient* was discussed in this chapter. This is the nonparametric version of the correlation coefficient, and it deals (like the Mann-Whitney U-Test) with ranked data. The Spearman r is denoted r_s.

Our optional section deals with the sampling distribution of r's. Here we meet the famous *Fisher z-transformation;* here also we test the null hypothesis, H_0: $\rho = 0$. "Rho" (ρ) is the population parameter for all population pairs of data being considered.

Can You Explain the Following?

1. linear correlation coefficient (r)
2. Pearson r
3. coefficient of determination
4. maximum and minimum values of r
5. Can you state a collection of data where $r = 0$?
6. Spearman r (r_s)

CUMULATIVE REVIEW

MISCELLANEOUS EXERCISES

14-25. Predict the linear correlation for each scatter diagram by inspection.

a. b.

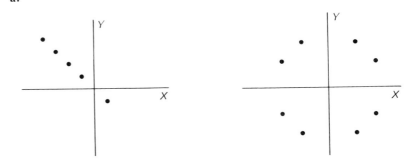

14-26. For each of the following indicate whether you would anticipate a positive correlation, a negative correlation, or zero correlation.

a. Overweight and incidence of heart attacks.
b. Weight of a baby at birth and length of the same baby at birth.
c. Age of your car and current market value.
d. Length of a jail term and incidence of crime.
e. Person's birthday and date of death.
f. Traveling at excessive speeds and accidents.
g. Speed limits and traffic accidents.
h. Study time for a course and course grade.

14-27. Find the value of the slope of the line of regression to a scatter diagram for which the value of $r = .85$, $B_y = .62$, and $B_x = .71$. [Hint: See Theorem 13-1.]

14-28. Express r in terms of $\hat{m}$, B_y, and B_x.

14-29. If $\hat{m} = .74$, $B_y = .62$, and $B_x = .71$, find r.

14-30. Given the value of r, find what percent of the variation in the dependent variable is explained by the line of regression.
(a) $r = .45$ (b) $r = 3/4$ (c) $r = -1$ (d) $r = .95$

14-31. You learn that the coefficient of determination (r^2) = 1. This means that the linear correlation could be either of what two values?

14-32. For the given value of r, classify the data pairs as varying directly, indirectly, or having no significant relationship.
(a) $r = .45$ (b) $r = .85$ (c) $r = 0$ (d) $r = -.93$

14-33. Find the correlation coefficient for:

y	−2	−1	3	2
x	−1	0	1	2

14-34. Find the value of r_s for

y	3	2	1	5	4
x	1	5	3	2	4

14-35. Ten employees of the Waldeck Pipe and Valve Company are trained as inspectors. The company has given each of these employees an aptitude test. They hope, in the future, to pick inspectors by using this aptitude-test score. However, they feel that they must first determine if there is a relationship between the aptitude-test score, x, and the average output, per shift, of each inspector. The data are given here. Compute the value of r^2 and r. Discuss the r^2 value.

y	70	60	24	38	45	30	35	50	25	20
x	80	80	45	70	95	20	50	90	25	50

14-36. Following are the weights and blood pressures of 10 male patients at St. Mary's Hospital. Rank the 10 weights (x) and the 10 blood pressures (y), and then compute r_s.

y	140	160	130	130	180	160	140	150	140	150
x	188	231	176	194	244	207	198	217	181	194

15 THE CHI-SQUARE DISTRIBUTION

15-1 INTRODUCTION

A national counseling service has been involved with marriage counseling for many years. Their experience has led these counselors to believe that there is a relationship between the degree of a couple's marital adjustment and their educational level. How do we establish this fact?

Workers at the registry of motor vehicles believe that if drivers are graded on accident reaction time versus blood alcohol content there will be a strong relationship. How do we test this belief?

A psychologist believes that he can accurately predict the probability of a minority group appearing at various levels of industrial management. A genetic researcher believes that she can predict the probability of certain colors occurring in a type of flower. How do we test to see whether or not the estimated probabilities are correct?

In the interest of answering such questions, this chapter discusses the splitting of a collection of data into k disjoint categories, called *cells;* compares *observed sample frequencies (O)* with *expected sample frequencies (E)*, and presents a null hypothesis that is a statement of probabilities for all cells. As for observed frequency and expected frequency, they mean exactly what they say. They are formally defined as follows:

Definition 15-1 The *observed frequency* in the ith cell, denoted O_i, is the number of pieces of sample data that we observe in the ith cell. The *expected frequency* in the ith cell, denoted E_i, is the theoretical cell frequency based upon H_0. $E_i = np_i$, where p_i (in H_0) is the probability associated with the ith cell.

For a sample of size n,

$$\sum_{i=1}^{k} O_i = \sum_{i=1}^{k} E_i = n$$

where we have k cells.

Karl Pearson, whom we already encountered, undertook the formulation of a method for measuring the difference between observed and expected frequencies. As a simple illustration, suppose a die is rolled 60 times and a record kept of the number of times each face comes up. If the die is honest, each face will have a probability of 1/6 of appearing on a single roll. Therefore, theoretically, we would expect each face to appear 10 times in an experiment of this kind. Suppose the experiment produced the results in the following table, where the row labeled O represents the observed frequencies from an actual sample of 60 rolls, and the row labeled E represents the expected frequencies.

$$E_1 = E_2 = E_3 = E_4 = E_5 = E_6 = (60)(1/6) = 10$$

Face	1	2	3	4	5	6	
O	15	7	5	10	7	16	$n = 60$
E	10	10	10	10	10	10	$n = 60$

Note, we have six observed cells and six expected cells.

How do we measure the difference between the observed frequencies and the corresponding expected frequencies? The principle of least squares motivated Pearson to begin with the summation

$$\sum_{i=1}^{k} c_i(O_i - E_i)^2$$

as an initial measure of deviation, the c_i's being, more or less, arbitrary constants. Such a summation is elegant due to its mere simplicity. It will equal zero when each observed frequency equals its corresponding expected frequency. This is a must. The greater the disparity between observed and expected frequencies, the greater the value of the summation. This feature is also desirable. Still, such a simple summation has a drawback. It can't be related to a probability curve.

Pearson overcame this problem by creating a new statistic, which he called chi-square (χ^2) because it is closely related to the famous chi-square probability curve. In fact, the approximation of the sampling distribution of the chi-square statistic by the chi-square probability curve is adequate whenever the number of pieces of data in each cell of the expected table is five or more.

We use identical symbolism to denote the chi-square statistic and the chi-square probability curve. This is a most unusual case in statistics; the context of the discussion indicates which interpretation of chi square is intended. Pearson created this new statistic by assigning a value to each constant, c_i, in his initial summation. He let

$$c_i = 1/np_i = 1/E_i$$

where p_i is the theoretical probability of a piece of data falling into the ith cell. Thus, Pearson arrived at a new statistic that offers us relative computational ease. It is defined formally in the following.

Definition 15-2 The *chi-square statistic* is denoted χ^2, and is equal to

$$\chi^2 = \sum_{i=1}^{k} \frac{(O_i - np_i)^2}{np_i} = \sum_{i=1}^{k} \frac{(O_i - E_i)^2}{E_i} \geq 0$$

where $E_i \neq 0$, $1 \leq i \leq k$

It is clear that this statistic can never be a negative number. The smallest value that it may assume is zero, which will occur if there is a perfect match between the observed and the expected frequencies. There is one term in the summation for each observed (or expected) cell in our table.

Example 15-1 Compute the chi-square statistic for the die illustration.

Solution

$$\chi^2 = \sum_{i=1}^{6} \frac{(O_i - E_i)^2}{E_i}$$

$$= \frac{25}{10} + \frac{9}{10} + \frac{25}{10} + 0 + \frac{9}{10} + \frac{36}{10}$$

$$= \frac{104}{10} \text{ or } 10.40$$

15-1 INTRODUCTION

The two main uses for the chi-square statistic are: (1) *goodness-of-fit*, which deals with making an inference about a population distribution in terms of the distribution observed in a sample, and (2) *independence of classification*, which deals with the question of determining, by use of sample data, whether two different classifications of the same sample are independent of each other. They are the topics of Sections 15-2 and 15-3, respectively. In Section 15-5 we show how this statistic serves as a tool in the interval estimation of σ^2 and σ. First, however, we would like you to become familiar with the important chi-square probability curve and its graph.

If $z_1, z_2, z_3, \ldots, z_n$ are standard normal variables, then $\sum_{i=1}^{n} z_i^2$ is a χ^2 probability variable with $\nu = n$. The n degrees of freedom ν follow immediately from the fact that we have n independent variables operating. There is a family of chi-square probability curves, and each curve depends upon the value of ν.

Table III presents probabilities for this family of curves. Each row of this table corresponds to a particular chi-square curve. Quoted probabilities are located under the extreme right of the curve, and are listed in columns labeled from .995 to .005, left to right. A specific curve is found by reading down the column headed ν until the desired ν value is found. For example, the critical point for the chi-square curve with $\nu = 20$ and a probability of .05 under the extreme right is 31.410. This critical point may be abbreviated, by using a double subscript, to: $\chi^2_{.05,20} = 31.410$. In similar fashion, $\chi^2_{.01,10}$ corresponds to the critical point of the chi-square curve with $\nu = 10$ and having a probability of .01 under the extreme right. The actual value is $\chi^2_{.01,10} = 23.209$. In general, the critical point corresponding to the curve with ν degrees of freedom and having α probability under its extreme right tail is denoted: $\chi^2_{\alpha,\nu}$. Figure 15-1 shows several members of the family of chi-square curves.

Figure 15-1

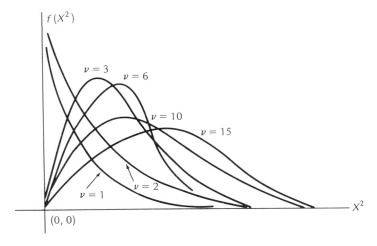

302 THE CHI-SQUARE DISTRIBUTION

With the exception of the chi-square probability curves associated with the values $\nu = 1$ and $\nu = 2$, all chi-square curves are similar in shape.

15-2 GOODNESS OF FIT

Here, we are going to make an inference about a population distribution based upon a sample distribution of data. Let $p_1, p_2, \ldots, p_k$ denote the expected values of the probabilities of our k cells. These probabilities are specified by the null hypothesis. Our problem is to determine whether or not they are consistent with the sample data.

In all of the following examples we will run a one-tail hypothesis test, because we don't want to reject H_0 when our chi-square statistic equals zero. This chi-square value indicates that $O_i = E_i$ for all values of i. A small chi-square value, such as $\chi^2 = 1$, indicates a small difference between corresponding observed and expected frequencies. This difference may be attributed to mere chance. We wish to reject H_0 only when our chi-square statistic is large, thereby indicating that a significant difference exists between corresponding observed and expected frequencies. For the two basic types of hypothesis testing covered in this chapter (goodness of fit and independence of classification), we always run a one-tail test.

Example 15-2 A sociologist wishes to determine whether attitudes toward legalized abortion are changing. A survey taken five years ago showed 60% opposed to legalized abortion, 25% indifferent, and 15% in favor. The sociologist takes a new sample of size 200 and the following data are recorded: 105 opposed, 40 indifferent, and 55 in favor. Use the chi-square goodness of fit test at the .05 level to determine if attitudes toward legalized abortion are changing.

Solution

H_0: $P(\text{opposed}) = .60$,
$P(\text{indifferent}) = .25$
$P(\text{for}) = .15$
H_a: At least one of these probabilities is false.

$E_1 = (200)(.60) = 120$, $E_2 = (200)(.25) = 50$, $E_3 = (200)(.15) = 30$

	Opposed	Indifferent	For	
O	105	40	55	n = 200
E	120	50	30	n = 200

$$\text{Chi-square test statistic} = \chi^2 = \sum_{i=1}^{3} \frac{(O_i - E_i)^2}{E_i}$$
$$= 1.87 + 2.00 + 20.83$$
$$= 24.70$$

Because there are three categories and our sample size $n = 200$ serves as a restriction in determining the expected frequencies, $\nu = 3 - 1$ or 2. Our critical point is $\chi^2_{.05,2} = 5.991$.

Figure 15-2

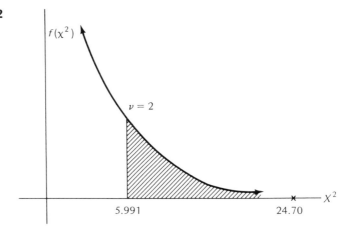

Conclusion Reject H_0 at the .05 level. The data indicate a significant change in attitude toward legalized abortion. See Fig. 15-2. There is little chance of making a type I error here because our test statistic lies so deeply inside the critical region.

Example 15-3 A fire department of a large city believes that the percentage of fires in different parts of the city is changing. Records of the past several years showed that 40% of fires occurred on the north side of the city, 20% on the south side, 30% on the east side, and 10% on the west side. This year, however, the following frequencies of fires occurred: north—110, south—50, east—80, west—60. Execute the chi-square test at the .01 level to see if there has been a shift in the different fire percentages.

Solution

H_0: $P(N) = .40$, $P(S) = .20$, $P(E) = .30$, $P(W) = .10$
H_a: At least one of these probabilities is false (the fire department's feelings).

	North	South	East	West	
O	110	50	80	60	$n = 300$
E	120	60	90	30	$n = 300$

$$\text{Chi-square test statistic} = \chi^2 = \sum_{i=1}^{4} \frac{(O_i - E_i)^2}{E_i}$$
$$= .83 + 1.67 + 1.11 + 30 = 33.61$$

Because there are four categories and our sample size $n = 300$ serves as a restriction in determining the expected frequencies, $\nu = 4 - 1 = 3$. Our critical point is $\chi^2_{.01,3} = 11.345$.

Conclusions Reject H_0. On the basis of these data there is a significant change in the percentages of fires within the four areas of the town. The fire department should consider the relocation of its fire equipment.

Steps in Conducting the Goodness of Fit Test

1. H_0: $P(\text{outcome one}) = p_1$ H_a: At least one of these
   $P(\text{outcome two}) = p_2$ probabilities is false.

   $P(\text{outcome } k) = p_k$
2. Obtain a sample of observed frequencies for the selected sample size, n.
3. On the basis of H_0, construct a second theoretical sample of n expected frequencies.
4. Determine the value of ν. It will be the number of categories k less the number of imposed restrictions.
5. Conduct a *one-tail* test using the chi-square statistic. Compare this test result with the value of the critical point in Table III.

15-2 GOODNESS OF FIT

Example 15-4 Two coins are tossed 80 times. We are interested in recording the number of heads. Run a chi-square goodness of fit test at the .01 level on the basis of the following to see if the coins are true.

Solution

H_0: $P(0H) = 1/4$
$P(1H) = 1/2$
$P(2H) = 1/4$
H_a: At least one probability is false.

	0H	1H	2H	
O	20	40	20	n = 80
E	20	40	20	n = 80

Conclusion Stop. $\chi^2 = 0$. The observed sample supports the conjecture in H_0 perfectly. The sample data provide no evidence that H_0 is untrue.

Example 15-5 A biologist believes the flowers she is breeding should appear: a white flower with green stigma, a white flower with red stigma, a red flower with green stigma, or a green flower with red stigma in the ratio of 9:3:3:1. In the same order, the observed frequencies in a sample of size 208 are 110, 48, 37, and 13. Test at the .05 level.

Solution H_0: $P(\text{first}) = 9/16$, $P(\text{second}) = 3/16$, $P(\text{third}) = 3/16$, $P(\text{fourth}) = 1/16$ (the biologist's theory)
H_a: At least one of the probabilities is false.
$E_1 = (208)(9/16) = 117$, $E_2 = (208)(3/16) = 39$, $E_3 = (208)(3/16) = 39$,
$E_4 = (208)(1/16) = 13$

	110	48	37	13	
O	110	48	37	13	n = 208
E	117	39	39	13	n = 208

$\chi^2 = .42 + 2.08 + .10 = 2.60 \not> \chi^2_{.05,3} = 7.815$

Conclusion There is insufficient evidence to reject H_0. The sample matches the biologist's theory.

Example 15-6 An educator conducted a survey of the marital status of students from five classes, each of which consisted of students from different major fields of learning. The following indicates the results of the survey.

	Class 1	Class 2	Class 3	Class 4	Class 5
Married	8	13	11	8	10
Single	28	31	21	28	42
Total	36	44	32	36	52

Test, at the .05 level, to see if the proportion of married students in each of these classes is the same.

Solution H_0: The proportion of married students in each class is the same.
H_a: The proportions are not equal.

We begin by computing the expected married frequency for each class (cell). To do this we must obtain an estimate of the true proportion of students who are married. Thus:

$$\frac{8 + 13 + 11 + 8 + 10}{36 + 44 + 32 + 36 + 52} = \frac{50}{200} = .25$$

It seems reasonable to estimate that .25 of the total student population is married. Out of 36 students in class 1, we would expect $(36)(.25) = 9$ to be married. We would expect 11, 8, 9, and 13 married students, respectively, in classes 2 to 5. In the next table, both observed and expected frequencies are within the same cell. The expected frequencies are within parentheses.

	Class 1	Class 2	Class 3	Class 4	Class 5
Married	8 (9)	13 (11)	11 (8)	8 (9)	10 (13)

$$\chi^2 = \frac{(8-9)^2}{9} + \frac{(13-11)^2}{11} + \frac{(11-8)^2}{8} + \frac{(8-9)^2}{9} + \frac{(10-13)^2}{13}$$
$$= .11 + .36 + 1.13 + .11 + .69 = 2.40 \not> \chi^2_{.05,4} = 9.488$$

Conclusion There is insufficient evidence to reject H_0.

EXERCISES A

15-1. Find the value of the following critical points in Table III.
  (a) $\chi^2_{.05,10}$ (b) $\chi^2_{.05,23}$ (c) $\chi^2_{.01,15}$

15-2. Find the value of the following critical points in Table III.
  (a) $\chi^2_{.025,14}$ (b) $\chi^2_{.99,19}$ (c) $\chi^2_{.005,25}$

15-3. Use Table III as an aid in completing the following.
  (a) $\chi^2_{.01,?} = 32.000$ (b) $\chi^2_{?,10} = 25.188$

15-4. Use Table III as an aid in completing the following.
  (a) $\chi^2_{.05,?} = 21.026$ (b) $\chi^2_{?,16} = 7.962$

15-5. (a) Explain why the chi-square statistic can't be negative.
  (b) Explain how the chi-square statistic can equal zero.

15-6. Compute the chi-square statistic on the basis of the following data. Is the test result significant at the .05 level? (Example 15-2)

k	1	2	3	4	5	
O	5	15	17	8	5	$n = 50$
E	10	10	10	10	10	$n = 50$

15-7. Compute the chi-square statistic on the basis of the following data. Is the test result significant at the .01 level? (Example 15-3)

k	1	2	3	4	
O	45	28	19	8	$n = 100$
E	50	25	15	10	$n = 100$

15-8. Compute the chi-square statistic on the basis of the following data. Is the test result significant at the .01 level?

k	1	2	3	4	
O	20	30	14	36	$n = 100$
E	25	25	25	25	$n = 100$

15-9. In the following replace the question mark by a greater- or less-than sign.
  (a) $\chi^2_{.95,15}$? $\chi^2_{.99,18}$
  (b) $\chi^2_{.025,16}$? $\chi^2_{.01,16}$

15-10. In the following replace the question mark by a greater- or less-than sign.
  (a) $\chi^2_{.05,20}$? $\chi^2_{.01,20}$
  (b) $\chi^2_{.05,20}$? $\chi^2_{.05,15}$

15-11. Is it possible for $E_i = (1/2)O_i$ for all i cells? Explain briefly.

15-12. If the sample size is not especially large, is it possible to obtain an observed cell frequency of zero? See the following. Will such a happening hinder the calculation of the chi-square statistic?

k	1	2	3	4	5	Total
O	9	10	10	0	11	$n = 40$
E	8	8	8	8	8	$n = 40$

15-13. It actually happened that a person inserted the data from a table, similar to the following one, into our computer. He was certain that he inserted the data correctly and yet he kept receiving an error message. Do you see why? Explain.

k	1	2	3	4	5	6	Total
O	12	16	10	11	10	11	$n = 60$
E	5	15	13	12	0	15	$n = 60$

15-14. To see if a die is honest, it is cast 120 times with the following observed frequencies for the six faces: 18, 23, 16, 21, 18, and 24. Test at the .05 level to see if the die is fair.

15-15. George thinks he has written a computer program to generate an infinite sequence of random decimals. He wishes to execute a chi-square goodness of fit test based on the run-out of the first 100 such decimals. It is decided to base the test upon the observed frequencies of the tenths digit. Carry out the test at the .05 level.

Tenths digit	0	1	2	3	4	5	6	7	8	9
Observed, O	8	13	7	15	10	8	7	15	5	12

15-16. It is believed that a certain lake contains 20% bass, 50% perch, and 30% catfish. For 50 fish netted, the following data are recorded: 12 bass, 22 perch, and 16 catfish. Are these results consistent with the stipulated beliefs? Test at the .05 level.

15-17. A sociologist is conducting a study of ethnic composition of people in managerial positions to see if there has been any significant change in the past 10 years. Several large companies are randomly selected and the breakdown by ethnic groups of people in managerial positions is recorded. From past records it is found that managerial positions were held by 1% blacks, 5% Orientals, 2% Chicanos, and 92% whites. In the present study, for 1000 managerial positions, the following data were obtained: 80 blacks, 60 Orientals, 25 Chicanos, and 835 whites. Test at the .05 level to see if there is a significant change.

CUMULATIVE REVIEW

 15-18. In a study of shopping habits, five different weeks were selected during which no holiday occurred and no special advertising promotions were planned. This was done so that any differences in the number of shoppers could be attributed to the night of the week on which the store was open and not to either of the other factors. The observed number of shoppers was

	M	T	W	Th	F	
O	175	205	195	210	215	Total 1000

Test the assumption that the number of shoppers is the same on all five nights. Test at the .05 level.

15-3 INDEPENDENCE OF CLASSIFICATION

Here we are interested in determining, by use of sample data, whether or not two different classifications of the same sample are independent of each other. Specifically, a sample might be partitioned into rows where each row could represent student height. This same sample might be partitioned into columns where each column could represent student weight.

As another example, we might partition a sample into rows representing various treatments for a known disease and into columns representing levels of recovery from this same disease. A sample of high school students may be placed into rows (A, B, C, D, E) by their English grammar grades and into columns $(1, 2, 3, 4, 5)$ according to their algebra I grade.

We may be comparing types of furniture defects (rows) versus types of furniture (columns), or we may be comparing the degree of approval of U.S. foreign policy with political affiliation. To make them more manageable, it is customary to record the data in a table of r rows and c columns.

Definition 15-3 A table recording the two classifications of the same sample first by r rows and then by c columns is called an $r \times c$ *contingency table*.

Definition 15-4 The intersection of the ith row and the jth column in an $r \times c$ contingency table is called the *ijth cell*. (Geometrically, it is a little rectangle.)

Figure 15-3

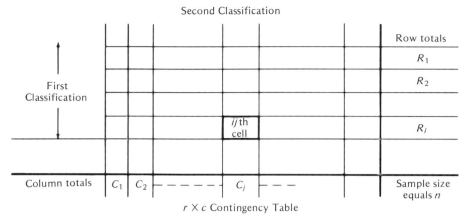

$r \times c$ Contingency Table

Clearly, an $r \times c$ contingency table has rc cells; see Fig. 15-3.

Example 15-7 A random sample of 200 people in a suburban community was interviewed to see whether they favored busing, were against busing, or had no real opinion. The data are recorded in a 3×2 table where rows represent busing opinion and columns represent sex of people interviewed.

	Men	Women	Row totals
Favor	17	43	60
Against	22	78	100
No opinion	11	29	40
Column totals	50	150	$n = 200$

Example 15-8. This example is a two-way breakdown of business dinner prices versus method of payment by businesspeople.

Payment Method

Dinner Price	Cash	Visa	Diner's Club	Totals
$8	200	70	130	400
$10	220	100	180	500
$12	190	80	130	400
$14	120	20	60	200
Totals	730	270	500	$n = 1500$

15-3 INDEPENDENCE OF CLASSIFICATION

The table tells us that 70 people paid $8 for dinner and used a Visa charge card, 180 people paid $10 for dinner and used Diner's Club, and 120 people paid $14 for dinner and paid cash. A total of 1500 people were interviewed.

For contingency tables: H_0: The two classifications are independent. H_a: the two classifications are dependent (not independent).

Intuitively, by independence, we mean that the proportion of each column that belongs in the ith row is the same for all columns and conversely. We realize that this interpretation is intuitive, and we will provide, in a later section, a more penetrating look at the meaning of independence.

If we denote the expected frequency of the ijth cell, E_{ij}, this intuitive meaning is

$$(1) \quad \frac{E_{ij}}{R_i} = \frac{C_j}{n} \quad \text{and} \quad (2) \quad \frac{E_{ij}}{C_j} = \frac{R_i}{n}$$

where R_i is the total of the ith row and C_j is the total of the jth column.

Both of these equations yield the expected cell frequency for the ijth cell, which is

$$\boxed{E_{ij} = \frac{R_i C_j}{n}}$$

Example 15-9 Return to Example 15-7 and compute the expected cell value E_{31}. This is the expected cell value that corresponds to the intersection of the third row (no opinion) and the first column (men).

Solution

$$E_{31} = \frac{(40)(50)}{200} = 10$$

Example 15-10 Return to Example 15-8 and evaluate E_{23}.

Solution $E_{23} = [(500)(500)]/1500 = 166.67$ or 167 to the nearest whole person.

As we have said, we will always execute a one-tail test of hypothesis. The one remaining ingredient needed to complete a test of independence of classification is the value of v. A formal definition of the value of v for an $r \times c$ contingency table is given next. The rationale for this definition will be apparent from the examples.

Definition 15-5 When dealing with $r \times c$ contingency tables, the value of $v = (r-1)(c-1)$.

Example 15-11 An item on a history exam is answered by 50 boys and 50 girls. The item is classified by the sex of the student and pass or fail. Test for independence of classification at the .05 level.

	Fail	Pass
Boy	25	25
Girl	45	5

HISTORY ITEM

Solution H_0: The two classifications are independent.
H_a: The two classifications are dependent.

We are given a 2×2 table of observed frequencies. We must construct another 2×2 table of expected frequencies:

$$E_{11} = \frac{R_1 C_1}{100} = \frac{(50)(70)}{100} = 35$$

Now we want our new 2×2 contingency table of expected frequencies to have exactly the same row and column totals as the given table of observed frequencies. With this one expected frequency tabulated we get

	Fail	Pass	Row total
Boy	(35)		50
Girl			50
Column total	70	30	$n = 100$

This table of expected frequencies has four cells and we have computed only one cell value. However, because of the row and column total restrictions, the other expected cell values are completely determined.

It remains only to fill in the remaining cell values: $E_{12} = 15$, $E_{21} = 35$, $E_{22} = 15$.
What all this means is that $\nu = 1$. Note that

$$(r - 1)(c - 1) = (2 - 1)(2 - 1) = 1$$

$$\text{Chi-square test statistic} = \chi^2 = \sum_{i=1}^{4} \frac{(O_i - E_i)^2}{E_i}$$

$$= 2.86 + 6.67 + 2.86 + 6.67$$

$$= 19.06 > \chi^2_{.05,1} = 3.841$$

Figure 15-4

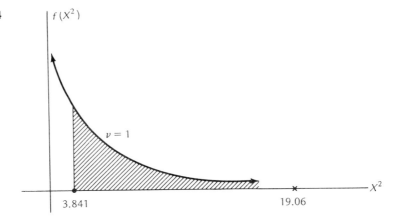

Conclusion Reject H_0; see Fig. 15-4. The two classifications are not independent.

Example 15-12a A sample of size $n = 300$ observed frequencies is recorded in a 3×4 table of observed frequencies. Construct the corresponding 3×4 contingency table of expected frequencies.

	Second classification				Row totals
First classification	25	45	15	15	100
	10	15	10	15	50
	25	60	35	30	150
Column totals	60	120	60	60	$n = 300$

Solution $E_{11} = [(100)(60)]/300 = 20$, $E_{12} = [(100)(120)]/300 = 40$, $E_{13} = 20$, $E_{21} = 10$, $E_{20} = 20$, $E_{23} = 10$

At this point, our 3×4 contingency table of expected frequencies looks like this.

	Second classification				Row totals
First classification	(20)	(40)	(20)		100
	(10)	(20)	(10)		50
					150
Column totals	60	120	60	60	$n = 300$

This table has 12 cells and we have computed only six of those cells. However, because of the row and column total restrictions, the other expected cell values are completely determined. The remaining expected cell values may be easily filled in by inspection. They are: $E_{14} = 20$, $E_{24} = 10$, $E_{31} = 30$, $E_{32} = 60$, $E_{33} = 30$, and $E_{34} = 30$. Here, $\nu = (3-1)(4-1) = 6$.

Example 15-12b Compute the chi-square test statistic for Example 15-12a.

$$\chi^2 = \sum_{i=1}^{12} \frac{(O_i - E_i)^2}{E_i}$$
$$= 25/20 + 25/40 + 25/20 + 25/20 + 0$$
$$+ 25/20 + 0 + 25/10 + 25/30 + 0 + 25/30 + 0$$
$$= 9.79$$

Steps in Conducting the Independence of Classification Test

1. H_0: Two classifications of the same sample are independent.
   H_a: The two classifications are dependent.
2. Record your sample of n observed frequencies in an $r \times c$ contingency table of observed frequencies.
3. Based upon H_0, construct a theoretical contingency table of r rows and c columns. The new $r \times c$ contingency table of expected frequencies must have exactly the same row and column totals as the original table of observed frequencies.
4. $\nu = (r-1)(c-1)$.
5. Conduct a *one-tail* test using the chi-square statistic. Compare this test result with the value of the critical point taken from Table III.

It is worth noting that the table of observed frequencies always precedes the table of expected frequencies.

Example 15-13 A random sample of 200 people in a suburban community is interviewed to see whether they favor busing, are against busing, or have no real opinion. The data are given in the following 3 × 2 table. Test to see if opinions on busing are independent of sex. Let $\alpha = .01$.

	Men	Women	Row totals
Favor	17 (15)	43 (45)	60
Against	22 (25)	78 (75)	100
No opinion	11 (10)	29 (30)	40
Column totals	50	150	$n = 200$

Solution H_0: The two classifications are independent.
H_a: The two classifications are dependent.

Here $\nu = 2$. That is, only two expected cell values need be computed, which we have circled. The remaining four cell values were inserted by inspection. Finally, we condensed the two tables into one table for conciseness.

$$\chi^2 = .27 + .09 + .36 + .12 + .10 + .03 = .97 \not> \chi^2_{.01,2} = 9.210$$

Conclusion There is insufficient evidence to reject H_0. On the basis of the data at hand, we can say that the two classifications are independent. In fact our extremely low chi-square test statistic makes reference to Table III unnecessary.

Here, let us take a closer look at ν. For an $r \times c$ contingency table of expected frequencies, we define $\nu = (r-1)(c-1)$. The cells to be filled by computation consist of one less row and one less column than the original table. Since our table has rc cells, less a row means less c cells, and less a column means less r cells. However, when we subtract c cells and then r cells from the total of rc cells, we discover that one cell has been subtracted twice. This necessitates an addition of one (+1) to obtain the correct number of cells that must be computed in our table of expected frequencies. Thus, $\nu = rc - c - r + 1 = (r-1)(c-1)$.

Next, we can combine the observed contingency table and the expected contingency table by placing each expected frequency within parentheses when recording cell frequencies. This method is recommended as a concise way of recording and presenting sample data.

Now look at the critical points in our chi-square table in the columns labeled .05, .025, and .01. Here, there are no critical point values less than or equal to one. At these levels, a chi-square value less than or equal to one will automatically mean there is insufficient evidence to reject H_0.

Example 15-14 A group of prominent psychologists has developed a marriage test, which, the psychologists believe, will indicate degree of marriage adjustment versus average educational background. To show that the test does establish such a relationship, 400 couples are obtained by random selection from across the nation, and the observed frequencies are recorded in a 3×4 contingency table. Test for independence of classification at the .05 level.

	Very low	Low	High	Very high	Row totals
College	20	30	70	80	200
High school	15	30	35	70	150
Grade school	15	10	15	10	50
Column totals	50	70	120	160	$n = 400$

MARRIAGE ADJUSTMENT RATING

Solution H_0: The two classifications are independent.
H_a: The two classifications are dependent.

The psychologists' personal feelings are contained in H_a: $\nu = (3 - 1)(4 - 1) = 6$. Here is the 3×4 contingency table of expected frequencies.

	Very low	Low	High	Very high	Row totals
College	25	35	60	80	200
High school	75/4	105/4	45	60	150
Grade school	25/4	35/4	15	20	50
Column totals	50	70	120	160	$n = 400$

MARRIAGE ADJUSTMENT RATING

By traveling across one row at a time, we obtain the chi-square test statistic:

$$\chi^2 = 1.00 + .71 + 1.67 + .75 + .54 + 2.22 + 1.67 + 12.25 + .18 + 5.00$$
$$= 25.99 > \chi^2_{.05,6} = 12.592$$

Conclusion Reject H_0. Accept H_a. The psychologists feel good. On the basis of the data, there is a dependency between the test rating and the educational level.

Example 15-15 In a medical study designed to establish the relationship between eye and hand laterality, 413 subjects were recorded as follows. Test for independence of classification at the .05 level.

	Left-eyed	Ambiocular	Right-eyed	Total
Left-handed	34 (35.4)	62 (58.5)	28 (30.0)	124
Ambidextrous	27 (21.4)	28 (35.4)	20 (18.2)	75
Right-handed	57 (61.1)	105 (101.0)	52 (51.8)	214
Total	118	195	100	$n = 413$

Solution

$$\chi^2 = \sum_{i=1}^{9} \frac{(O_i - E_i)^2}{E_i}$$
$$= .055 + .209 + .133 + 1.465 + 1.547 + .178 + .275 + .158 + .001$$
$$\chi^2 = 4.021 < \chi^2_{.05,4} = 9.488$$

Conclusion Eye and hand laterality are independent.

15-4 YATES' CORRECTION

If χ^2 is computed in the ordinary way for $\nu = 1$ and $\nu = 2$, the result in practice is that a type I error is made more often than it should be. In an attempt to correct this situation, we introduce a correction factor that makes

Rothamsted Experimental Station
Dr. Frank Yates

our chi-square statistic smaller. This, in turn, reduces the probability of a type I error so that it is not more than the chosen value of α. Of course, such a correction will increase the probability of a type II error.

One correction used is simply to bring the observed values slightly closer to what is expected by adding .5 to those values that are less than what is expected and by subtracting .5 from those that are larger than the corresponding expected values. This is called *Yates' correction* for continuity. Note that there is an absolute value in this definition.

Definition 15-6 *Yates' correction* for the chi-square statistic applies when $\nu = 1$ or $\nu = 2$. It is a modification of Definition 15-2.

$$\chi^2_{(\text{corrected})} = \sum_{i=1}^{k} \frac{[|O_i - E_i| - .5]^2}{E_i}$$

Example 15-16 Redo Example 15-2 using Yates' correction.

Solution

$$\chi^2_{(\text{corrected})} = \sum_{i=1}^{3} \frac{(|O_i - E_i| - .5)^2}{E_i}$$
$$= \frac{(14.5)^2}{120} + \frac{(9.5)^2}{50} + \frac{(24.5)^2}{30}$$
$$= 1.75 + 1.81 + 20.01$$
$$= 23.57$$

Our goal for using Yates' correction has been achieved. Our corrected statistic is less than the previous answer in Example 15-2.

Example 15-17 Redo Example 15-11 using Yates' correction.

Solution

$$\chi^2_{(\text{corrected})} = \frac{(9.5)^2}{35} + \frac{(9.5)^2}{15} + \frac{(9.5)^2}{35} + \frac{(9.5)^2}{15}$$
$$= 2.58 + 6.02 + 2.58 + 6.02$$
$$= 17.20, \text{ which is less than the uncorrected answer in Example 15-11.}$$

EXERCISES A

15-19. State the number of degrees of freedom for the following contingency tables.
(a) 3 × 5 (b) 6 × 5 (c) 4 × 8

15-20. We have a square contingency table and $\nu = 81$. How many rows does this table have?

15-21. Here is a 3 × 4 contingency table of expected frequencies. Fill in the missing cell values.

3		2		12
	4		7	20
6	5			18
15	15	10	10	$n = 50$

15-22. Here is a 2 × 3 contingency table of expected frequencies. Fill in the missing cell values.

25	35		75
			75
50	50	50	$n = 150$

15-23. Here is a 3 × 3 contingency table of expected frequencies. Fill in the missing cell values.

55			100
	60	40	120
	15		80
110	90	100	$n = 300$

15-24. We are dealing with a contingency table such that $\nu = 42$ and $r = 8$. This implies that $c = $ what? How many cells does this table have?

15-25. We are given a 3 × 3 contingency table of observed frequencies. Construct the corresponding 3 × 3 contingency table of expected frequencies. How many cells does this table have? State the value of ν. Compute the X^2 statistic. (Example 15-7)

	Second classification			Totals
	1	2	3	6
First classification	2	1	1	4
	3	1	6	10
Totals	6	4	10	$n = 20$

   15-26. A marketing research firm surveyed a sample of 100 males and 100 females. A preference for brand A over brand B was stated by 33 males and 18 females. Run a chi-square test, at the .05 level, to see if there is any relationship between brand and sex of the person surveyed. (Example 15-11)

   15-27. Redo Exercise 15-26 using Yates' correction. (Example 15-16)

15-28. Forty-seven female physical education majors filled out a questionnaire on smoking versus coffee drinking. Test at the .05 level to see if the two classifications are independent.

	Smoke	Don't smoke
Drink coffee	12	18
Don't drink coffee	2	15

15-29. Redo Exercise 15-28 using Yates' correction.

   15-30. A turkey grower feeds different rations to three different groups of turkey chicks. Test the hypothesis that the mortality rate is homogeneous (the same) across the three rations. Let $\alpha = .05$. Use the data in the following table. (Example 15-15)

	Number of chicks that lived	Number of chicks that died	Totals
Ration A	84	16	100
Ration B	97	3	100
Ration C	93	7	100
Totals	274	26	$n = 300$

15-4 YATES' CORRECTION

15-31. Following is a table of frequencies that were observed for a sample of 300 adults. One classification is by ethnic group and the other is by political preference. Are the two classifications independent at the .05 level?

	Democratic	Republican	Liberal	Conservative	Totals
Black	36	8	14	2	60
White	84	72	18	26	200
Spanish-speaking	18	8	10	4	40
Totals	138	88	42	32	$n = 300$

15-32. A sample of employees of an industrial organization was asked to indicate a preference for one of three pension plans. Is there reason to believe that the employees' preferences depend on their job classification? Let $\alpha = .01$.

	Plan I	Plan II	Plan III
Supervisors	9	13	20
Clerical	39	50	19
Labor	52	57	41

15-33. Some recent studies have indicated the possibility that there might be a relationship between month of birth and intelligence. To determine whether there is such a relationship, J. Orme reported in the *British Journal of Medical Psychology* (Vol. 35, page 235) the results of intelligence tests conducted on two groups of intellectual subnormals (less than 70 IQ). The results are given here. Test for independence of classification at the .05 level.

	Summer	Autumn	Winter	Spring	Totals
55–69 IQ	29	19	12	18	78
40–54 IQ	13	17	20	20	70
Totals	42	36	32	38	$n = 148$

Hint: Your results will be significant.

322 THE CHI-SQUARE DISTRIBUTION

 15-34. Suppose that we ask 100 men and 50 women whether they are early risers, late risers, or risers with no preference. Conduct a chi-square test of independence of classification at the .05 level.

	Early risers	Late risers	No preference	Totals
Male	10	50	40	100
Female	10	30	10	50
Totals	20	80	50	$n = 150$

B
CUMULATIVE REVIEW

 15-35. The data in the following table were obtained from a random sampling of reports of shoplifting in a certain chain of department stores. Test the hypothesis that the number of shoplifters is independent of the day of the week. Let $\alpha = .05$.

	Fewer than two shoplifters	Between two and four (inclusive) shoplifters	More than four shoplifters	Totals
Monday	10	7	11	28
Tuesday	14	10	19	43
Wednesday	23	13	10	46
Thursday	13	18	9	40
Friday	7	14	21	42
Saturday	6	8	5	19
Total	73	70	75	$n = 218$

15-5 CONFIDENCE INTERVALS FOR σ^2 AND σ

In previous chapters we obtained confidence interval estimates for μ and p. Here we discuss how we obtain confidence interval estimates for σ^2 and σ.

We assume that our population is normally distributed. Take the variance of one random sample, multiply this by the sample size less one $(n-1)$, and then divide the product by the population variance (σ^2). The result of doing this repeatedly is the creation of an important sampling distribution based on the statistic $[(n-1)s^2]/\sigma^2$ that possesses a chi-square distribution with $\nu = n - 1$.

As our sample size increases to $n = 30$ (becomes large), the family of chi-square curves based upon the statistic $[(n-1)s^2]/\sigma^2$ yields a sequence of curves such that any two successive curves are closer to each other (vertical-distance). This family of curves approaches the chi-square curve associated with $n = 30$. Because of this, table III will stop tabulating with the value $\nu = 30$ ($n = 31$). To compute confidence intervals for σ^2 and σ, for large sample sizes, use the bottom row of table III.

From the knowledge that the sampling distribution of this statistic is chi-square, we immediately move to the following theorem.

Theorem 15-1

$$\frac{(n-1)s^2}{\chi^2_{\alpha/2, n-1}} \leq \sigma^2 \leq \frac{(n-1)s^2}{\chi^2_{1-\alpha/2, n-1}}$$

Proof

1. $\chi^2_{1-(\alpha/2), n-1} \leq \dfrac{(n-1)s^2}{\sigma^2} \leq \chi^2_{(\alpha/2), n-1}$

2. $\dfrac{1}{\chi^2_{1-(\alpha/2), n-1}} \geq \dfrac{\sigma^2}{(n-1)s^2} \geq \dfrac{1}{\chi^2_{(\alpha/2), n-1}}$

3. $\dfrac{(n-1)s^2}{\chi^2_{(\alpha/2), n-1}} \leq \sigma^2 \leq \dfrac{(n-1)s^2}{\chi^2_{1-(\alpha/2), n-1}}$

This theorem (line 3) represents the $100(1 - \alpha)\%$ confidence interval for σ^2. The corresponding interval for σ is obtained by taking the square root across line 3.

Example 15-18 Compute the 95% confidence interval estimate of the standard deviation of typing speeds for the population of typists in the city of New York.

Do this computation on the basis of a sample of 15 typists randomly selected in New York. The mean of this sample is 60 words per minute and the variance is 5.

Solution

$$\frac{14(5)}{26.12} \leq \sigma^2 \leq \frac{14(5)}{5.63}$$

$$2.68 \leq \sigma^2 \leq 12.43$$

$$1.64 \leq \sigma \leq 3.53$$

Example 15-19 A variety of wheat gives five yields: 82, 90, 110, 78, and 85. The data are in tons. Find the 95% confidence interval for σ^2 and σ.

Solution sample mean = 89

$$s^2 = (1/4)[7^2 + 1^2 + 21^2 + 11^2 + 4^2]$$
$$= 157$$

$$\frac{(4)(157.00)}{11.14} \leq \sigma^2 \leq \frac{(4)(157.00)}{.48}$$

$$56.37 \leq \sigma^2 \leq 1308.33$$

$$\sqrt{56.37} \leq \sigma \leq \sqrt{1308.33}$$

$$7.51 \leq \sigma \leq 36.17$$

EXERCISES A

In Exercises 15-36 through 15-38, for the given sample data, find the requested confidence interval for σ^2 and σ.

	n	s^2	Interval Size, %
15-36.	101	2	95
15-37.	51	3.5	99
15-38.	166	1.38	95

15-39. In order to determine a 95% confidence interval for the population variance of weights of packages of grass seed, a sample of 10 packages was randomly selected and it was found that $s = .28$. Compute this interval. (Example 15-18)

B
15-40. A random sample of 20 cigarettes from a normal population has an average nicotine content of 18.6 mg and a standard deviation of 2.4 mg. Compute the 99% confidence interval for the population variance.

CUMULATIVE REVIEW

15-41. An experimenter wanted to check the variability of equipment designed to measure the volume of an audio source. Four independent measurements recorded by his equipment for the same sound were 4.1, 5.2, 8.5, and 10.2. Estimate σ^2 with a 95% confidence interval. Estimate σ with a 95% confidence interval.

15-6 MORE ABOUT CHI-SQUARE

This section will deal with two topics: (1) an alternate formula for computing the chi-square statistic; and (2) a more penetrating look at ijth expected cell value, E_{ij}.

The Alternate Formula

An alternate formula for computing the chi-square (χ^2) statistic of Definition 15-2 is derived as follows:

1.
$$\chi^2 = \sum_{i=1}^{k} \frac{(O_i - E_i)^2}{E_i}$$

Recall:
$$\sum_{i=1}^{k} O_i = \sum_{i=1}^{k} E_i = n.$$

2.
$$\chi^2 = \sum_{i=1}^{k} \frac{O_i^2}{E_i} - 2\sum_{i=1}^{k} \frac{O_i E_i}{E_i} + \sum_{i=1}^{k} \frac{E_i^2}{E_i}$$

Theorem 15-2

3.
$$\boxed{\chi^2 = \sum_{i=1}^{k} \frac{O_i^2}{E_i} - n}$$

where n is the sample size and k the number of cells.

Example 15-20 Compute the chi-square statistic of Example 15-2 by using the alternate version of our formula for computing the chi-square statistic just derived in Theorem 15-2.

Solution

$$\chi^2 = \sum_{i=1}^{3} \frac{O_i^2}{E_i} - 200$$
$$= \left[\frac{105^2}{120} + \frac{40^2}{50} + \frac{55^2}{30}\right] - 200$$
$$= 91.88 + 32.00 + 100.83 - 200.00$$
$$\chi^2 = 24.71$$

Example 15-21 Compute the chi-square statistic of Example 15-3 by using the alternate version of the formula for the chi-square statistic derived in Theorem 15-2.

Solution

$$\chi^2 = \left[\frac{110^2}{120} + \frac{50^2}{60} + \frac{80^2}{90} + \frac{60^2}{30}\right] - 300$$
$$= 100.83 + 41.67 + 71.11 + 120.00 - 300.00$$
$$\chi^2 = 33.61$$

The ijth Expected Cell Value

Let's take a more penetrating look at the expected cell value for the ijth cell, E_{ij}. The ijth expected cell frequency is really based upon our old definition of two independent events (Definition 4-6). Recall that events A and B are independent if, and only if, $P(A \cap B) = P(A)P(B)$.

An example will illustrate how this probability definition of independence enters into the derivation of the formula for computing E_{ij}. Suppose that we wish to study the affiliation between geographical region and religion. Two groups of people are chosen at random, one from the East Coast and one from the West Coast of the United States. Each person is classified as Protestant, Catholic, or Jewish. The resulting observed frequencies are presented in a 2 × 3 table.

	Protestant	Catholic	Jewish	Row totals
East Coast	182	215	203	600
West Coast	154	136	110	400
Column totals	336	351	313	$n = 1000$

The row and column totals are called *marginal frequencies*. To test our null hypothesis of independence of classification, we assume the two classifications are independent.

15-7 THE CHI-SQUARE TEST FOR POPULATION NORMALITY (optional)

Let P be Protestant, C be Catholic, J be Jewish, E be East Coast, W be West Coast. By using our marginal frequencies we may estimate the following probabilities:

$P(P) = 336/1000 \quad P(C) = 351/1000 \quad P(J) = 313/1000$
$\quad\quad\quad\quad\quad\quad\quad\quad P(E) = 600/1000 \quad P(W) = 400/1000$

Now, if H_0 is true and the two variables are independent:

$P(E \cap P) = (600/1000)(336/1000)$
$P(W \cap P) = (400/1000)(336/1000)$
$P(E \cap C) = (600/1000)(351/1000)$
$P(W \cap C) = (400/1000)(351/1000)$
$P(E \cap J) = (600/1000)(313/1000)$
$P(W \cap J) = (400/1000)(313/1000)$

> Thus: $P_{ij} = \dfrac{R_i}{n} \cdot \dfrac{C_j}{n}$; P_{ij} being the probability of the ijth cell

Being reasonable people, we believe the expected cell value to be a product of the cell probability times the sample size. For example, E_{12} will be

$$E_{12} = [(600/1000)(351/1000)][1000] = \frac{(600)(351)}{1000}$$

When this thinking is generalized to each cell in our $r \times c$ table of expected frequencies, we see that the sample size always cancels and we, in effect, are left with our old formula. For:

$$E_{ij} = (P_{ij})(n) = \left(\frac{R_i}{n}\right)\left(\frac{C_j}{n}\right) n = \frac{R_i C_j}{n} \quad n \neq 0.$$

15-7 THE CHI-SQUARE TEST FOR POPULATION NORMALITY (optional)

In this section we deal with an important applied problem—that of testing the common demand that a population be normally distributed. Our problem is resolved by conducting a goodness of fit test. We accomplish this by fitting to our sample a normal probability curve with a mean equal to our sample mean and a variance equal to our sample variance. The use of these two values results in two restrictions against the number of degrees of freedom (ν) used in evaluating our chi-square statistic. Our choice of sample size serves as a third and final restriction. If our sample is a good fit to this normal probability curve, it lends immediate credence to the belief that the population from which this sample was selected is normally distributed.

By inspecting the histogram of Fig. 15-5 (Fig. 1-1 of Chapter 1), we obtain a strong, intuitive indication that the data from this sample may be successfully fitted to a normal distribution with mean equal to the sample mean and variance equal to the sample variance.

Figure 15-5

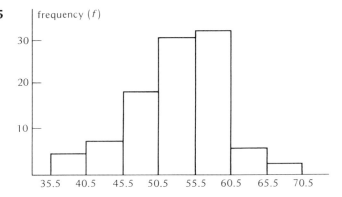

A more penetrating look at the normality of the sample of 100 driver's speeds on U.S. Route 95 in Massachusetts (See the beginning of section 1-8) is obtained by using the chi-square statistic in a goodness-of-fit test. For testing purposes, our value of ν = the number of classes selected less three. We lose three degrees of freedom because we select the sample size and use the sample mean and sample variance. For the case of k classes, $\nu = k - 3$.

Our plan of attack is to transform the normal distribution to which we are fitting our sample into the corresponding standard normal distribution. Once this is done, probabilities expected within classes may be easily found in Table I. Once we know the expected probabilities within each class, we will be able to compute the corresponding expected frequency of each class (E_i). For the case of k classes, $E_i = np_i$, $1 \leq i \leq k$. Example 15-22 illustrates this technique.

Example 15-22 A random sample of 80 pieces of data, drawn randomly from a population we believe to be normally distributed, is recorded in a frequency table. Conduct the chi-square goodness of fit test at the .05 level.

15-7 THE CHI-SQUARE TEST FOR POPULATION NORMALITY (optional)

Solution

H_0: The sample data are normally distributed.
H_a: The sample data are not normally distributed.

Recall that we assume all data within a class to be located at the class mark. The mean is 67.50, $s^2 = 126.33$, and $s = 11.24$.

Classes	O_i	Upper Class Boundary	z Score of Boundary	Probability Below	Probability Within	E_i
90–94	1	94.5	2.40	.9918	.0168	1.34
85–89	4	89.5	1.96	.9750	.0405	3.24
80–84	7	84.5	1.51	.9345	.0768	6.14
75–79	10	79.5	1.07	.8577	.1253	10.02
70–74	13	74.5	.62	.7324	.1610	12.88
65–69	15	69.5	.18	.5714	.1778	14.22
60–64	11	64.5	−.27	.3936	.1457	12.38
55–59	8	59.5	−.71	.2389	.1159	9.27
50–54	6	54.5	−1.16	.1230	.0682	5.46
45–49	3	49.5	−1.60	.0548	.0346	2.77
40–44	2	44.5	−2.05	.0202	*.0202	1.62
Totals	80					79.34

* Duplicating the below probability in the within column will serve as a reasonable approximation here.

We selected 11 classes so $\nu = 11 - 3 = 8$. Our critical point is $\chi^2_{.05,8} = 15.507$.

$$\text{Chi-square test statistic} = \sum_{i=1}^{11} \frac{(O_i - E_i)^2}{E_i}$$
$$= .09 + .18 + .12 + .000 + .001 + .04$$
$$+ .15 + .17 + .05 + .02 + .09$$
$$= .91$$

Conclusion There is insufficient evidence to reject H_0. In fact, the chi-square analysis seems to indicate that our population may be safely considered to be normally distributed. See Fig. 15-6.

Figure 15-6

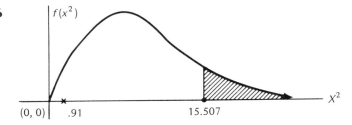

CUMULATIVE REVIEW EXERCISES B

 15-42. Use the chi-square goodness of fit test at the .05 level to see if the sample data of Exercise 1-59 of Chapter 1 may be reasonably assumed to be part of a normal population. [Hint: Use your previous work from Exercise 1-59 of Chapter 1.]

The following data represent the number of accidents per year at 48 intersections in a large city.

33	22	32	15	5	10	13	31	11	25	12	20
14	11	13	21	15	15	4	34	22	16	17	8
17	18	23	19	21	24	26	7	23	18	16	15
28	28	16	26	14	20	20	19	19	17	25	11

 15-43. The following frequency distribution table represents the seasonal rainfall in Sacramento, Calif., for a 120-day period. Conduct a chi-square goodness of fit test at the .01 level to see if we may reasonably assume that rainfall in Sacramento is normally distributed.

Class	Observed Frequency
7.5–10.5	14
10.5–13.5	16
13.5–16.5	23
16.5–19.5	25
19.5–22.5	13
22.5–25.5	18
25.5–28.5	5
28.5–31.5	1
31.5–34.5	4
34.5–37.5	1

SUMMARY

In this chapter we meet another important probability curve (distribution) the *chi-square distribution*. There are a family of chi-square distributions recorded by rows in Table III. Like the t distributions, each chi-square distribution is determined by specifying a number of degrees of freedom.

We discussed *goodness of fit*. Here we are given a collection of data that we believe yields (when graphed) a specific distribution such as a uniform distribution (perfectly level) or a normal distribution. Our job is to use chi-square analysis to see if our collection of data does indeed conform to the theorized distribution.

We discussed *independence of classification*. Here sample data is classified in two ways. One classification is by rows and the other classification is by columns. For purposes of analysis, we set up a table of r rows and c columns called an $r \times c$ *contingency table*. Our job is to determine whether or not the two classifications of the same sample are independent.

Both goodness of fit and independence of classification depend on the chi-square statistic: $\chi^2 = \sum_{i=1}^{k} (O_i - E_i)^2/E_i$ where all $E_i > 0$. In the formula for the chi-square statistic, O_i denotes the *observed frequency*, E_i denotes the *expected frequency*, and $E_{ij} = R_i C_j / n$.

Chi-square distributions associated with $\nu = 1$ and $\nu = 2$ are quite different, in shape, from other members of the family of chi-square distributions. When $\nu = 1$ or $\nu = 2$, we find that we encounter an error of type I more frequently than is desirable. This is overcome by introducing *Yates' Correction* into our formula for computing the chi-square statistic.

In this chapter we discussed the sampling distribution of the statistic $\frac{(n-1)s^2}{\sigma^2}$. The sampling distribution of this statistic is chi-square with $\nu = n - 1$. Based on this sampling distribution, we are able to calculate confidence interval estimates for σ^2 and σ. Recall s is the point estimate of σ.

The optional section shows us how to use chi-square analysis to determine whether or not it is reasonable to believe that a collection of data was drawn from a normal population. In other words, we are dealing with an important goodness of fit test.

Can You Explain the Following?

1. observed frequency (O)
2. expected frequency (E)
3. cell
4. the formula for the chi-square statistic
5. goodness of fit

6. independence of classification
7. Yates' Correction
8. contingency table
9. the value of E_{ij} for a contingency table
10. the value of ν for a contingency table
11. marginal frequencies

CUMULATIVE REVIEW

MISCELLANEOUS EXERCISES

15-44. **a.** In a 9×10 contingency table what does ν equal?
b. How many columns are in a contingency table of 12 rows if $\nu = 154$?

15-45. Explain in your own words how it is possible for $\chi^2 = 0$.

15-46. Replace the question mark by a greater- or less-than symbol in the following.

$$\chi^2_{.05,15} \quad ? \quad \chi^2_{.01,15}$$

15-47. We are given the following distribution and asked to test it for uniformity (all expected cell values are equal) at the 5% level. Conduct a chi-square goodness of fit test.

k	1	2	3	4	5	Total
o	17	23	23	17	20	$n = 100$

15-48. Following is a 3×3 table of observed frequencies (O_i). **(a)** State the value of ν. **(b)** Construct the corresponding table of expected frequencies (E_i). The answers may be left in fractional form.

11/3	10/3	3	10
11/6	10/6	3/2	5
11/2	5	9/2	15
11	10	9	$n = 30$

15-49. A sample of 100 students was randomly selected to test the suitability of an English question for a standardized test. The students were classified by sex and pass-or-fail in the following table. Execute a chi-square test at the 5% level (without Yates' correction).

	Fail	Pass
Boys	22	28
Girls	28	22

15-50. Redo Exercise 15-49 using Yates' correction.

15-51. A political analyst believes that religious beliefs have no bearing on the way people vote (Democratic or Republican). It is decided to test this feeling by running a chi-square test at the 5% level. Is the analyst correct on the basis of the following data?

	Catholic	Jewish	Protestant	Totals
Democratic	25	15	20	60
Republican	10	10	20	40
Totals	35	25	40	$n = 100$

15-52. A campus poll was taken as to whether or not left-wing speakers should be allowed on campus. Test at the 1% level to see if the response is independent of the academic year of the student responding.

	Freshmen	Sophomores	Juniors	Seniors	
Should be allowed	99	98	98	97	392
Should not be allowed	121	112	92	83	408
	220	210	190	180	$n = 800$

15-53. Compute the 95% confidence interval for σ^2 based on a sample of size 20, in which $s^2 = 2.00$.

15-54. Ace Security Company is interested in estimating the variance of times needed to patrol a certain size warehouse. A sample of 40 patrols yields $s^2 = 8.52$ minutes. Compute a 99% confidence interval estimate of σ^2. Company security plans demand that this estimate be obtained quickly. [Hint: Use the last line of the chi-square table.]

16 THE SAMPLING DISTRIBUTION OF F RATIOS

16-1 INTRODUCTION

Two reading specialists are trying out elementary school reading programs. They would like to compare their testing results to see which program has the most variation.

A major automobile manufacturer wishes to automate the production of a part that he has been having hand-crafted. He speculates, on the basis of samples received, that the automated product and the hand-crafted product both have the same average life expectancy. The manufacturer wonders whether giving this important part over to automation will result in increased variation of the dimensions.

What our reading specialists and our auto manufacturer are faced with is the problem of comparing the variances of two populations. This is our next topic. Specifically, we are interested in testing the null hypothesis, H_0: $\sigma_1^2 = \sigma_2^2$, *with each variance coming from a normal population*.

In order to test this null hypothesis, we select a random sample from each of the two normal populations in question. We assume that the two samples are selected independently of each other and that their respective sizes are n_1 and n_2. How should we decide whether or not the two sample variances lend credence to H_0? Certainly, if the sample variances are equal they match our conjecture in H_0. But if the two sample variances aren't equal, should we record the difference value and begin an investigation of the sampling distribution of differences between sample variances? We have already encountered the sampling distribution of the difference between sample means.

As interesting as it may seem, our decision to reject or not to reject H_0 is based on the ratio of our two sample variances, not on the magnitude of their difference. If $s_1^2/s_2^2 = 1$, our two sample variances provide no evidence to allow us to consider rejecting H_0. If s_1^2/s_2^2 is nearly equal to one, we still have little reason to reject H_0. However, if this ratio of sample variances were very large or very small, we would be forced to reject H_0 as unreasonable. To determine just how large or small this ratio should be before we reject H_0, we must look at the sampling distribution of all such ratios.

16-2 THE F RATIO

Definition 16-1 Given two samples of sizes n_1 and n_2 drawn independently from two normal populations, the ratio of the two point estimates s_1^2/s_2^2, $s_2^2 \neq 0$ is called an F ratio.

Definition 16-2 For fixed sample sizes n_1 and n_2, the sampling distribution of s_1^2/s_2^2, $s_2^2 \neq 0$ is called an F distribution.

Before discussing this sampling distribution, we should point out that: (1) an F ratio may be created by using other point estimates of σ_1^2 and σ_2^2 besides s_1^2 and s_2^2; (2) a special case results when we draw independent samples from the same normal population; and (3) Definition 16-2 refers to fixed sample sizes n_1 and n_2. This correctly implies that there is more than one F distribution. Indeed, there is a family of F distributions.

Because each sample variance is nonnegative, we can never have a negative F ratio. The smallest possible F ratio is $F = 0$, which occurs when $s_1^2 = 0$ and $s_2^2 \neq 0$. In the unlikely event that $s_2^2 = 0$, we simply forget this potential ratio and move on. Our F ratios tend to cluster in a small interval about $F = +1$. In fact, should the null hypothesis H_0: $\sigma_1^2 = \sigma_2^2$ be true, an F ratio of exactly $+1$ is possible. We might begin to think intuitively that there are just as many F ratios below as above $+1$ but, in general, this is only approximately true. It is true for the case $n_1 = n_2$.

Should the numerator estimate be large and the denominator estimate small, we obtain a large F ratio. For example, if $s_1^2 = 10$ and $s_2^2 = 1/10$, then $F = 100$—but such a large F ratio would occur very infrequently. The larger the F ratio, the more unlikely is its occurrence. Every F distribution starts at $(0, 0)$, rises to a modal peak near the value $F = +1$, and then steadily declines toward the horizontal F-axis. The distribution approaches the positive portion of the F-axis as the value of F increases in magnitude.

The particular F distribution with which we are dealing depends on the numerator sample size n_1 and the denominator sample size n_2. Each sample variance serves as a restriction. Thus, our numerator really depends on $n_1 - 1$ values and our denominator on $n_2 - 1$ values. We express this by saying that each F distribution has two degrees of freedom: $\nu_1 = n_1 - 1$ for the numerator and $\nu_2 = n_2 - 1$ for the denominator. This is an unusual situation. It is our first encounter with a statistic that demands two degrees of freedom.

Our sampling distribution of F ratios is a discrete distribution. What we need, and have, is a probability curve that will approximate our sampling distribution. This is the topic of the next section.

16-3 TESTING THE EQUALITY OF POPULATION VARIANCES

One of the most important probability curves in applied statistics is the F distribution. Theoretically, the F variable is defined as the ratio of two independent chi-square variables, each divided by its degrees of freedom.

Definition 16-3 The F variable is

$$F = \frac{U/\nu_1}{V/\nu_2}$$

where U and V are independent random variables having chi-square distributions with $\nu_1 \neq 0$ and $\nu_2 \neq 0$ degrees of freedom, respectively.

There are two F distribution tables—Table IV, which deals only with the .05 level of significance, and Table V, which deals only with the .01 level of significance. Both tables represent a tabulation of alpha, α, probability beneath the extreme right of the distribution; see Fig. 16-1. Observe the triple subscript notation used to denote the critical point. The value of ν_1 is always written before the value of ν_2.

Figure 16-1

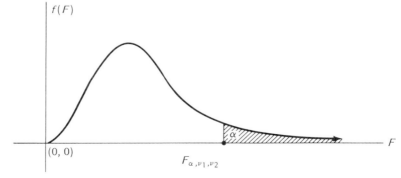

16-3 TESTING THE EQUALITY OF POPULATION VARIANCES

We believe that reading Table IV or V is more or less straightforward. Locate ν_1 by reading across the top of the table and locate ν_2 by reading down the leftmost column, which is entitled ν_2. As a check, locate the critical points $F_{.05,8,20}$ and $F_{.05,20,25}$; their respective values are 2.45 and 2.01.

While we are looking at the tables, we can make another interesting observation. For the special case that our two samples are drawn independently from the same normal population, as $n_1 \to \infty$ and $n_2 \to \infty$, what should we expect the F ratio s_1^2/s_2^2 to approach? The answer is that as both point estimates are approaching the same population variance, they must be approaching each other. Thus, $F \to +1$. Tables IV and V show that this is exactly what happens. We have seen the two tables compressed into one by the use of two different type settings.

Definition 16-3 defines the F statistic as the ratio of two chi-square variables, each divided by its own number of degrees of freedom. Further, recall that in Section 15-5 we discussed the fact that the statistic $(n-1)s^2/\sigma^2$ possessed a chi-square distribution with $\nu = n - 1$. It follows immediately from our theory, then, that the quotient of two such statistics, divided by their sample size less one, should produce an F statistic (from Definition 16-3).

$$F = \frac{[(n_1 - 1)s_1^2/\sigma_1^2]/(n_1 - 1)}{[(n_2 - 1)s_2^2/\sigma_2^2]/(n_2 - 1)} = \frac{\sigma_2^2}{\sigma_1^2} \frac{s_1^2}{s_2^2}$$

and as a result of H_0: $\sigma_1^2 = \sigma_2^2$, we obtain the result

$$F = \frac{s_1^2}{s_2^2}$$

This is the motivation for Definition 16-1.

We can't help but notice that the construction of the two F-distribution tables seems to favor one-tail testing. For example, if we wish to conduct a two-tail test using Table IV, we must execute the test at the .10 level. In order to use Table V in a two-tail test situation, we must execute our test at the .02 level.

We are at liberty to label either of the two populations as population I. If the population with the larger sample variance is designated as population II, then our F ratio will be less than one. We will be forced to deal with a rejection region under the lower-left portion of our F distribution. *We can avoid this difficulty by always designating the population with the larger sample variance as population I. This will result in an F-test statistic that will always be greater than or equal to +1.* We will adopt this convention and thus be concerned only with upper-tail (right-side) rejection. Once we realize that this is an accepted convention, it becomes clear why a table with its entire region of rejection under the extreme right is so desirable.

Example 16-1 Our two reading specialists, Makokian and Mailloux, select random samples from students participating in their respective reading programs. Makokian selects a random sample of 21 test results and finds the sample variance to be 6. Mailloux selects a random sample of 31 participants from his program and, using the same standard test, finds his sample variance to be 3. Test H_0: $\sigma_1^2 = \sigma_2^2$ versus H_a: $\sigma_1^2 > \sigma_2^2$. Let $\alpha = .05$.

Solution Because Makokian's sample variance is larger, we will designate it s_1^2

$$s_1^2 = 6, \qquad s_2^2 = 3, \qquad \nu_1 = 21 - 1 = 20, \qquad \nu_2 = 31 - 1 = 30$$

F-test statistic $= 6/3 = 2 > F_{.05,20,30} = 1.93$. See Fig. 16-2.

Figure 16-2

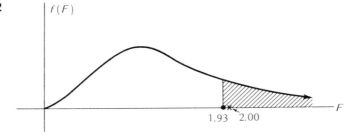

Conclusion Reject H_0. Be careful of making a type I error here. Accept H_a. The test results from Makokian's participants seem to be more variable. However, the results are close.

Example 16-2 A civic organization wishes to study statements attributed to politicians pertaining to what they consider a living wage for a family of four. For 61 Democrats selected, the variance of their quoted amounts is $2.20; and for 61 Republicans selected, the variance of their quoted amounts is $1.80. Try to establish $\sigma_D^2 > \sigma_R^2$ at the .01 level.

Solution We will conduct a one-tail test.

$$H_0: \sigma_D^2 = \sigma_R^2 \qquad H_a: \sigma_D^2 > \sigma_R^2$$

16-3 TESTING THE EQUALITY OF POPULATION VARIANCES

Let the Democrat sample variance be $s_1^2 = \$2.20$.
Our F test statistic is $\$2.20/\$1.80 = 1.22 \not> F_{.01,60,60} = 1.84$.

Conclusion There is insufficient evidence to reject H_0.

Example 16-3 On the first day of every month for a year, the highest temperature of the day was recorded in Boston and in Brockton, Mass. The data collected showed that Boston had a mean high temperature of 70.2°F with $s = 10.8$ and Brockton had a mean high temperature of 74.1°F with $s = 15.3$. Test the hypothesis that the high temperatures of the two cities are equally variable. Execute a two-tail test at the .10 level of significance.

Solution

$$H_0: \sigma_1^2 = \sigma_2^2$$
$$H_a: \sigma_1^2 \neq \sigma_2^2$$
$$\nu_1 = 12 - 1 = 11, \; \nu_2 = 12 - 1 = 11$$
$$F\text{-test statistic} = (15.3)^2/(10.8)^2 = 2.01 < F_{.05,11,11} = 2.82$$

We averaged two table values to obtain the foregoing critical point.

Conclusion There is insufficient evidence to reject H_0. The difference in the variability of the high temperatures of the two cities is not significant.

EXERCISES A

16-1. (a) $F_{.05,5,24} = ?$ (b) $F_{.01,10,20} = ?$
16-2. (a) $F_{.05,8,18} = ?$ (b) $F_{.01,7,25} = ?$
16-3. (a) $F_{.05,4,?} = 2.87$ (b) $F_{.05,?,28} = 2.36$
16-4. In the following, substitute a correct inequality for the question mark.
$F_{.05,\nu_1,\nu_2}$? $F_{.01,\nu_1,\nu_2}$
16-5. It can be shown that for $\nu_2 > 2$, the mean of the F distribution is $\nu_2/(\nu_2 - 2)$. When $\nu_2 = 10$, what is the value of the mean?
16-6. In Example 16-1 our two reading specialists are in some danger of making an error of which type?

16-7. Execute a two-tail test of the equality of variance hypothesis at the .01 level on the basis of the following two sample sets of IQs. Test $\sigma_1^2 > \sigma_2^2$. (Example 16-1)
I: 92, 85, 104, 91, 80, 109
II: 96, 112, 103, 118, 92, 111

16-8. A manufacturer believes he has two types of tires whose variability in wear is the same. If a sample of 25 tires of the first type produced a variance of 2000 miles and a sample of 20 tires of the second type produced a variance of 1600 miles, is the manufacturer correct? Test at the .02 level. Hint: $H_a: \sigma_1^2 \neq \sigma_2^2$. (Example 16-3)

16-9. Suppose a random sample of $n_1 = n_2 = 11$ cans from two brands are examined to determine the weight of the contents, in ounces. The data are
$\bar{x}_1 = 11.70 \quad \bar{x}_2 = 11.60$
$s_1^2 = .06 \quad s_2^2 = .022$
Do these data present sufficient evidence for us to believe there is variability in the contents? Test at the .02 level.

16-10. In a gasoline economy study, 10 one-gallon samples of a particular brand of gasoline were used. Two cars, A and B, both averaged approximately the same mileage (per gallon). $s_a^2 = 3.56$ and $s_b^2 = .87$. Test $\sigma_A^2 > \sigma_B^2$ at the .05 level. (Example 16-2)

B **16-11.** In an experiment to test the effect of a large amount of lime on marigolds, Angus Taylor obtained the following numbers of plants per plot for control and treated plots.
Control: 140, 142, 36, 129, 49, 37, 114, 125
Treated: 117, 137, 137, 143, 130, 112, 130, 121
Test for equality of variances at the .10 level.

CUMULATIVE REVIEW

16-12. A very interesting and true relationship with regard to F distributions is the equality: $F_{1-\alpha,\nu,\nu_2} = 1/F_{\alpha,\nu_2,\nu_1}$. *Hint:* The number of degrees of freedom of the numerator is always written first by convention.
Using this relationship and the tables at the end of the text find:
(a) $F_{.95,6,8}$ (b) $F_{.99,12,10}$

16-4 A HISTORICAL NOTE

SIR RONALD FISHER

The letter F which we have been using is in honor of Sir Ronald Fisher who pioneered much of the work in this area in the 1920's. Fisher, who was a friend of Gosset, is said to have looked at the student's t-distribution and realized that its principle could be extended to several groups.

The first actual use of the letter F was by another well-known statistician, George Snedecor, in the 1930's. Snedecor, incidentally, is credited with the formulation of the first F table. The creation of this table in precomputer times required the herculean task of setting up sampling distributions of ratios of variances for all combinations of degrees of freedom in the table.

GEORGE SNEDECOR

Fisher was a giant of modern statistics. He was the principal leader in bringing about the use of the correct number of degrees of freedom in testing independence from contingency tables. He pioneered the well known *method of maximum likelihood*. This method is today accepted, with high regard, as a general method of estimation. He did work on the topic of sufficient statistics, on better methods of curve fitting, on exact distributions of small samples, and on partial and multiple correlation coefficients. He introduced valid and efficient experimental designs through the principle of randomization.

He made valuable contributions toward the improvement of computational techniques in statistics. And he contributed to the method of finding an inverse of a matrix.

When Fisher died in 1962 the world of statistics suffered a great loss.

16-5 ONE-WAY ANALYSIS OF VARIANCE

An executive of a large firm is considering five competing computerized storage systems for her company's office records, and would like to know whether they are equally efficient. An agricultural seed company interested in developing multipurpose seed would like to know whether a new variety of corn will produce similar or differing yields in different geographical regions. A sociologist is interested in determining if there is a difference in income for immigrants to the United States from five countries during their first year of residence. A product-testing company wishes to compare the writing lifetime of four brands of similarly priced ballpoint pens. We wish to study the average toilet-training age in children from three different socioeconomic levels. What all of these examples have in common is an applied comparison of means from three or more groups.

Since we have already discussed the comparison of means (the sampling distribution of the difference between means) for two groups, we could compare three or more groups in all possible ways two at a time. However, such a comparison is a time-consuming job, and may cause us to lose considerable insight into the analysis of our several groups. For example, let's assume that we are interested in some specific feature of five different ingredients under consideration for a mixture of exterior house paint. By concentrating all of our attention on only two ingredients at a time, we may completely overlook the most desirable mixture of all five ingredients.

We wish to analyze the means of several groups that differ in exactly one way. Our subject is called *one-way analysis of variance* because we will execute a test based upon two different estimates of σ^2. Our test statistic will be a quotient of these two variance estimates. Our test statistic is an F ratio. When we consider two distinct aspects of each of our groups, we say we are dealing with two-way analysis of variance—which is beyond the scope of this work.

Assume we have k normal populations and all k populations possess a common variance σ^2. We would like to test the following null hypothesis.

H_0: $\mu_1 = \mu_2 = \mu_3 = \cdots \cdots = \mu_k$
H_a: At least one of the population means is different from the others.

The selection of k random samples that support H_0, is equivalent to saying that all k normal populations are really the same normal population. We begin our testing procedure by selecting k random samples of equal size, one from each of the k normal populations. For the time being, we assume that all k samples are of size n; see Fig. 16-3. If $\bar{x}_1 = \bar{x}_2 = \bar{x}_3 = \cdots = \bar{x}_k$, our samples will lend credence to the conjecture of H_0. Sample means that differ only slightly in magnitude still tend to support H_0, since mere chance fluctuation would account for some difference among the k sample means. Only when one or more of the sample means varies widely in value must we seriously doubt the truth of H_0. We can measure the spread of these k sample means by computing their variance. Remember that the variance of any set of data is measured in reference to the data's mean. We now define the mean of the k sample means.

Figure 16-3

k random samples

One-way analysis of variance

sample X_1	sample X_2	sample X_j	sample X_k
x_{11}	x_{12}	x_{1j}	x_{1k}
x_{21}	x_{22}	x_{2j}	x_{2k}
x_{31}	x_{32}	x_{3j}	x_{3k}
—	—	—	—
—	—	—	—
x_{n1}	x_{n2}	x_{nj}	x_{nk}

Definition 16-4 The mean of our k sample means is called the *grand mean* and will be denoted $\bar{x}$. (Here, $\bar{x}$ is not a sample mean.)

$$\bar{x} = 1/k \sum_{j=1}^{k} \bar{x}_j$$

All by itself, the variance of our k means goes quite far. If this variance equals zero, all the sample means must be equal and we then have no reason to doubt H_0. When the variance is small, it tends to indicate that the k means in the set are close in value. Mere chance alone would lead us to expect some fluctuation between sample means. Thus, a small variance among sample means would not seem to encourage rejection of H_0.

However, it is difficult to determine at just what value the variance of our k sample means ceases being small and becomes a rejection value. We refine the testing process by computing an F-test ratio. The numerator of this ratio is an estimate of σ^2 called the *between variance estimate* (BVE). The denominator of this ratio is an estimate of σ^2 called the *within variance estimate* (WVE). The rationale for the between variance estimate stems from the relationship $\sigma_{\bar{x}}^2 = \sigma^2/n$, which implies that $n \cdot \sigma_{\bar{x}}^2 = \sigma^2$. Thus, if we can use our k sample means to estimate $\sigma_{\bar{x}}^2$, this approximation times n will serve as an approximation for σ^2. For the present, all sample sizes are assumed equal.

Definition 16-5 The *between variance estimate* of the population variance, σ^2, is given by the formula:

$$\mathrm{BVE} = [n/(k-1)] \sum_{j=1}^{k} (\bar{x}_j - \bar{x})^2, \quad k \neq 1$$

The rationale of the within variance estimate is simple. Each of the k sample variances is a point estimate of σ^2. Thus, an average of these k point estimates should also be an estimate of σ^2. The within variance estimate possesses two properties that the between variance estimate doesn't have: (1) it is relatively free of the actual values of the sample means, and (2) it tends to be relatively constant.

Definition 16-6 The *within variance estimate* of the population variance, σ^2, is given by the formula:

$$\mathrm{WVE} = [1/k(n-1)] \sum_{j=1}^{k} \sum_{i=1}^{n} (x_{ij} - \bar{x}_j)^2 = 1/k \sum_{j=1}^{k} s_j^2$$

By using both of these new point estimates of σ^2, we obtain our F-test statistic:

> *Case of Equal Sample Sizes*
>
> $$F = \frac{BVE}{WVE} = \frac{kn(n-1)\sum_{j=1}^{k}(x_j - \bar{x})^2}{(k-1)\sum_{j=1}^{k}\sum_{i=1}^{n}(x_{ij} - \bar{x}_j)^2} \qquad k \neq 1$$
>
> Here, $\nu_1 = k - 1$; we lose one degree of freedom because we know $\bar{x}$.
> Here, $\nu_2 = k(n-1) = kn - k$. We lose k degrees of freedom because we know the mean of each sample.

In actual practice, it may be easier to compute the numerator and denominator of our F-test ratio separately.

Example 16-4 Test the null hypothesis of equality of means for the following three-column samples. Let α be .05.

Sample X_1	Sample X_2	Sample X_3
1	6	2
2	3	4
3	0	6

Solution H_0: $\mu_1 = \mu_2 = \mu_3$
H_a: At least one of these means doesn't equal the others.

$$k = 3, n = 3, \text{the grand mean; } \bar{x} = 3, \nu_1 = 2, \nu_2 = (3)(2) = 6$$
$$s_1^2 = 1, s_2^2 = 9, s_3^2 = 4 \qquad \bar{x}_1 = 2, \bar{x}_2 = 3, \bar{x}_3 = 4$$
$$BVE = 3[1/2][(-1)^2 + 1^2] = 3$$
$$WVE = [1/3][s_1^2 + s_2^2 + s_3^2] = 14/3$$

Therefore,

$$F\text{-test statistic} = 9/14 < F_{.05,2,6} = 5.14$$

Conclusion There is insufficient evidence to reject H_0.

With regard to the previous example, we have two comments. First, our F-test ratio is less than +1. Any F-test ratio less than +1 automatically implies an inability of the sample data to reject H_0. Observe that all critical

Figure 16-4

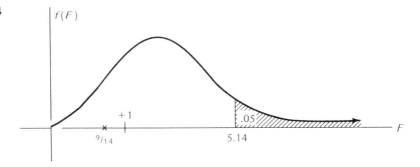

point values in Tables IV and V are greater than or equal to +1. Second, we executed a one-tail test. This is always the case when we are testing the hypothesis of equality of means. Why? Assume we were to run a two-tail test. This would mean that an F-test ratio of zero would result in the rejection of H_0. However, F can be zero only when the BVE is zero, which, in turn, implies that all of our sample means are equal and we have no reason to consider rejecting H_0. Clearly, this would leave us in a ludicrous situation. We reject H_0 by obtaining a significantly large F-test ratio.

Steps in Conducting a One-Way Analysis of Variance Test for Equal Sample Sizes

1. H_0: $\mu_1 = \mu_2 = \mu_3 = \cdots = \mu_k$.
   H_a: At least one of the sample means does not equal the others.
2. Calculate: $\bar{x}_1, \bar{x}_2, \bar{x}_3, \ldots \bar{x}_k$.
3. Average the k sample means of step 2. This yields the grand mean, $\bar{x}$.
4. Determine the variance of these same k sample means, and multiply this answer by n. This is the value of BVE.
5. Calculate: $s_1^2, s_2^2, s_3^2, \ldots s_k^2$.
6. Average the k sample variances of step 5. This is the value of WVE.
7. Divide BVE by WVE. This is the value of the F-test statistic.
8. $\nu_1 = k - 1$, $\nu_2 = k(n - 1)$.
9. Conduct a one-tail test using the F-test statistic of step 7. Compare this test statistic with the value of the critical point F_α, ν_1, ν_2 from the table.

Example 16-5 From a normal population of freshman biology grades, three random samples are drawn. Each of the three samples represents a different major field of concentration. Test the null hypothesis of equality of means at the .01 level.

X_1 First Field of Concentration	X_2 Second Field of Concentration	X_3 Third Field of Concentration
40	50	60
30	50	60
40	50	60
60	60	70
30	40	50

Solution H_0: $\mu_1 = \mu_2 = \mu_3$
H_a: At least one of the means is different from the others.

$$\bar{x}_1 = 40, \bar{x}_2 = 50, \bar{x}_3 = 60 \quad \nu_1 = 2, \nu_2 = 12, \bar{x} = 50$$
$$s_1^2 = 150, s_2^2 = 50, s_3^2 = 50 \quad n = 5, k = 3$$
$$\text{BVE} = 5[1/2][(-10)^2 + 10^2] = 500$$
$$\text{WVE} = 1/3[s_1^2 + s_2^2 + s_3^2] = 83.33$$

Therefore,

$$F\text{-test ratio} = 6.00 < F_{.01,2,12} = 6.93$$

Conclusion There is insufficient evidence to reject H_0. (We would have rejected H_0 at the .05 level.)

Example 16-6 A manufacturer is concerned about coffee breaks at the plant. The manufacturer wonders whether or not the amount of caffeine consumed affects production. In order to relieve the manufacturer's anxiety, five random samples of five people representing various levels of caffeine consumption are obtained. Test the hypothesis of equality of means at the .05 level. The following table gives the number of parts assembled.

X_1 Very Large Consumption	X_2 Large Consumption	X_3 Moderate Consumption	X_4 Small Consumption	X_5 No Coffee at All
16	16	16	16	14
16	14	14	14	13
13	13	12	12	13
14	13	10	12	10
12	10	10	12	10

16-5 ONE-WAY ANALYSIS OF VARIANCE

Solution H_0: $\mu_1 = \mu_2 = \mu_3 = \mu_4 = \mu_5$
H_a: At least one of the means doesn't equal the others.

$$\bar{x}_1 = 14.20, \ \bar{x}_2 = 13.20, \ \bar{x}_3 = 12.40, \ \bar{x}_4 = 13.20, \ \bar{x}_5 = 12.00$$

$$\bar{x} = 13.00, \ v_1 = 4, \ v_2 = 20, \ s_1^2 = 3.20, \ s_2^2 = 4.70, \ s_3^2 = 6.80,$$

$$s_4^2 = 3.20, \ s_5^2 = 3.50$$

$$\text{BVE} = (5)(1/4)[1.44 + .04 + .36 + .04 + 1.00] = 3.60$$

$$\text{WVE} = 1/5[3.20 + 4.70 + 6.80 + 3.20 + 3.50] = 4.28$$

Therefore,

$$F\text{-test ratio} = .84$$

Conclusion   Because $.84 < 1$ there is insufficient evidence to reject H_0.

What do we do for an F-test ratio in the more likely situation where sample sizes are not all equal? For example, we may use unequal size samples for convenience sake. Perhaps we began our experiment with equal sample sizes and dropouts have left us with unequal sizes. Perhaps it is a desire to use all available data that dictates our use of unequal sample sizes. We will state here, without proof, the formulas for computing BVE and WVE for this case. The interested reader will find an explanation of these formulas in an optional section at the end of this chapter.

The Case of Unequal Sample Sizes

$$\text{BVE} = \frac{\sum_{i=1}^{k} n_i \bar{x}_i^2 - \frac{\left[\sum_{i=1}^{k} n_i \bar{x}_i\right]^2}{N}}{k - 1} \qquad v_1 = k - 1, \ k \neq 1$$

where n_1 is our first sample size, n_2 is our second sample size, etc.

where $N = n_1 + n_2 + n_3 + \cdots + n_k$

where k still represents the number of groups

$$\text{WVE} = \frac{(n_1 - 1)s_1^2 + (n_2 - 1)s_2^2 + (n_3 - 1)s_3^2 + \cdots + (n_k - 1)s_k^2}{n_1 + n_2 + n_3 + \cdots + n_k - k}$$

where $v_2 = (n_1 + n_2 + n_3 + \cdots + n_k) - k$ and $k \neq N$

Example 16-7 Redo the calculation of the F-test ratio for Example 16-4 using the unequal sample size formulas for BVE and WVE.

Solution

$$\text{BVE} = \frac{[3 \cdot 2^2 + 3 \cdot 3^2 + 3 \cdot 4^2] - [3 \cdot 2 + 3 \cdot 3 + 3 \cdot 4]^2/9}{2}$$

$$= \frac{87 - 729/9}{2}$$

$$= 3$$

$$\text{WVE} = \frac{2 \cdot 1 + 2 \cdot 9 + 2 \cdot 4}{3 + 3 + 3 - 3} = 14/3$$

Therefore,

$$F\text{-test statistic} = 9/14$$

Example 16-8 Redo the calculation of the F-test ratio for Example 16-5 using the unequal sample size formulas for BVE and WVE.

Solution

$$\text{BVE} = \frac{[5 \cdot 40^2 + 5 \cdot 50^2 + 5 \cdot 60^2] - [5 \cdot 40 + 5 \cdot 50 + 5 \cdot 60]^2/15}{2}$$

$$= \frac{[8000 + 12{,}500 + 18{,}000] - [200 + 250 + 300]^2/15}{2}$$

$$= \frac{38{,}500 - 37{,}500}{2} = 500$$

$$\text{WVE} = \frac{4(150) + 4(50) + 4(50)}{5 + 5 + 5 - 3} = 83.33$$

Therefore,

$$F = 6.00$$

Example 16-9 Three different makes of electronic calculators to be used for classroom programming instruction are being evaluated for volume purchase by a university. A standard set of problems is created that is representative of the uses to which the calculators will be put. It is estimated that 10 hours will be required to complete the set of problems. Because of variations in the times manufacturers could make demonstrators available, only 12 operators of 15 assigned to the three calculators were able to complete the standard set of problems. The completion times, rounded to the nearest hour, are

	Calculator	
A	B	C
7	8	14
6	6	8
8	9	12
	10	10
	7	

At the .05 level, test the null hypothesis that the mean completion times are the same for all three calculators.

Solution H_0: $\mu_1 = \mu_2 = \mu_3$
H_a: At least one of the means is different from the others.

$\bar{a} = 7, \bar{b} = 8, \bar{c} = 11 \qquad k = 3, n_1 = 3, n_2 = 5, n_3 = 4, N = 12$

$s_a^2 = 1, s_b^2 = 5/2, s_c^2 = 20/3 \qquad \nu_1 = 2, \nu_2 = 9$

$$\text{BVE} = \frac{[3 \cdot 7^2 + 5 \cdot 8^2 + 4 \cdot 11^2] - [3 \cdot 7 + 5 \cdot 8 + 4 \cdot 11]^2/12}{2}$$

$$= \frac{951 - 11{,}025/12}{2} = 16.13$$

$$\text{WVE} = \frac{2(1) + 4(5/2) + 3(20/3)}{9} = 32/9 = 3.56$$

F-test ratio $= 16.13/3.56 = 4.54 > F_{.05,2,9} = 4.26$

Conclusion Reject H_0. If we return to the data, inspection shows calculator C is poor.

Example 16-10 Four groups of students were subjected to different teaching techniques and tested at the end of the semester. As a result of dropouts due to sickness and transfer, the number of students being tested varied from group to group. Test at the .05 level. Do the data shown present sufficient evidence to indicate a difference in the mean achievement for the four teaching techniques?

I	II	III	IV
65	75	59	94
87	69	78	89
73	83	67	80
79	81	62	88
81	72	83	
69	79	76	
	90		

Solution H_0: $\mu_1 = \mu_2 = \mu_3 = \mu_4$
H_a: At least one of the means differs from the rest.

$$n_1 = 6, n_2 = 7, n_3 = 6, n_4 = 4 \quad k = 4, N = 23$$
$$\bar{x}_1 = 75.67, \bar{x}_2 = 78.43, \bar{x}_3 = 70.83, \bar{x}_4 = 87.75$$
$$\text{BVE} = 237.53 \quad\quad \bar{x} = 78.17$$
$$\text{WVE} = 62.98$$

Therefore,

$$F\text{-test ratio} = 3.77 > F_{.05,3,19} = 3.13$$

Conclusion Reject H_0. The evidence is sufficient to conclude that there is a difference in the mean achievement for the four teaching techniques.

Example 16-11 An experiment was carried out to study the effect of a small lesion on a structure in a rat's brain. The experiment involved testing the rat's ability to discriminate. A first group of rats had the lesion on the left side of the structure. A second group had the lesion on the right side of the structure. A third group, acting as a control group, had no lesion. The dependent variable score was the number of trials it took each rat to learn a task to a specified level. Each group started with 10 rats but, due to death or postoperative incapacity, the final group sizes were 6, 7, and 5, respectively. Use the given data to test the null hypothesis that the three populations are identical in their average ability to learn the task. Test at the .01 level of significance.

H_0: $\mu_1 = \mu_2 = \mu_3$
H_a: At least one of the means differs from the others.

	Group	
I (left)	II (right)	III (no lesion)
21	21	20
17	23	18
22	18	27
19	22	19
25	27	21
22	18	
	18	

Solution

$$n_1 = 6, \, n_2 = 7, \, n_3 = 5$$
$$\bar{x}_1 = 21, \, \bar{x}_2 = 21, \, \bar{x}_3 = 21$$

Conclusion Stop the testing process. Because all the sample means are equal they support the conjecture in H_0.

Before we can leave this section, we must consider the fundamental question of one-way analysis of variance. What do we do after we have rejected H_0? True, we now know that at least one of the sample means has a significantly different value from that of the other means. But which mean is causing the rejection? Is it a combination of two or more than two means that is causing the rejection? A statement of rejection implies that our work is only partly done.

Mere inspection of the values of the sample means will often provide us with a degree of insight into our problem. However, common-sense inspection just isn't adequate for research purposes.

The problem of making several comparisons after an F test ratio has been rejected has long been a troublesome problem. Certainly it is not a trivial problem. It is called the problem of *Post-Hoc Comparisons*.

To begin with let's discuss a way out of our dilemma which has the features of being both fast and readily available to us. It will be our way, although not ideal, of resolving the problem. We will immediately use our one-way analysis of variance test (or test for the difference between sample means) for the case $k = 2$ to analyze key sets of two sample means or possibly all sets of two sample means. This last idea is often practical when we are working with computer support facilities. We do have such facilities and we will put them to good use in our next example. Of course, one immediate drawback in our resolution of the problem is an inability to compare sets of three or more sample means simultaneously.

There are good established post-hoc comparison techniques. Without going into any detail, we would simply like to mention what we believe is one of the most popular of these techniques. Any in-depth investigation will be left to the reader.

Perhaps the most popular post-hoc method of comparison is that of Henry Scheffe of the University of California at Berkeley. It was published in the late 1950's. Scheffe's method has the advantage of relative simplicity and may be applied to groups of unequal sizes. The method is relatively insensitive to departures from normality and lack of homogeneity (equality) of population variances.

Example 16-12   The following are five consecutive Saturday earnings (in dollars) of three salesmen employed by R. Quindley, Inc.

X_1: Mr. Jones	X_2: Mr. Smith	X_3: Mr. Brown
172	203	161
185	172	149
165	187	183
194	183	156
212	179	144

a. Use a .05 level of significance to test the hypothesis of equality of means.
b. We will discover that the F-test ratio will be rejected. Follow this up by making all possible comparisons of two means. At each step, state "reject" or "insufficient evidence to reject."

Solution H_0: $\mu_1 = \mu_2 = \mu_3$
H_a: At least one of the means doesn't equal the others.

$$n_1 = n_2 = n_3 = 5 \quad k = 3, N = 15, \nu_1 = 2, \nu_2 = 12$$
$$\bar{x}_1 = 185.6, \quad \bar{x}_2 = 184.8, \bar{x}_3 = 158.6 \quad \bar{x} = 176.33$$
$$s_1^2 = 344.3, \, s_2^2 = 134.2, \, s_3^2 = 228.3$$
$$\text{BVE} = 1{,}180.07$$
$$\text{WVE} = 235.60$$
$$F\text{-test ratio} = 5.01 > F_{.05, 2, 12} = 3.8$$

Conclusion Reject H_0.

The critical point for all three pairs is $F_{.05,1,8} = 5.32$. Our *post-hoc* comparisons:
1. Jones versus Smith: Insufficient evidence to reject H_0; $F = .0067$
2. Jones versus Brown: Reject H_0; $F = 6.37$
3. Smith versus Brown: Reject H_0; $F = 9.47$

Conclusion Brown alone is causing our initial F-test ratio to be rejected. Inspection of the sample means reveals that his average dollar sales is significantly lower than that of Jones or Smith.

Finally, because we used computer facilities, the entire *post-hoc* comparison took only minutes to do.

EXERCISES A

16-13. For the case of unequal sample sizes, show that the formula for computing v_2 reduces to the formula for v_2 when it is stipulated that all samples are equal in size.

The next three exercises are intended primarily for computational expertise. As such, they will violate the condition that all sample sizes be five or larger.

16-14. The quantity of demand for a product according to price is given in the following. Test the null hypothesis of equality of means at the .05 level. (Example 16-4)

X_1: Low Price	X_2: Medium Price	X_3: High Price
12	9	6
10	8	5
8	7	7

16-15. Below are three consecutive weeks' earnings (in dollars) for three part-time salesmen. Test, at the .05 level, to see if there is a significant difference among their average weekly earnings. (Example 16-5)

Mr. L	Mr. M	Mr. N
152	181	160
175	171	130
180	203	124

16-16. Test the null hypothesis $H_0: \mu_1 = \mu_2 = \mu_3 = \mu_4$ for the following data. Test at the .01 level. (Example 16-9)

X_1	X_2	X_3	X_4
2	3	1	0
3	4	2	1
4	5	6	2
	6	7	4
	7		5
			6

 16-17. Three farmers, A, B, and C, all raise watermelons as an extra crop on their farms. Because all three have been life-long competitors, they now wonder whether or not they have the same average ability to grow watermelons. The following data represent the number of watermelons per harvest for each farmer for the last five years.

A	B	C
320	370	490
380	430	520
360	380	505
340	420	485
400	400	500

Test at the .05 level to see if $\mu_a = \mu_b = \mu_c$. If the answer is reject, push further and try to ascertain who has the best average, who has the lowest average, etc. (Example 16-12)

 16-18. Suppose data on daily productivity of three machine operators have been calculated for five days as follows: operator A—100, 110, 92, 95, and 108 parts; operator B—94, 97, 90, 101, and 98 parts; operator C—98, 104, 113, 97, and 103 parts. Determine at the .05 level if the three operators are producing at the same average daily rate.

 16-19. A group of psychologists was interested in studying the effect of anxiety upon learning as measured by student performance on a series of tests. On the basis of a pretest, 27 participants were placed into three groups. Group I was composed of those students who scored very low on the anxiety pretest. Group III was composed of those students who scored very high on the anxiety pretest. The same battery of tests was given to all 27 students. The following is a summary of the results. Use the sample results to test that the average score was identical. Test at the .05 level of significance. (Example 16-9)

Group I: $n_1 = 6, \bar{x}_1 = 88, s_1^2 = 10.1$
Group II: $n_2 = 12, \bar{x}_2 = 82, s_2^2 = 14.8$
Group III: $n_3 = 9, \bar{x}_3 = 78, s_3^2 = 13.9$

CUMULATIVE REVIEW

16-20. Fourteen children with hearing disabilities were matched on the basis of their unaided speech-reception thresholds, and then randomly assigned to wear one of three types of hearing aid. The table shows their aided speech-reception thresholds. Do these data provide sufficient evidence to indicate a difference among types of hearing aids? Let $\alpha = .05$.

Type of Hearing Aid		
A	B	C
15	17	10
17	18	12
16	20	14
15	22	11
13	21	

16-21. These data represent the scores obtained from independent random samples of smokers and nonsmokers. All scores are anxiety levels. Do the data provide sufficient evidence to believe that there is a difference with respect to mean anxiety of the four groups? Test at the .01 level.

	Smokers		
Nonsmokers	Light	Moderate	Heavy
24	36	30	56
17	49	33	55
13	41	27	72
18	37	29	73
13	26	30	74
15	45	38	71
24	29		

16-22. A psychologist wants to find out whether porpoises, monkeys, and rats learn at the same rate of speed. She places one animal at a time in a T-maze where the left exit contains an electric shock punishment and the right exit contains food. There were 6 porpoises, 9 monkeys, and 11 rats available for testing purposes. Following is a record of the number of times each animal took to learn to go to the right. At the .05 level, can we conclude that all species of animals have the same average learning time?

Porpoises: 3, 7, 9, 15, 6, 2
Monkeys: 18, 14, 7, 5, 11, 9, 3, 11, 12
Rats: 4, 21, 11, 16, 19, 23, 7, 10, 17, 6, 8

HISTORICAL NOTE

WILLIAM KRUSKAL

WILSON ALLEN WALLIS

William Kruskal is associated with the Department of Statistics, University of Chicago, Chicago, Illinois.

W. Allen Wallis, Chancellor of the University of Rochester, was born November 5, 1912, in Philadelphia. When he was three years old his family moved to Berkeley, where his father was an instructor in anthropology at the University of California.

Wallis entered the University of Minnesota in 1928. His major subject was psychology and his minor sociology.

At Minnesota, Wallis was president of the Chi Phi fraternity, an editorial writer for the college paper for two years, and a member of Phi Beta Kappa. He received his degree magna cum laude in 1932 at the age of 19. He was a graduate student in economics at the University of Minnesota 1932–33.

Wallis's first teaching appointment, in 1937, was at Yale University as instructor in Political Economy. In the fall of 1938, he joined the Department of Economics at Stanford University, an association that continued, with extensive interruptions, until 1946. At Stanford he taught intermediate economic theory, mathematical economics, and advanced statistics, and occasionally lectured on introductory economics.

In 1956, Wallis was appointed dean of Chicago's Graduate School of Business, which, under his leadership, came to be recognized as one of the nation's best.

He became President (later Chancellor) of the University of Rochester in 1962.

16-6 THE KRUSKAL-WALLIS TEST

The Kruskal-Wallis test is the nonparametric alternative to the one-way analysis of variance test. It may be employed usefully when the assumption of normality is inappropriate. Like the Spearman r and the Mann-Whitney U-test, which preceded it, it too is a rank order test.

This test may be correctly thought of as an extension of the Mann-Whitney U-test.

H_0: The k independent samples are from the same population.
H_a: The k independent samples aren't from the same population.

We select k independent samples from our k populations. Let

$$N = \sum_{i=1}^{k} n_i$$

where n_i, $1 \leq i \leq k$ is the size of the ith sample. All the data from the k samples are pooled into one large sample and ranked. The lowest value is assigned the rank of 1, the next lowest 2, and so on. If there are no ties in the last positions to be ranked, our last rank value will be N. Ties are handled as they have been in previous rank order tests. Tied data values are assigned the mean rank position of the positions they occupy.

Let R_i, $1 \leq i \leq k$ be the sum of the ranks associated with the ith sample. A statistic H is now calculated from these rank values. It can be shown that the sampling distribution of this statistic has a chi-square distribution with $\nu = k - 1$ when all sample sizes have five or more pieces of data. Kruskal and Wallis introduced this test statistic in 1952. We conjecture that the use of the letter H is in honor of Hoetelling, a well-known statistician.

The *Kruskal-Wallis H statistic* is defined formally for emphasis.

Definition 16-7   The Kruskal-Wallis H statistic is as follows:

$$H = \left[\frac{12}{N(N+1)} \sum_{i=1}^{k} \frac{R_i^2}{n_i} \right] - 3(N+1)$$

If decimal values are used in the calculation of the factor $12/N(N + 1)$, in H, experience has shown that the number of decimal places used in our numerical values makes a big difference. We suggest a minimum of three decimal places (or fractions) be used to improve arithmetic accuracy.

Example 16-13 Three different methods of weather prediction—astrology, meteorology, and Chinese weather prediction—are to be compared. Six forecasters from each category are selected and their number of day-to-day correct predictions for a one-year period are as given in the table. Use the Kruskal-Wallis H test, at the .05 level, to test that there is no difference in accuracy among the methods of weather prediction.

Astrologers	Meteorologists	Chinese Weather Predictors
115	198	284
121	160	109
48	201	162
168	173	212
150	81	224
79	105	155

Solution H_0: There is no difference in the methods.
             H_a: There is a difference in the methods.
             Ranks of the weather predictors are as follows:

Astrologers	Meteorologists	Chinese
6	14	18
7	10	5
1	15	11
12	13	16
8	3	17
2	4	9
$R_1 = 36$	$R_2 = 59$	$R_3 = 76$

$$H\text{-test statistic} = \frac{12}{(18)(19)}\left[\frac{36^2 + 59^2 + 76^2}{6}\right] - 3(19)$$
$$= (.035)(10{,}553/6) - 57$$
$$= 4.56 \not> \chi^2_{.05,2} = 5.991$$

Conclusion   There is insufficient evidence to reject H_0.

Should ties occur as we rank the data, the H-test statistic tends to be too small. To compensate for ties, we divide our basic H-test statistic (Definition 16-7) by a correction term.

Definition 16-8 For the case of ties in the Kruskal-Wallis H test, a correction term C is divided into our basic H-test statistic.

$$C = 1 - \frac{\left[\sum_{i=1}^{L}(t_i^3 - t_i)\right]}{N^3 - N}$$

where L is the number of sets of tied observations and t_i is the number tied in any set i.

Example 16-14 Apply the Kruskal-Wallis test to the data of Example 16-4. Maintain the .05 level of significance. This example is designed as an aid to computational expertise. We realize that the sample sizes are too small to make the testing really effective.

Solution H_0: The three samples are from the same population.
H_a: The three samples aren't from the same population.

Pooled sample: → 0   1   2   2   3   3   4   6   6
                 ↕   ↕   ↕   ↕   ↕   ↕   ↕   ↕   ↕
Ranks: →   1   2 3.5 3.5 5.5 5.5 7 8.5 8.5

Ranks

Sample X_1	Sample X_2	Sample X_3
2.0	8.5	3.5
3.5	5.5	7.0
5.5	1.0	8.5
$R_1 = 11$	$R_2 = 15$	$R_3 = 19$

There are three sets of tied observations; $L = 3$. Each set of ties has two members so all $t_i = 2$ here.

Therefore,

$$C = 1 - \frac{3[2^3 - 2]}{720} = .97$$

$$H\text{-test statistic} = \frac{\frac{12}{(9)(10)}\left[\frac{11^2 + 15^2 + 19^2}{3}\right] - 3(10)}{.97}$$

$$= 1.42/.97 = 1.46 \not> \chi^2_{.05,2} = 5.991$$

Conclusion There is insufficient evidence to reject H_0. Observe in this computational example and in Example 16-4 that we arrived at the same conclusion. In neither case was the conclusion close. Also, notice here that, because of the tie in the last two places, our largest rank value is 8.5 and not 9.

Example 16-15 We wish to test the null hypothesis of equality of distributions for the arithmetic reasoning scores of three different school districts. Use the Kruskal-Wallis H test at the .01 level.

	School District	
I	II	III
62	40	83
51	53	74
70	75	62
33	61	65
76	42	71
67	59	56
60	77	79
43	59	68
52		76
57		63
		67
		71

Solution

	Ranks	
I	*II*	*III*
14.5	2	30
5	7	24
21	25	14.5
1	13	17
26.5	3	22.5
18.5	10.5	8
12	28	29
4	10.5	20
6		26.5
9		16
		18.5
		22.5
$R_1 = 117.5$	$R_2 = 99$	$R_3 = 248.5$

$L = 5$, each $t_i = 2$, and

$$C = 1 - \frac{30}{26{,}970} = .999$$

$$H\text{-test statistic} = \frac{\left[\frac{12}{(30)(31)}\left(\frac{117.5^2}{10} + \frac{99^2}{8} + \frac{248.5^2}{12}\right)\right] - 3(31)}{.999}$$

$$= \frac{(.013)(1380.63 + 1225.13 + 5146.02) - 93}{.999}$$

$$= \frac{7.02}{.999}$$

$$= 7.03 > \chi^2_{.05,2} = 5.991$$

Conclusion Reject H_0. Observe how close our correction term, C, is to 1. Because it is so close to 1, it may be omitted without injuring the conclusion.

The rationale behind rejecting H when it is large enough comes from the fact, which can be shown, that a large H value will occur when the assigned ranks are not spread evenly throughout the k groups.

EXERCISES A

 16-23. Apply the Kruskal-Wallis test to the data of Exercise 16-14. (Example 16-14) Hint: C = .98.

 16-24. Apply the Kruskal-Wallis test to the data of Exercise 16-16.

 16-25. Apply the Kruskal-Wallis test to the data of Exercise 16-17. Hint: Drop C = .999 (Example 16-15)

 16-26. Apply the Kruskal-Wallis test to the data of Exercise 16-22.

 16-27. In Exercise 16-24 we returned to Exercise 16-16, and executed the Kruskal-Wallis H test for the data. Run the Kruskal-Wallis H test for the data of Exercise 16-16 again with the following change. Let the largest piece of data have the rank value 1, the second largest piece of data the rank value 2, and so on. [Hint: The answer you obtain should be the same as the answer obtained in Exercise 16-23. It is immaterial whether the ranks run lowest to highest or highest to lowest.]

 16-28. Execute the Kruskal-Wallis test, at the .01 level, for the following data.
Sample I: 3, 7, 11, 16, 22, 29, 31, 36
Sample II: 3, 4, 7, 18, 19, 32
Sample III: 22, 38, 46, 47, 47, 50, 53, 54, 56

16-7 SOME THEORY (optional)

On the derivation of the unequal-size formulas for BVE and WVE

The unequal-size formulas for BVE and WVE do a job that needs to be done and this, in itself, is not an unimportant fact. In addition, they have two properties that strongly recommend them:

I. If all of our sample sizes are equal ($n_1 = n_2 = \cdots = n_k = n$) these formulas reduce to the formulas for equal sample sizes of Definitions 16-5 and 16-6. Observe:

i. $\text{WVE} = \dfrac{(n_1 - 1)s_1^2 + (n_2 - 1)s_2^2 + \cdots + (n_k - 1)s_k^2}{n_1 + n_2 + n_3 + \cdots + n_k - k}$

$= \dfrac{(n - 1)[s_1^2 + s_2^2 + \cdots + s_k^2]}{kn - k}$

$= 1/k \sum_{j=1}^{k} s_j^2$ (See Definition 16-6.)

ii. $\text{BVE} = \dfrac{\sum_{i=1}^{k} n_i \bar{x}_i^2 - \dfrac{\left(\sum_{i=1}^{k} n_i \bar{x}_i\right)^2}{N}}{k-1}$

$= \dfrac{n \sum_{i=1}^{k} \bar{x}_i^2 - \dfrac{n^2 \left[\sum_{i=1}^{k} \bar{x}_i\right]^2}{kn}}{k-1}$

$= n/(k-1) \left[\sum_{i=1}^{k} \bar{x}_i^2 - \dfrac{\left(\sum_{i=1}^{k} \bar{x}_i\right)^2}{k} \right]$

$= n/(k-1) \sum_{i=1}^{k} (\bar{x}_i - \bar{x})^2 \qquad$ (See Definition 16-5.)

II. $t^2 = F$, an established result.

Proof

$$t^2 = \dfrac{(\bar{x}_2 - \bar{x}_1)^2}{\left[\dfrac{(n_1 - 1)s_1^2 + (n_2 - 1)s_2^2}{n_1 + n_2 - 2}\right][1/n_1 + 1/n_2]}$$

We immediately recognize

$$\dfrac{(n_1 - 1)s_1^2 + (n_2 - 1)s_2^2}{n_1 + n_2 - 2}$$

as WVE for the case $k = 2$. It remains then only to show that

$$\dfrac{(\bar{x}_2 - \bar{x}_1)^2}{1/n_1 + 1/n_2}$$

is equal to BVE. This we will now do:

$\dfrac{(\bar{x}_2 - \bar{x}_1)^2}{1/n_1 + 1/n_2} = \dfrac{(\bar{x}_2 - \bar{x}_1)^2 n_1 n_2}{n_1 + n_2}$

$= \dfrac{n_1 n_2 \bar{x}_1^2 - 2 n_1 n_2 \bar{x}_1 \bar{x}_2 + n_1 n_2 \bar{x}_2^2}{n_1 + n_2}$

$= \dfrac{[n_1^2 \bar{x}_1^2 + n_1 n_2 \bar{x}_2^2 + n_1 n_2 \bar{x}_1^2 + n_2^2 \bar{x}_2^2] + [-n_1^2 \bar{x}_1^2 - 2 n_1 n_2 \bar{x}_1 \bar{x}_2 - n_2^2 \bar{x}_2^2]}{n_1 + n_2}$

$= \dfrac{\sum_{i=1}^{k=2} n_i \bar{x}_i^2 - \dfrac{\left(\sum_{i=1}^{k=2} n_i \bar{x}_i\right)^2}{n_1 + n_2}}{2 - 1}$

$= \text{BVE}$

Finally, we wish to say, in conclusion, that we hope this text has served to give you a good start toward learning applied statistics.

SUMMARY

This chapter introduces the important *F probability curve* (distribution) which is named in honor of Ronald A. Fisher. There are a family of F distributions recorded in two tables. Table IV is constructed for work at the 5% test-level; Table V is constructed for work at the 1% test-level.

You learned how to test the null hypothesis, H_0: $\sigma_1^2 = \sigma_2^2$. This was accomplished by using an F ratio as a test statistic. The understanding in the computation is that the larger sample variance will always be placed in the numerator of the test statistic.

One-way analysis of variance (sometimes abbreviated ANOVA) was discussed which allows us to test the equality of three or more population means at one time. Our test statistic is an F ratio. The numerator of the ratio is called the *between variance estimate* of σ^2 (B.V.E.); the denominator of the ratio is called the *within variance estimate* of σ^2 (W.V.E.). In doing the work here, it is assumed that we have k normal populations and all k populations possess a common variance. The numerator has ν_1 degrees of freedom and the denominator has ν_2 degrees of freedom. Thus, we are dealing with a test statistic that demands two degrees of freedom. We began with a one-way analysis of variance formula built on equal sample sizes and ended with a one-way analysis of variance formula for unequal sample sizes.

The *Kruskal-Wallis H-Test* is the nonparametric "match" for the parametric one-way analysis of variance test. This test is another example of a rank-order test.

Can You Explain the Following?

1. F ratio
2. F distribution
3. Ronald A. Fisher
4. between variance estimate (B.V.E.)
5. within variance estimate (W.V.E.)
6. grand mean
7. H_0 for one-way analysis of variance
8. Kruskal-Wallis H-test

CUMULATIVE REVIEW

MISCELLANEOUS EXERCISES

16-29. Explain the following inequality: smallest sample mean $\leq$ grand mean $\leq$ largest sample mean; k samples.

16-30. In your own words explain why: **(a)** BVE ≥ 0 **(b)** WVE ≥ 0

16-31. It is possible for WVE = 0. Explain. Clearly, we could not compute the F-test statistic in such a case, however, there is no real problem as the answer is predetermined. Explain.

16-32. A weather forecaster who wants to compare the variability in rainfall between two cities, A and B, finds (from two samples) that $s_a^2 = 2.76$ and $s_b^2 = .85$. On the basis of these two samples is there sufficient evidence to reject the hypothesis that rainfall of the two cities is equally variable? Test at the 10% level. $n_a = 31$ and $n_b = 41$

16-33. It is desired to compare the variability in potency of two vitamin supplements. The established brand, A, yields a test sample with $s_a^2 = 2.47$. The new experimental brand, B, yields a test sample with $s_b^2 = 1.03$. Test H_a: $\sigma_A^2 > \sigma_B^2$ at the 1% level. $n_a = n_b = 61$.

16-34. A national educator believes that physical dexterity of seventh, eighth, and ninth grade boys is significantly different. A 10-second manipulative skill test is administered to three random samples of boys from grades 7, 8, and 9. Use one-way analysis of variance to determine if there is a significant difference between the means of the three groups.

Physical Dexterity Test Results

Seventh Grade (A)	Eighth Grade (B)	Ninth Grade (C)
20	16	22
12	17	18
15	18	23
13	19	17
20	20	20

16-35. Test the null hypothesis H_0: $\mu_1 = \mu_2 = \mu_3$ for the following data. Test at the .01 level. (A computation practice exercise.)

X_1	X_2	X_3
2	5	1
4	10	3
6	15	5
8		7
		9

16-36. An experiment is conducted to show that the practice of writing simple English paragraphs promotes understanding. A class of 30 students is randomly divided into three groups of 10 each. Because the class makeup is homogeneous, it is anticipated that random grouping will not alter the test result. Group A receives the standardized test being used at the high school. Groups B and C are given the same exam with paragraphs modified in such a way (it is hoped) as to promote understanding. Test to see if H_0: $\mu_1 = \mu_2 = \mu_3$ can be rejected at the .05 level.

A: 38, 54, 39, 52, 63, 54, 47, 52, 46, 25

B: 58, 44, 63, 94, 72, 42, 89, 68, 53, 47

C: 76, 51, 83, 84, 51, 67, 40, 89, 76, 53

 16-37. Conduct the Kruskal-Wallace H test to determine if there is any difference in the effects of sales in the basis of three different commercial displays of Beautiful Cosmetics. Each test display was placed in 10 different department stores. The data show units sold in these stores. Test at the 1% level.

Display 1	Display 2	Display 3
15	26	9
21	22	15
24	14	12
10	29	18
14	22	22
16	14	17
12	18	14
19	25	14
22	30	16
17	20	13

 16-38. Suppose that in a large company three groups of technical sales trainees are taught a certain trouble-shooting procedure by different methods, A, B, and C. At the end of the instruction, the subjects are given performance tests, and we want to decide whether, in general, the mean performance for the three groups is the same. Conduct the Kruskal-Wallace H test at the 5% level.

Method A: 33, 30, 32, 25, 30, 34

Method B: 29, 28, 27, 28, 21, 24, 28

Method C: 31, 22, 24, 26, 23

TABLES

TABLE I THE STANDARD NORMAL

An entry in the table is the area under the entire curve which is between $z = 0$ and a positive value of z. Areas for negative values of z are obtained by symmetry.

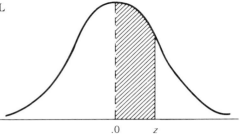

z	.00	.01	.02	.03	.04	.05	.06	.07	.08	.09
0.0	.0000	.0040	.0080	.0120	.0160	.0199	.0239	.0279	.0319	.0359
0.1	.0398	.0438	.0478	.0517	.0557	.0596	.0636	.0675	.0714	.0753
0.2	.0793	.0832	.0871	.0910	.0948	.0987	.1026	.1064	.1103	.1141
0.3	.1179	.1217	.1255	.1293	.1331	.1368	.1406	.1443	.1480	.1517
0.4	.1554	.1591	.1628	.1664	.1700	.1736	.1772	.1808	.1844	.1879
0.5	.1915	.1950	.1985	.2019	.2054	.2088	.2123	.2157	.2190	.2224
0.6	.2257	.2291	.2324	.2357	.2389	.2422	.2454	.2486	.2517	.2549
0.7	.2580	.2611	.2642	.2673	.2703	.2734	.2764	.2794	.2823	.2852
0.8	.2881	.2910	.2939	.2967	.2995	.3023	.3051	.3078	.3106	.3133
0.9	.3159	.3186	.3212	.3238	.3264	.3289	.3315	.3340	.3365	.3389
1.0	.3413	.3438	.3461	.3485	.3508	.3531	.3554	.3577	.3599	.3621
1.1	.3643	.3665	.3686	.3708	.3729	.3749	.3770	.3790	.3810	.3830
1.2	.3849	.3869	.3888	.3907	.3925	.3944	.3962	.3980	.3997	.4015
1.3	.4032	.4049	.4066	.4082	.4099	.4115	.4131	.4147	.4162	.4177
1.4	.4192	.4207	.4222	.4236	.4251	.4265	.4279	.4292	.4306	.4319
1.5	.4332	.4345	.4357	.4370	.4382	.4394	.4406	.4418	.4429	.4441
1.6	.4452	.4463	.4474	.4484	.4495	.4505	.4515	.4525	.4535	.4545
1.7	.4554	.4564	.4573	.4582	.4591	.4599	.4608	.4616	.4625	.4633
1.8	.4641	.4649	.4656	.4664	.4671	.4678	.4686	.4693	.4699	.4706
1.9	.4713	.4719	.4726	.4732	.4738	.4744	.4750	.4756	.4761	.4767
2.0	.4772	.4778	.4783	.4788	.4793	.4798	.4803	.4808	.4812	.4817
2.1	.4821	.4826	.4830	.4834	.4838	.4842	.4846	.4850	.4854	.4857
2.2	.4861	.4864	.4868	.4871	.4875	.4878	.4881	.4884	.4887	.4890
2.3	.4893	.4896	.4898	.4901	.4904	.4906	.4909	.4911	.4913	.4916
2.4	.4918	.4920	.4922	.4925	.4927	.4929	.4931	.4932	.4934	.4936
2.5	.4938	.4940	.4941	.4943	.4945	.4946	.4948	.4949	.4951	.4952
2.6	.4953	.4955	.4956	.4957	.4959	.4960	.4961	.4962	.4963	.4964
2.7	.4965	.4966	.4967	.4968	.4969	.4970	.4971	.4972	.4973	.4974
2.8	.4974	.4975	.4976	.4977	.4977	.4978	.4979	.4979	.4980	.4981
2.9	.4981	.4982	.4982	.4983	.4984	.4984	.4985	.4985	.4986	.4986
3.0	.4987	.4987	.4987	.4988	.4988	.4989	.4989	.4989	.4990	.4990

Reprinted from P. Hoel, *Elementary Statistics*, John Wiley & Sons, by permission of the publisher.

TABLE II THE STUDENT-t DISTRIBUTION

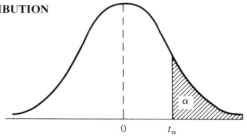

ν	t_α				
	0.10	0.05	0.025	0.01	0.005
1	3.078	6.314	12.706	31.821	63.657
2	1.886	2.920	4.303	6.965	9.925
3	1.638	2.353	3.182	4.541	5.841
4	1.533	2.132	2.776	3.747	4.604
5	1.476	2.015	2.571	3.365	4.032
6	1.440	1.943	2.447	3.143	3.707
7	1.415	1.895	2.365	2.998	3.499
8	1.397	1.860	2.306	2.896	3.355
9	1.383	1.833	2.262	2.821	3.250
10	1.372	1.812	2.228	2.764	3.169
11	1.363	1.796	2.201	2.718	3.106
12	1.356	1.782	2.179	2.681	3.055
13	1.350	1.771	2.160	2.650	3.012
14	1.345	1.761	2.145	2.624	2.977
15	1.341	1.753	2.131	2.602	2.947
16	1.337	1.746	2.120	2.583	2.921
17	1.333	1.740	2.110	2.567	2.898
18	1.330	1.734	2.101	2.552	2.878
19	1.328	1.729	2.093	2.539	2.861
20	1.325	1.725	2.086	2.528	2.845
21	1.323	1.721	2.080	2.518	2.831
22	1.321	1.717	2.074	2.508	2.819
23	1.319	1.714	2.069	2.500	2.807
24	1.318	1.711	2.064	2.492	2.797
25	1.316	1.708	2.060	2.485	2.787
26	1.315	1.706	2.056	2.479	2.779
27	1.314	1.703	2.052	2.473	2.771
28	1.313	1.701	2.048	2.467	2.763
29	1.311	1.699	2.045	2.462	2.756
inf.	1.282	1.645	1.960	2.326	2.576

Reprinted from Fisher and Yates: *Statistical Tables for Biological, Agricultural, and Medical Research*, Oliver and Boyd Lt., Edinburgh, by permission of authors and publishers.

TABLE III THE CHI-SQUARE DISTRIBUTION

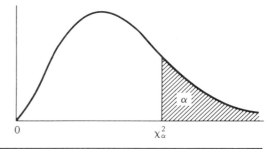

ν	α							
	0.995	0.99	0.975	0.95	0.05	0.025	0.01	0.005
1	0.0^4393	0.0^3157	0.0^3982	0.0^2393	3.841	5.024	6.635	7.879
2	0.0100	0.0201	0.0506	0.103	5.991	7.378	9.210	10.597
3	0.0717	0.115	0.216	0.352	7.815	9.348	11.345	12.838
4	0.207	0.297	0.484	0.711	9.488	11.143	13.277	14.860
5	0.412	0.554	0.831	1.145	11.070	12.832	15.086	16.750
6	0.676	0.872	1.237	1.635	12.592	14.449	16.812	18.548
7	0.989	1.239	1.690	2.167	14.067	16.013	18.475	20.278
8	1.344	1.646	2.180	2.733	15.507	17.535	20.090	21.955
9	1.735	2.088	2.700	3.325	16.919	19.023	21.666	23.589
10	2.156	2.558	3.247	3.940	18.307	20.483	23.209	25.188
11	2.603	3.053	3.816	4.575	19.675	21.920	24.725	26.757
12	3.074	3.571	4.404	5.226	21.026	23.337	26.217	28.300
13	3.565	4.107	5.009	5.892	22.362	24.736	27.688	29.819
14	4.075	4.660	5.629	6.571	23.685	26.119	29.141	31.319
15	4.601	5.229	6.262	7.261	24.996	27.488	30.578	32.801
16	5.142	5.812	6.908	7.962	26.296	28.845	32.000	34.267
17	5.697	6.408	7.564	8.672	27.587	30.191	33.409	35.718
18	6.265	7.015	8.231	9.390	28.869	31.526	34.805	37.156
19	6.844	7.633	8.907	10.117	30.144	32.852	36.191	38.582
20	7.434	8.260	9.591	10.851	31.410	34.170	37.566	39.997
21	8.034	8.897	10.283	11.591	32.671	35.479	38.932	41.401
22	8.643	9.542	10.982	12.338	33.924	36.781	40.289	42.796
23	9.260	10.196	11.689	13.091	35.172	38.076	41.638	44.181
24	9.886	10.856	12.401	13.848	36.415	39.364	42.980	45.558
25	10.520	11.524	13.120	14.611	37.652	40.646	44.314	46.928
26	11.160	12.198	13.844	15.379	38.885	41.923	45.642	48.290
27	11.808	12.879	14.573	16.151	40.113	43.194	46.963	49.645
28	12.461	13.565	15.308	16.928	41.337	44.461	48.278	50.993
29	13.121	14.256	16.047	17.708	42.557	45.722	49.588	52.336
30	13.787	14.953	16.791	18.493	43.773	46.979	50.892	53.672

* Abridged from Table 8 of *Biometrika Tables for Statisticians*, Vol. I, by permission of the Biometrika Trustees.

TABLE IV THE F-DISTRIBUTION
($\alpha = 0.5$)

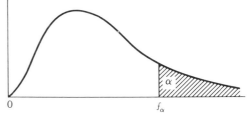

$$f_{0.05}(\nu_1, \nu_2)$$

ν_2	ν_1 = 1	2	3	4	5	6	7	8	9
1	161.4	199.5	215.7	224.6	230.2	234.0	236.8	238.9	240.5
2	18.51	19.00	19.16	19.25	19.30	19.33	19.35	19.37	19.38
3	10.13	9.55	9.28	9.12	9.01	8.94	8.89	8.85	8.81
4	7.71	6.94	6.59	6.39	6.26	6.16	6.09	6.04	6.00
5	6.61	5.79	5.41	5.19	5.05	4.95	4.88	4.82	4.77
6	5.99	5.14	4.76	4.53	4.39	4.28	4.21	4.15	4.10
7	5.59	4.74	4.35	4.12	3.97	3.87	3.79	3.73	3.68
8	5.32	4.46	4.07	3.84	3.69	3.58	3.50	3.44	3.39
9	5.12	4.26	3.86	3.63	3.48	3.37	3.29	3.23	3.18
10	4.96	4.10	3.71	3.48	3.33	3.22	3.14	3.07	3.02
11	4.84	3.98	3.59	3.36	3.20	3.09	3.01	2.95	2.90
12	4.75	3.89	3.49	3.26	3.11	3.00	2.91	2.85	2.80
13	4.67	3.81	3.41	3.18	3.03	2.92	2.83	2.77	2.71
14	4.60	3.74	3.34	3.11	2.96	2.85	2.76	2.70	2.65
15	4.54	3.68	3.29	3.06	2.90	2.79	2.71	2.64	2.59
16	4.49	3.63	3.24	3.01	2.85	2.74	2.66	2.59	2.54
17	4.45	3.59	3.20	2.96	2.81	2.70	2.61	2.55	2.49
18	4.41	3.55	3.16	2.93	2.77	2.66	2.58	2.51	2.46
19	4.38	3.52	3.13	2.90	2.74	2.63	2.54	2.48	2.42
20	4.35	3.49	3.10	2.87	2.71	2.60	2.51	2.45	2.39
21	4.32	3.47	3.07	2.84	2.68	2.57	2.49	2.42	2.37
22	4.30	3.44	3.05	2.82	2.66	2.55	2.46	2.40	2.34
23	4.28	3.42	3.03	2.80	2.64	2.53	2.44	2.37	2.32
24	4.26	3.40	3.01	2.78	2.62	2.51	2.42	2.36	2.30
25	4.24	3.39	2.99	2.76	2.60	2.49	2.40	2.34	2.28
26	4.23	3.37	2.98	2.74	2.59	2.47	2.39	2.32	2.27
27	4.21	3.35	2.96	2.73	2.57	2.46	2.37	2.31	2.25
28	4.20	3.34	2.95	2.71	2.56	2.45	2.36	2.29	2.24
29	4.18	3.33	2.93	2.70	2.55	2.43	2.35	2.28	2.22
30	4.17	3.32	2.92	2.69	2.53	2.42	2.33	2.27	2.21
40	4.08	3.23	2.84	2.61	2.45	2.34	2.25	2.18	2.12
60	4.00	3.15	2.76	2.53	2.37	2.25	2.17	2.10	2.04
120	3.92	3.07	2.68	2.45	2.29	2.17	2.09	2.02	1.96
∞	3.84	3.00	2.60	2.37	2.21	2.10	2.01	1.94	1.88

* Reproduced from Table 18 of *Biometrika Tables for Statisticians*, Vol. I, by permission of the Biometrika Trustees.

TABLE IV   THE F-DISTRIBUTION ($\alpha = .05$) (Continued)

$$f_{0.05}(\nu_1, \nu_2)$$

ν_2	\multicolumn{9}{c}{ν_1}
	10	12	15	20	24	30	40	60	120	∞
1	241.9	243.9	245.9	248.0	249.1	250.1	251.1	252.2	253.3	254.3
2	19.40	19.41	19.43	19.45	19.45	19.46	19.47	19.48	19.49	19.50
3	8.79	8.74	8.70	8.66	8.64	8.62	8.59	8.57	8.55	8.53
4	5.96	5.91	5.86	5.80	5.77	5.75	5.72	5.69	5.66	5.63
5	4.74	4.68	4.62	4.56	4.53	4.50	4.46	4.43	4.40	4.36
6	4.06	4.00	3.94	3.87	3.84	3.81	3.77	3.74	3.70	3.67
7	3.64	3.57	3.51	3.44	3.41	3.38	3.34	3.30	3.27	3.23
8	3.35	3.28	3.22	3.15	3.12	3.08	3.04	3.01	2.97	2.93
9	3.14	3.07	3.01	2.94	2.90	2.86	2.83	2.79	2.75	2.71
10	2.98	2.91	2.85	2.77	2.74	2.70	2.66	2.62	2.58	2.54
11	2.85	2.79	2.72	2.65	2.61	2.57	2.53	2.49	2.45	2.40
12	2.75	2.69	2.62	2.54	2.51	2.47	2.43	2.38	2.34	2.30
13	2.67	2.60	2.53	2.46	2.42	2.38	2.34	2.30	2.25	2.21
14	2.60	2.53	2.46	2.39	2.35	2.31	2.27	2.22	2.18	2.13
15	2.54	2.48	2.40	2.33	2.29	2.25	2.20	2.16	2.11	2.07
16	2.49	2.42	2.35	2.28	2.24	2.19	2.15	2.11	2.06	2.01
17	2.45	2.38	2.31	2.23	2.19	2.15	2.10	2.06	2.01	1.96
18	2.41	2.34	2.27	2.19	2.15	2.11	2.06	2.02	1.97	1.92
19	2.38	2.31	2.23	2.16	2.11	2.07	2.03	1.98	1.93	1.88
20	2.35	2.28	2.20	2.12	2.08	2.04	1.99	1.95	1.90	1.84
21	2.32	2.25	2.18	2.10	2.05	2.01	1.96	1.92	1.87	1.81
22	2.30	2.23	2.15	2.07	2.03	1.98	1.94	1.89	1.84	1.78
23	2.27	2.20	2.13	2.05	2.01	1.96	1.91	1.86	1.81	1.76
24	2.25	2.18	2.11	2.03	1.98	1.94	1.89	1.84	1.79	1.73
25	2.24	2.16	2.09	2.01	1.96	1.92	1.87	1.82	1.77	1.71
26	2.22	2.15	2.07	1.99	1.95	1.90	1.85	1.80	1.75	1.69
27	2.20	2.13	2.06	1.97	1.93	1.88	1.84	1.79	1.73	1.67
28	2.19	2.12	2.04	1.96	1.91	1.87	1.82	1.77	1.71	1.65
29	2.18	2.10	2.03	1.94	1.90	1.85	1.81	1.75	1.70	1.64
30	2.16	2.09	2.01	1.93	1.89	1.84	1.79	1.74	1.68	1.62
40	2.08	2.00	1.92	1.84	1.79	1.74	1.69	1.64	1.58	1.51
60	1.99	1.92	1.84	1.75	1.70	1.65	1.59	1.53	1.47	1.39
120	1.91	1.83	1.75	1.66	1.61	1.55	1.50	1.43	1.35	1.25
∞	1.83	1.75	1.67	1.57	1.52	1.46	1.39	1.32	1.22	1.00

TABLE V THE F-DISTRIBUTION ($\alpha = .01$)

$$f_{0.01}(\nu_1, \nu_2)$$

ν_2	ν_1								
	1	2	3	4	5	6	7	8	9
1	4052	4999.5	5403	5625	5764	5859	5928	5981	6022
2	98.50	99.00	99.17	99.25	99.30	99.33	99.36	99.37	99.39
3	34.12	30.82	29.46	28.71	28.24	27.91	27.67	27.49	27.35
4	21.20	18.00	16.69	15.98	15.52	15.21	14.98	14.80	14.66
5	16.26	13.27	12.06	11.39	10.97	10.67	10.46	10.29	10.16
6	13.75	10.92	9.78	9.15	8.75	8.47	8.26	8.10	7.98
7	12.25	9.55	8.45	7.85	7.46	7.19	6.99	6.84	6.72
8	11.26	8.65	7.59	7.01	6.63	6.37	6.18	6.03	5.91
9	10.56	8.02	6.99	6.42	6.06	5.80	5.61	5.47	5.35
10	10.04	7.56	6.55	5.99	5.64	5.39	5.20	5.06	4.94
11	9.65	7.21	6.22	5.67	5.32	5.07	4.89	4.74	4.63
12	9.33	6.93	5.95	5.41	5.06	4.82	4.64	4.50	4.39
13	9.07	6.70	5.74	5.21	4.86	4.62	4.44	4.30	4.19
14	8.86	6.51	5.56	5.04	4.69	4.46	4.28	4.14	4.03
15	8.68	6.36	5.42	4.89	4.56	4.32	4.14	4.00	3.89
16	8.53	6.23	5.29	4.77	4.44	4.20	4.03	3.89	3.78
17	8.40	6.11	5.18	4.67	4.34	4.10	3.93	3.79	3.68
18	8.29	6.01	5.09	4.58	4.25	4.01	3.84	3.71	3.60
19	8.18	5.93	5.01	4.50	4.17	3.94	3.77	3.63	3.52
20	8.10	5.85	4.94	4.43	4.10	3.87	3.70	3.56	3.46
21	8.02	5.78	4.87	4.37	4.04	3.81	3.64	3.51	3.40
22	7.95	5.72	4.82	4.31	3.99	3.76	3.59	3.45	3.35
23	7.88	5.66	4.76	4.26	3.94	3.71	3.54	3.41	3.30
24	7.82	5.61	4.72	4.22	3.90	3.67	3.50	3.36	3.26
25	7.77	5.57	4.68	4.18	3.85	3.63	3.46	3.32	3.22
26	7.72	5.53	4.64	4.14	3.82	3.59	3.42	3.29	3.18
27	7.68	5.49	4.60	4.11	3.78	3.56	3.39	3.26	3.15
28	7.64	5.45	4.57	4.07	3.75	3.53	3.36	3.23	3.12
29	7.60	5.42	4.54	4.04	3.73	3.50	3.33	3.20	3.09
30	7.56	5.39	4.51	4.02	3.70	3.47	3.30	3.17	3.07
40	7.31	5.18	4.31	3.83	3.51	3.29	3.12	2.99	2.89
60	7.08	4.98	4.13	3.65	3.34	3.12	2.95	2.82	2.72
120	6.85	4.79	3.95	3.48	3.17	2.96	2.79	2.66	2.56
∞	6.63	4.61	3.78	3.32	3.02	2.80	2.64	2.51	2.41

Reproduced from *Biometrika Tables for Statisticians*, Vol. I, by permission of the Biometrika Trustees.

TABLE V THE F-DISTRIBUTION ($\alpha = .01$) (Continued)

$$f_{0.01}(\nu_1, \nu_2)$$

ν_2	ν_1 10	12	15	20	24	30	40	60	120	∞
1	6056	6106	6157	6209	6235	6261	6287	6313	6339	6366
2	99.40	99.42	99.43	99.45	99.46	99.47	99.47	99.48	99.49	99.50
3	27.23	27.05	26.87	26.69	26.60	26.50	26.41	26.32	26.22	26.13
4	14.55	14.37	14.20	14.02	13.93	13.84	13.75	13.65	13.56	13.46
5	10.05	9.89	9.72	9.55	9.47	9.38	9.29	9.20	9.11	9.02
6	7.87	7.72	7.56	7.40	7.31	7.23	7.14	7.06	6.97	6.88
7	6.62	6.47	6.31	6.16	6.07	5.99	5.91	5.82	5.74	5.65
8	5.81	5.67	5.52	5.36	5.28	5.20	5.12	5.03	4.95	4.86
9	5.26	5.11	4.96	4.81	4.73	4.65	4.57	4.48	4.40	4.31
10	4.85	4.71	4.56	4.41	4.33	4.25	4.17	4.08	4.00	3.91
11	4.54	4.40	4.25	4.10	4.02	3.94	3.86	3.78	3.69	3.60
12	4.30	4.16	4.01	3.86	3.78	3.70	3.62	3.54	3.45	3.36
13	4.10	3.96	3.82	3.66	3.59	3.51	3.43	3.34	3.25	3.17
14	3.94	3.80	3.66	3.51	3.43	3.35	3.27	3.18	3.09	3.00
15	3.80	3.67	3.52	3.37	3.29	3.21	3.13	3.05	2.96	2.87
16	3.69	3.55	3.41	3.26	3.18	3.10	3.02	2.93	2.84	2.75
17	3.59	3.46	3.31	3.16	3.08	3.00	2.92	2.83	2.75	2.65
18	3.51	3.37	3.23	3.08	3.00	2.92	2.84	2.75	2.66	2.57
19	3.43	3.30	3.15	3.00	2.92	2.84	2.76	2.67	2.58	2.49
20	3.37	3.23	3.09	2.94	2.86	2.78	2.69	2.61	2.52	2.42
21	3.31	3.17	3.03	2.88	2.80	2.72	2.64	2.55	2.46	2.36
22	3.26	3.12	2.98	2.83	2.75	2.67	2.58	2.50	2.40	2.31
23	3.21	3.07	2.93	2.78	2.70	2.62	2.54	2.45	2.35	2.26
24	3.17	3.03	2.89	2.74	2.66	2.58	2.49	2.40	2.31	2.21
25	3.13	2.99	2.85	2.70	2.62	2.54	2.45	2.36	2.27	2.17
26	3.09	2.96	2.81	2.66	2.58	2.50	2.42	2.33	2.23	2.13
27	3.06	2.93	2.78	2.63	2.55	2.47	2.38	2.29	2.20	2.10
28	3.03	2.90	2.75	2.60	2.52	2.44	2.35	2.26	2.17	2.06
29	3.00	2.87	2.73	2.57	2.49	2.41	2.33	2.23	2.14	2.03
30	2.98	2.84	2.70	2.55	2.47	2.39	2.30	2.21	2.11	2.01
40	2.80	2.66	2.52	2.37	2.29	2.20	2.11	2.02	1.92	1.80
60	2.63	2.50	2.35	2.20	2.12	2.03	1.94	1.84	1.73	1.60
120	2.47	2.34	2.19	2.03	1.95	1.86	1.76	1.66	1.53	1.38
∞	2.32	2.18	2.04	1.88	1.79	1.70	1.59	1.47	1.32	1.00

TABLE VI SQUARES AND SQUARE ROOTS

n	n^2	$\sqrt{n}$	n	n^2	$\sqrt{n}$	n	n^2	$\sqrt{n}$
			35	1 225	5.916 08	70	4 900	8.366 60
1	1	1.000 00	36	1 296	6.000 00	71	5 041	8.426 15
2	4	1.414 21	37	1 369	6.082 76	72	5 184	8.485 28
3	9	1.732 05	38	1 444	6.164 41	73	5 329	8.544 00
4	16	2.000 00	39	1 521	6.245 00	74	5 476	8.602 33
5	25	2.236 07	40	1 600	6.324 56	75	5 625	8.660 25
6	36	2.449 49	41	1 681	6.403 12	76	5 776	8.717 80
7	49	2.645 75	42	1 764	6.480 74	77	5 929	8.774 96
8	64	2.828 43	43	1 849	6.557 44	78	6 084	8.831 76
9	81	3.000 00	44	1 936	6.633 25	79	6 241	8.888 19
10	100	3.162 28	45	2 025	6.708 20	80	6 400	8.944 27
11	121	3.316 63	46	2 116	6.782 33	81	6 561	9.000 00
12	144	3.464 10	47	2 209	6.855 66	82	6 724	9.055 39
13	169	3.605 55	48	2 304	6.928 20	83	6 889	9.110 43
14	196	3.741 66	49	2 401	7.000 00	84	7 056	9.165 15
15	225	3.872 98	50	2 500	7.071 07	85	7 225	9.219 54
16	256	4.000 00	51	2 601	7.141 43	86	7 396	9.273 62
17	289	4.123 11	52	2 704	7.211 10	87	7 569	9.327 38
18	324	4.242 64	53	2 809	7.280 11	88	7 744	9.380 83
19	361	4.358 90	54	2 916	7.348 47	89	7 921	9.433 98
20	400	4.472 14	55	3 025	7.416 20	90	8 100	9.486 83
21	441	4.582 58	56	3 136	7.483 32	91	8 281	9.539 39
22	484	4.690 42	57	3 249	7.549 83	92	8 464	9.591 66
23	529	4.795 83	58	3 364	7.615 77	93	8 649	9.643 65
24	576	4.898 98	59	3 481	7.618 15	94	8 836	9.695 36
25	625	5.000 00	60	3 600	7.745 97	95	9 025	9.746 79
26	676	5.099 02	61	3 721	7.810 25	96	9 216	9.797 96
27	729	5.196 15	62	3 844	7.874 01	97	9 409	9.848 86
28	784	5.291 50	63	3 969	7.937 25	98	9 604	9.899 50
29	841	5.385 17	64	4 096	8.000 00	99	9 801	9.949 87
30	900	5.477 23	65	4 225	8.062 26	100	10 000	10.00000
31	961	5.567 76	66	4 356	8.124 04	101	10 201	10.04998
32	1 024	5.656 85	67	4 489	8.185 35	102	10 404	10.09950
33	1 089	5.744 56	68	4 624	8.246 21	103	10 609	10.14889
34	1 156	5.830 95	69	4 761	8.306 62	104	10 816	10.19804

TABLE VI SQUARES AND SQUARE ROOTS (Continued)

n	n^2	$\sqrt{n}$	n	n^2	$\sqrt{n}$	n	n^2	$\sqrt{n}$
105	11 025	10.24695	145	21 025	12.04159	185	34 225	13.60147
106	11 236	10.29563	146	21 316	12.08305	186	34 596	13.63818
107	11 449	10.34408	147	21 609	12.12436	187	34 969	13.67479
108	11 664	10.39230	148	21 904	12.16553	188	35 344	13.71131
109	11 881	10.44031	149	22 201	12.20656	189	35 721	13.74773
110	12 100	10.48809	150	22 500	12.24745	190	36 100	13.78405
111	12 321	10.53565	151	22 801	12.28821	191	36 481	13.82027
112	12 544	10.58301	152	23 104	12.32883	192	36 864	13.85641
113	12 769	10.63015	153	23 409	12.36932	193	37 249	13.89244
114	12 996	10.67708	154	23 716	12.40967	194	37 636	13.92839
115	13 225	10.72381	155	24 025	12.44990	195	38 025	13.96424
116	13 456	10.77033	156	24 336	12.49000	196	38 416	14.00000
117	13 689	10.81665	157	24 649	12.52996	197	38 809	14.03567
118	13 924	10.86278	158	24 964	12.56981	198	39 204	14.07125
119	14 161	10.90871	159	25 281	12.60952	199	39 601	14.10674
120	14 400	10.95445	160	25 600	12.64911	200	40 000	14.14214
121	14 641	11.00000	161	25 921	12.68858	201	40 401	14.17745
122	14 884	11.04536	162	26 244	12.72792	202	40 804	14.21267
123	15 129	11.09054	163	26 569	12.76715	203	41 209	14.24781
124	15 376	11.13553	164	26 806	12.80625	204	41 616	14.28286
125	15 625	11.18034	165	27 225	12.84523	205	42 025	14.31782
126	15 876	11.22497	166	27 556	12.88410	206	42 436	14.35270
127	16 129	11.26943	167	27 889	12.92285	207	42 849	14.38749
128	16 384	11.31371	168	28 224	12.96148	208	43 264	14.42221
129	16 641	11.35782	169	28 561	13.00000	209	43 681	14.45683
130	16 900	11.40175	170	28 900	13.03840	210	44 100	14.49138
131	17 161	11.44552	171	29 241	13.07670	211	44 521	14.52584
132	17 424	11.48913	172	29 584	13.11488	212	44 944	14.56022
133	17 689	11.53256	173	29 929	13.15295	213	45 369	14.59452
134	17 956	11.57584	174	30 276	13.19091	214	45 796	14.62874
135	18 225	11.61895	175	30 625	13.22876	215	46 225	14.66288
136	18 496	11.66190	176	30 976	13.26650	216	46 656	14.69694
137	18 769	11.70470	177	31 329	13.30413	217	47 089	14.73092
138	19 044	11.74734	178	31 684	13.34166	218	47 524	14.76482
139	19 321	11.78983	179	32 041	13.37909	219	47 961	14.79865
140	19 600	11.83216	180	32 400	13.41641	220	48 400	14.83240
141	19 881	11.87434	181	32 761	13.45362	221	48 841	14.86607
142	20 164	11.91638	182	33 124	13.49074	222	49 284	14.89966
143	20 449	11.95826	183	33 489	13.52775	223	49 729	14.93318
144	20 736	12.00000	184	33 856	13.56466	224	50 176	14.96663

TABLE VI SQUARES AND SQUARE ROOTS (Continued)

n	n^2	$\sqrt{n}$	n	n^2	$\sqrt{n}$	n	n^2	$\sqrt{n}$
225	50 625	15.00000	265	70 225	16.27882	305	93 025	17.46425
226	51 076	15.03330	266	70 756	16.30951	306	93 636	17.49286
227	51 529	15.06652	267	71 289	16.34013	307	94 249	17.52142
228	51 984	15.09967	268	71 824	16.37071	308	94 864	17.54993
229	52 441	15.13275	269	72 361	16.40122	309	95 481	17.57840
230	52 900	15.16575	270	72 900	16.43168	310	96 100	17.60682
231	53 361	15.19868	271	73 441	16.46208	311	96 721	17.63519
232	53 824	15.23155	272	73 984	16.49242	312	97 344	17.66352
233	54 289	15.26434	273	74 529	16.52271	313	97 969	17.69181
234	54 756	15.29706	274	75 076	16.55295	314	98 596	17.72005
235	55 225	15.32971	275	75 625	16.58312	315	99 225	17.74824
236	55 696	15.36229	276	76 176	16.61235	316	99 856	17.77639
237	56 169	15.39480	277	76 729	16.64332	317	100 489	17.80449
238	56 644	15.42725	278	77 284	16.67333	318	101 124	17.83255
239	57 121	15.45962	279	77 841	16.70329	319	101 761	17.86057
240	57 600	15.49193	280	78 400	16.73320	320	102 400	17.88854
241	58 081	15.52417	281	78 961	16.76305	321	103 041	17.91647
242	58 564	15.55635	282	79 524	16.79286	322	103 684	17.94436
243	59 049	15.58846	283	80 089	16.82260	323	104 329	17.97220
244	59 536	15.62050	284	80 656	16.85230	324	104 976	18.00000
245	60 025	15.65248	285	81 225	16.88194	325	105 625	18.02776
246	60 516	15.68439	286	81 796	16.91153	326	106 276	18.05547
247	61 009	15.71623	287	82 369	16.94107	327	106 929	18.08314
248	61 504	15.74902	288	82 944	16.97056	328	107 584	18.11077
249	62 001	15.77973	289	83 521	17.00000	329	108 241	18.13836
250	62 500	15.81139	290	84 100	17.02939	330	108 900	18.16590
251	63 001	15.84298	291	84 681	17.05872	331	109 561	18.19341
252	63 504	15.87451	292	85 264	17.08801	332	110 224	18.22087
253	64 009	15.90597	293	85 849	17.11724	333	110 889	18.24829
254	64 516	15.93738	294	86 436	17.14643	334	111 556	18.27567
255	65 025	15.96872	295	87 025	17.17556	335	112 225	18.30301
256	65 536	16.00000	296	87 616	17.20465	336	112 896	18.33030
257	66 049	16.03122	297	88 209	17.23369	337	113 569	18.35756
258	66 564	16.06238	298	88 804	17.26268	338	114 244	18.38478
259	67 081	16.09348	299	89 401	17.29162	339	114 921	18.41195
260	67 600	16.12452	300	90 000	17.32051	340	115 600	18.43909
261	68 121	16.15549	301	90 601	17.34935	341	116 281	18.46619
262	68 644	16.18641	302	91 204	17.37815	342	116 964	18.49324
263	69 169	16.21727	303	91 809	17.40690	343	117 649	18.52026
264	69 696	16.24808	304	92 416	17.43560	344	118 336	18.54724

TABLE VI SQUARES AND SQUARE ROOTS (Continued)

n	n^2	$\sqrt{n}$	n	n^2	$\sqrt{n}$	n	n^2	$\sqrt{n}$
345	119 025	18.57418	385	148 225	19.62142	425	180 625	20.61553
346	119 716	18.60108	386	148 996	19.64688	426	181 476	20.63977
347	120 409	18.62794	387	149 769	19.67232	427	182 329	20.66398
348	121 104	18.65476	388	150 544	19.69772	428	183 184	20.68816
349	121 801	18.68154	389	151 321	19.72308	429	184 041	20.71232
350	122 500	18.70829	390	152 100	19.74842	430	184 900	20.73644
351	123 201	18.73499	391	152 881	19.77372	431	185 761	20.76054
352	123 904	18.76166	392	153 664	19.79899	432	186 624	20.78461
353	124 609	18.78829	393	154 449	19.82423	433	187 489	20.80865
354	125 316	18.81489	394	155 236	19.84943	434	188 356	20.83267
355	126 025	18.84144	395	156 025	19.87461	435	189 225	20.85665
356	126 736	18.86796	396	156 816	19.89975	436	190 096	20.88061
357	127 449	18.89444	397	157 609	19.92486	437	190 969	20.90454
358	128 164	18.92089	398	158 404	19.94994	438	191 844	20.92845
359	128 881	18.94730	399	159 201	19.97498	439	192 721	20.95233
360	129 600	18.97367	400	160 000	20.00000	440	193 600	20.97618
361	130 321	19.00000	401	160 801	20.02498	441	194 481	21.00000
362	131 044	19.02630	402	161 604	20.04994	442	195 364	21.02380
363	131 769	19.05256	403	162 409	20.07486	443	196 249	21.04757
364	132 496	19.07878	404	163 216	20.09975	444	197 136	21.07131
365	133 225	19.10497	405	164 025	20.12461	445	198 025	21.09502
366	133 956	19.13113	406	164 836	20.14944	446	198 916	21.11871
367	134 689	19.15724	407	165 649	20.17424	447	199 809	21.14237
368	135 424	19.18333	408	166 464	20.19901	448	200 704	21.16601
369	136 161	19.20937	409	167 281	20.22375	449	201 601	21.18962
370	136 900	19.23538	410	168 100	20.24864	450	202 500	21.21320
371	137 641	19.26136	411	168 921	20.27313	451	203 401	21.23676
372	138 384	19.28730	412	169 744	20.29778	452	204 304	21.26029
373	139 129	19.31321	413	170 569	20.32240	453	205 209	21.28380
374	139 876	19.33908	414	171 396	20.34699	454	206 116	21.30728
375	140 625	19.36492	415	172 225	20.37155	455	207 025	21.33073
376	141 376	19.39072	416	173 056	20.39608	456	207 936	21.35416
377	142 129	19.41649	417	173 889	20.42058	457	208 849	21.37756
378	142 884	19.44222	418	174 724	20.44505	458	209 764	21.40093
379	143 641	19.46792	419	175 561	20.46949	459	210 681	21.42429
380	144 400	19.49359	420	176 400	20.49390	460	211 600	21.44761
381	145 161	19.51922	421	177 241	20.51828	461	212 521	21.47091
382	145 924	19.54482	422	178 084	20.54264	462	213 444	21.49419
383	146 689	19.57039	423	178 929	20.56696	463	214 369	21.51743
384	147 456	19.59592	424	179 776	20.59126	464	215 296	21.54066

TABLE VI SQUARES AND SQUARE ROOTS (Continued)

n	n^2	$\sqrt{n}$	n	n^2	$\sqrt{n}$	n	n^2	$\sqrt{n}$
465	216 225	21.56386	505	255 025	22.47221	545	297 025	23.34524
466	217 156	21.58703	506	256 036	22.49444	546	298 116	23.36664
467	218 089	21.61018	507	257 049	22.51666	547	299 209	23.38803
468	219 024	21.63331	508	258 064	22.53886	548	300 304	23.40940
469	219 961	21.65641	509	259 081	22.56103	549	301 401	23.43075
470	220 900	21.67948	510	260 100	22.58318	550	302 500	23.45208
471	221 841	21.70253	511	261 121	22.60531	551	303 601	23.47339
472	222 784	21.72556	512	262 144	22.62742	552	304 704	23.49468
473	223 729	21.74856	513	263 169	22.64950	553	305 809	23.51595
474	224 676	21.77154	514	264 196	22.67157	554	306 916	23.53720
475	225 625	21.79449	515	265 225	22.69361	555	308 025	23.55844
476	226 576	21.81742	516	266 256	22.71563	556	309 136	23.57965
477	227 529	21.84033	517	267 289	22.73763	557	310 249	23.60085
478	228 484	21.86321	518	268 324	22.75961	558	311 364	23.62202
479	229 441	21.88607	519	269 361	22.78157	559	312 481	23.64318
480	230 400	21.90890	520	270 400	22.80351	560	313 600	23.66432
481	231 361	21.93171	521	271 441	22.82542	561	314 721	23.68544
482	232 324	21.95450	522	272 484	22.84732	562	315 844	23.70654
483	233 289	21.97726	523	273 529	22.86919	563	316 969	23.72762
484	234 256	22.00000	524	274 576	22.89105	564	318 096	23.74868
485	235 225	22.02272	525	275 625	22.91288	565	319 225	23.76973
486	236 196	22.04541	526	276 676	22.93469	566	320 356	23.79075
487	237 169	22.06808	527	277 729	22.95648	567	321 489	23.81176
488	238 144	22.09072	528	278 784	22.97825	568	322 624	23.83275
489	239 121	22.11334	529	279 841	23.00000	569	323 761	23.85372
490	240 100	22.13594	530	280 900	23.02173	570	324 900	23.87467
491	241 081	22.15852	531	281 961	23.04344	571	326 041	23.89561
492	242 064	22.18107	532	283 024	23.06513	572	327 184	23.91652
493	243 049	22.20360	533	284 089	23.08679	573	328 329	23.93742
494	244 036	22.22611	534	285 156	23.10844	574	329 476	23.95830
495	245 025	22.24860	535	286 225	23.13007	575	330 625	23.97916
496	246 016	22.27106	536	287 296	23.15167	576	331 776	24.00000
497	247 009	22.29350	537	288 369	23.17326	577	332 929	24.02082
498	248 004	22.31591	538	289 444	23.19483	578	334 084	24.04163
499	249 001	22.33831	539	290 521	23.21637	579	335 241	24.06242
500	250 000	22.36068	540	291 600	23.23790	580	336 400	24.08319
501	251 001	22.38303	541	292 681	23.25941	581	337 561	24.10394
502	252 004	22.40536	542	293 764	23.28089	582	338 724	24.12468
503	253 009	22.42766	543	294 849	23.30236	583	339 889	24.14539
504	254 016	22.44994	544	295 936	23.32381	584	341 056	24.16609

TABLE VI SQUARES AND SQUARE ROOTS (Continued)

n	n^2	$\sqrt{n}$	n	n^2	$\sqrt{n}$	n	n^2	$\sqrt{n}$
585	342 225	24.18677	625	390 625	25.00000	665	442 225	25.78759
586	343 396	24.20744	626	391 876	25.01999	666	443 556	25.80698
587	344 569	24.22808	627	393 129	25.03997	667	444 889	25.82634
588	345 744	24.24871	628	394 384	25.05993	668	446 224	25.84570
589	346 921	24.26932	629	395 641	25.07987	669	447 561	25.86503
590	348 100	24.28992	630	396 900	25.09980	670	448 900	25.88436
591	349 281	24.31049	631	398 161	25.11971	671	450 241	25.90367
592	350 464	24.33105	632	399 424	25.13961	672	451 584	25.92296
593	351 649	24.35159	633	400 689	25.15949	673	452 929	25.94224
594	352 836	24.37212	634	401 956	25.17936	674	454 276	25.96151
595	354 025	24.39262	635	403 225	25.19921	675	455 625	25.98076
596	355 216	24.41311	636	404 496	25.21904	676	456 976	26.00000
597	356 409	24.43358	637	405 769	25.23886	677	458 329	26.01922
598	357 604	24.45404	638	407 044	25.25866	678	459 684	26.03843
599	358 801	24.47448	639	408 321	25.27845	679	461 041	26.05763
600	360 000	24.49490	640	409 600	25.29822	680	462 400	26.07681
601	361 201	24.51530	641	410 881	25.31798	681	463 761	26.09598
602	362 404	24.53569	642	412 164	25.33772	682	465 124	26.11513
603	363 609	24.55606	643	413 449	25.35744	683	466 489	26.13427
604	364 816	24.57641	644	414 736	25.37716	684	467 856	26.15339
605	366 025	24.59675	645	416 025	25.39685	685	469 225	26.17250
606	367 236	24.61707	646	417 316	25.41653	686	470 596	26.19160
607	368 449	24.63737	647	418 609	25.43619	687	471 969	26.21068
608	369 664	24.65766	648	419 904	25.45584	688	473 344	26.22975
609	370 881	24.67793	649	421 201	25.47548	689	474 721	26.24881
610	372 100	24.69818	650	422 500	25.49510	690	476 100	26.26785
611	373 321	24.71841	651	423 801	25.51470	691	477 481	26.28688
612	374 544	24.73863	652	425 104	25.53429	692	478 864	26.30589
613	375 769	24.75884	653	426 409	25.55386	693	480 249	26.32489
614	376 996	24.77902	654	427 716	25.57342	694	481 636	26.34388
615	378 225	24.79919	655	429 025	25.59297	695	483 025	26.36285
616	379 456	24.81935	656	430 336	25.61250	696	484 416	26.38181
617	380 689	24.83948	657	431 649	25.63201	697	485 809	26.40076
618	381 924	24.85961	658	432 964	25.65151	698	487 204	26.41969
619	383 161	24.87971	659	434 281	25.67100	699	488 601	26.43861
620	384 400	24.89980	660	435 600	25.69047	700	490 000	26.45751
621	385 641	24.91987	661	436 921	25.70992	701	491 401	26.47640
622	386 884	24.93993	662	438 244	25.72936	702	492 804	26.49528
623	388 129	24.95997	663	439 569	25.74879	703	494 209	26.51415
624	389 376	24.97999	664	440 896	25.76820	704	495 616	26.53300

TABLE VI SQUARES AND SQUARE ROOTS (Continued)

n	n^2	$\sqrt{n}$	n	n^2	$\sqrt{n}$	n	n^2	$\sqrt{n}$
705	497 025	26.55184	745	555 025	27.29469	785	616 225	28.01785
706	498 436	26.57066	746	556 516	27.31300	786	617 796	28.03569
707	499 849	26.58947	747	558 009	27.33130	787	619 369	28.05352
708	501 264	26.60827	748	559 504	27.34959	788	620 944	28.07134
709	502 681	26.62705	749	561 001	27.36786	789	622 521	28.08914
710	504 100	26.64583	750	562 500	27.38613	790	624 100	28.10694
711	505 521	26.66458	751	564 001	27.40438	791	625 681	28.12472
712	506 944	26.68333	752	565 504	27.42262	792	627 264	28.14249
713	508 369	26.70206	753	567 009	27.44085	793	628 849	28.16026
714	509 796	26.72078	754	568 516	27.45906	794	630 436	28.17801
715	511 225	26.73948	755	570 025	27.47726	795	632 025	28.19574
716	512 656	26.75818	756	571 536	27.49545	796	633 616	28.21347
717	514 089	26.77686	757	573 049	27.51363	797	635 209	28.23119
718	515 524	26.79552	758	574 564	27.53180	798	636 804	28.24889
719	516 961	26.81418	759	576 081	27.54995	799	638 401	28.26659
720	518 400	26.83282	760	577 600	27.56810	800	640 000	28.28472
721	519 841	26.85144	761	579 121	27.58623	801	641 601	28.30194
722	521 284	26.87006	762	580 644	27.60435	802	643 204	28.31960
723	522 729	26.88866	763	582 169	27.62245	803	644 809	28.33725
724	524 176	26.90725	764	583 696	27.64055	804	646 416	28.35489
725	525 625	26.92582	765	585 225	27.65863	805	648 025	28.37252
726	527 076	26.94439	766	586 756	27.67671	806	649 636	28.39014
727	528 529	26.96294	767	588 289	27.69476	807	651 249	28.40775
728	529 984	26.98148	768	589 824	27.71281	808	652 864	28.42534
729	531 441	27.00000	769	591 361	27.73085	809	654 481	28.44293
730	532 900	27.01851	770	592 900	27.74887	810	656 100	28.46050
731	534 361	27.03701	771	594 441	27.76689	811	657 721	28.47806
732	535 824	27.05550	772	595 984	27.78489	812	659 344	28.49561
733	537 289	27.07397	773	597 529	27.80288	813	660 969	28.51315
734	538 756	27.09243	774	599 076	27.82086	814	662 596	28.53069
735	540 225	27.11088	775	600 625	27.83882	815	664 225	28.54820
736	541 696	27.12932	776	602 176	27.85678	816	665 856	28.56571
737	543 169	27.14774	777	603 729	27.87472	817	667 489	28.58321
738	544 644	27.16616	778	605 284	27.89265	818	669 124	28.60070
739	546 121	27.18455	779	606 841	27.91057	819	670 761	28.61818
740	547 600	27.20294	780	608 400	27.92848	820	672 400	28.63564
741	549 081	27.22132	781	609 961	27.94638	821	674 041	28.65310
742	550 564	27.23968	782	611 524	27.96426	822	675 684	28.67054
743	552 049	27.25803	783	613 089	27.98214	823	677 329	28.68798
744	553 536	27.27636	784	614 656	28.00000	824	678 976	28.70540

TABLE VI SQUARES AND SQUARE ROOTS (Continued)

n	n^2	$\sqrt{n}$	n	n^2	$\sqrt{n}$	n	n^2	$\sqrt{n}$
825	680 625	28.72281	865	748 225	29.41088	905	819 025	30.08322
826	682 726	28.74022	866	749 956	29.42788	906	820 836	30.09983
827	683 929	28.75761	867	751 689	29.44486	907	822 649	30.11644
828	685 584	28.77499	868	753 424	29.46184	908	824 464	30.13304
829	687 241	28.79236	869	755 161	29.47881	909	826 281	30.14963
830	688 900	28.80972	870	756 900	29.49576	910	828 100	30.16621
831	690 561	28.82707	871	758 641	29.51271	911	829 921	30.18278
832	692 224	28.84441	872	760 384	29.52965	912	831 744	30.19934
833	693 889	28.86174	873	762 129	29.54657	913	833 569	30.21589
834	695 556	28.87906	874	763 876	29.56349	914	835 396	30.23243
835	697 225	28.89637	875	765 625	29.58040	915	837 225	30.24897
836	698 896	28.91366	876	767 376	29.59730	916	839 056	30.26549
837	700 569	28.93095	877	769 129	29.61419	917	840 889	30.28201
838	702 244	28.94823	878	770 884	29.63106	918	842 724	30.29851
839	703 921	28.96550	879	772 641	29.64793	919	844 561	30.31501
840	705 600	28.98275	880	774 400	29.66479	920	846 400	30.33150
841	707 281	29.00000	881	776 161	29.68164	921	848 241	30.34798
842	708 964	29.01724	882	777 924	29.69848	922	850 084	30.36445
843	710 649	29.03446	883	779 689	29.71532	923	851 929	30.38092
844	712 336	29.05168	884	781 456	29.73214	924	853 776	30.39737
845	714 025	29.06888	885	783 225	29.74895	925	855 625	30.41381
846	715 716	29.08608	886	784 996	29.76575	926	857 476	30.43025
847	717 409	29.10326	887	786 769	29.78255	927	859 329	30.44667
848	719 104	29.12044	888	788 544	29.79933	928	861 184	30.46309
849	720 801	29.13760	889	790 321	29.81610	929	863 041	30.47950
850	722 500	29.15476	890	792 100	29.83287	930	864 900	30.49590
851	724 201	29.17190	891	793 881	29.84962	931	866 761	30.51229
852	725 904	29.18904	892	795 664	29.86637	932	868 624	30.52868
853	727 609	29.20616	893	797 449	29.88311	933	870 489	30.54505
854	729 316	29.22328	894	799 236	29.89983	934	872 356	30.56141
855	731 025	29.24038	895	801 025	29.91655	935	874 225	30.57777
856	732 736	29.25748	896	802 816	29.93326	936	876 096	30.59412
857	734 449	29.27456	897	804 609	29.94996	937	877 969	30.61046
858	736 164	29.29164	898	806 404	29.96665	938	879 844	30.62679
859	737 881	29.30870	899	808 201	29.98333	939	881 721	30.64311
860	739 600	29.32576	900	810 000	30.00000	940	883 600	30.65942
861	741 321	29.34280	901	811 801	30.01666	941	885 481	30.67572
862	743 044	29.35984	902	813 604	30.03331	942	887 364	30.69202
863	744 769	29.37686	903	815 409	30.04996	943	889 249	30.70831
864	746 496	29.39388	904	817 216	30.06659	944	891 136	30.72458

TABLE VI   SQUARES AND SQUARE ROOTS (Continued)

n	n^2	$\sqrt{n}$	n	n^2	$\sqrt{n}$	n	n^2	$\sqrt{n}$
945	893 025	30.74085	965	931 225	31.06445	985	970 225	31.38471
946	894 916	30.75711	966	933 156	31.08054	986	972 196	31.40064
947	896 809	30.77337	967	935 089	31.09662	987	974 169	31.41656
948	898 704	30.78961	968	937 024	31.11270	988	976 144	31.43247
949	900 601	30.80584	969	938 961	31.12876	989	978 121	31.44837
950	902 500	30.82207	970	940 900	31.14482	990	980 100	31.46427
951	904 401	30.83829	971	942 841	31.16087	991	982 081	31.48015
952	906 304	30.85450	972	944 784	31.17691	992	984 064	31.49603
953	908 209	30.87070	973	946 729	31.19295	993	986 049	31.51190
954	910 116	30.88689	974	948 676	31.20897	994	988 036	31.52777
955	912 025	30.90307	975	950 625	31.22499	995	990 025	31.54362
956	913 936	30.91925	976	952 576	31.24100	996	992 016	31.55947
957	915 849	30.93542	977	954 529	31.25700	997	994 009	31.57531
958	917 764	30.95158	978	956 484	31.27299	998	996 004	31.59114
959	919 681	30.96773	979	958 441	31.28898	999	998 001	31.60696
960	921 600	30.98387	980	960 400	31.30495	1000	1000 000	31.62278
961	923 521	31.00000	981	962 361	31.32092			
962	925 444	31.01612	982	964 324	31.33688			
963	927 369	31.03224	983	966 289	31.35283			
964	929 296	31.04835	984	968 256	31.36877			

Computed by J. Huang, Department of Statistics, University of Florida.

TABLE VII   RANDOM NUMBERS

48611	62866	33963	14045	79451	04934	45576
78812	03509	78673	73181	29973	18664	04555
19472	63971	37271	31445	49019	49405	46925
51266	11569	08697	91120	64156	40365	74297
55806	96275	26130	47949	14877	69594	83041
77527	81360	18180	97421	55541	90275	18213
77680	58788	33016	61173	93049	04694	43534
15404	96554	88265	34537	38526	67924	40474
14045	22917	60718	66487	46346	30949	03173
68376	43918	77653	04127	69930	43283	35766
93385	13421	67957	20384	58731	53396	59723
09858	52104	32014	53115	03727	98624	84616
93307	34116	49516	42148	57740	31198	70336
04794	01534	92058	03157	91758	80611	45357
86265	49096	97021	92582	61422	75890	86442
65943	79232	45702	67055	39024	57383	44424
90038	94209	04055	27393	61517	23002	96560
97283	95943	78363	36498	40662	94188	18202
21913	72958	75637	99936	58715	07943	23748
41161	37341	81838	19389	80336	46346	91895
23777	98392	31417	98547	92058	02277	50315
59973	08144	61070	73094	27059	69181	55623
82690	74099	77885	23813	10054	11900	44653
83854	24715	48866	65745	31131	47636	45137
61980	34997	41825	11623	07320	15003	56774
99915	45821	97702	87125	44488	77613	56823
48293	86847	43186	42951	37804	85129	28993
33225	31280	41232	34750	91097	60752	69783
06846	32828	24425	30249	78801	26977	92074
32671	45587	79620	84831	38156	74211	82752
82096	21913	75544	55228	89796	05694	91552
51666	10433	10945	55306	78562	89630	41230
54044	67942	24145	42294	27427	84875	37022
66738	60184	75679	38120	17640	36242	99357
55064	17427	89180	74018	44865	53197	74810
89599	60264	84549	78007	88450	06488	72274
64756	87759	92354	78694	63638	80939	98644
80817	74533	68407	55862	32476	19326	95558
39847	96884	84657	33697	39578	90197	80532
90401	41700	95510	61166	33757	23279	85523
78227	90110	81378	96659	37008	04050	04228
87240	52716	87697	79433	16336	52862	69149
08486	10951	26832	39763	02485	71688	90936
39338	32169	03713	93510	61244	73774	01245
21188	01850	69689	49426	49128	14660	14143
13287	82531	04388	64693	11934	35051	68576
53609	04001	19648	14053	49623	10840	31915
87900	36194	31567	53506	34304	39910	79630
81641	00496	36058	75899	46620	70024	88753
19512	50277	71508	20116	79520	06269	74173

TABLE VII RANDOM NUMBERS (Continued)

24418	26508	91507	76455	54941	72711	39408
57404	73678	08272	62941	02349	71389	45605
77644	98489	86268	73652	98210	44546	27174
68366	65614	01443	07607	11826	91326	29664
64472	72294	95432	53555	96810	17100	35066
88205	37913	98633	81009	81060	33449	68055
98455	78685	71250	10329	56135	80647	51406
48977	36794	56054	59243	57361	65304	93258
93077	72941	92779	23581	24548	56415	61927
84533	26564	91583	83411	66504	02036	02922
11338	12903	14514	27585	45068	05520	56321
23853	68500	92274	87026	99717	01542	72990
94096	74920	25822	98026	05394	61840	83089
83160	82362	09350	98536	38155	42661	02363
97425	47335	69709	01386	74319	04318	99387
83951	11954	24317	20345	18134	90062	10761
93085	35203	05740	03206	92012	42710	34650
33762	83193	58045	89880	78101	44392	53767
49665	85397	85137	30496	23469	42846	94810
37541	82627	80051	72521	35342	56119	97190
22145	85304	35348	82854	55846	18076	12415
27153	08662	61078	52433	22184	33998	87436
00301	49425	66682	25442	83668	66236	79655
43815	43272	73778	63469	50083	70696	13558
14689	86482	74157	46012	97765	27552	49617
16680	55936	82453	19532	49988	13176	94219
86938	60429	01137	86168	78257	86249	46134
33944	29219	73161	46061	30946	22210	79302
16045	67736	18608	18198	19468	76358	69203
37044	52523	25627	63107	30806	80857	84383
61471	45322	35340	35132	42163	69332	98851
47422	21296	16785	66393	89240	51463	95963
24133	39719	14484	58613	88717	29289	77360
67253	67064	10748	16006	16767	57345	42285
62382	76941	01635	35829	77516	98468	51686
98011	16503	09201	03523	87192	66483	55649
37366	24386	20654	85117	74078	64120	04643
73587	83993	54176	05221	94119	20108	78101
33583	68291	50547	96085	62180	27453	18567
02878	33223	39199	49536	56199	05993	71201
91498	41673	17195	33175	04994	09879	70337
91127	19815	30219	55591	21725	43827	78862
12997	55013	18662	81724	24305	37661	18956
96098	13651	15393	69995	14762	69734	89150
97627	17837	10472	18983	28387	99781	52977
40064	47981	31484	76603	54088	91095	00010
16239	68743	71374	55863	22672	91609	51514
58354	24913	20435	30965	17453	65623	93058
52567	65085	60220	84641	18273	49604	47418
06236	29052	91392	07551	83532	68130	56970

ANSWERS AND HINTS TO SELECTED EXERCISES

Chapter 1
Section 1-5

1-1. a. For example, when your population consists of fire insurance claim data for an entire year.
b. When your research deals only with fire insurance claims for the month of November.

1-3. Here are 4 statistics: (1) the mean height, (2) the median height, (3) the modal height, and (4) the smallest height

1-5. 15

1-7. 110

1-9. 30

1-11. 10

1-13. 32. Hint: when i = 1 we get $1^1 = 1$

1-15. $\sum_{i=1}^{10} x^i$

1-17. $\sum_{k=2}^{10} 5k$

1-19. for $n = 3$: $\sum_{k=1}^{3} k = 1 + 2 + 3 = 6 = \frac{(3)(4)}{2}$

for $n = 5$: $\sum_{k=1}^{5} k = \frac{(5)(6)}{2}$

for $n = 10$: $\sum_{k=1}^{10} k = \frac{(10)(11)}{2}$

1-20. a. $\sum_{i=1}^{3} x_i = 2 + 4 + 5 = 11$. (This is the whole solution)
b. 45

1-22. a. Let $t_1 = 4$, $t_2 = 2$, $t_3 = 5$, $t_4 = 3$, $t_5 = 2$, $t_6 = 7$.
b. $\sum_{i=1}^{6} (t_i + 3) = 7 + 5 + 8 + 6 + 5 + 10 = 41$
c. 137

1-24. $5k$

1-26. n [Hint: see the text examples.]

1-28. $3a + 3b - 3c$ [Hint: $(a + b - c)$ is just a big constant.]

1-30. 21

Section 1-7

1-31. a. discrete
b. continuous
c. discrete
d. continuous

1-33. mean = 4, mode = 2, median = 3.5, range = 5

1-35. $29,477.86

1-37. 96 [Hint: If the average of three history grades is 80, the total is 240.]

1-39. mean = 47.60, median = 47.50, mode = none, range = 20

1-41. 3.75 lb. (gained)

1-43. $12,600; $12,508 [Hint: Add $708.]

1-45. mean = (1/12)($485.47) = $40.46

1-47. mean = $1.03 [Hint: The sum of the values is $20.53.]
median = $1.01
mode = $1.09
range = $.52

1-49. mean = $45.00, median = $48.51, range = $56.92

1-51. [Hint: There are six negative numbers in the summation.]

Section 1-8

1-52.

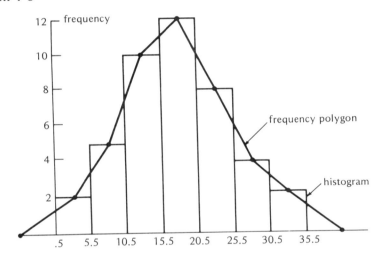

1-54.

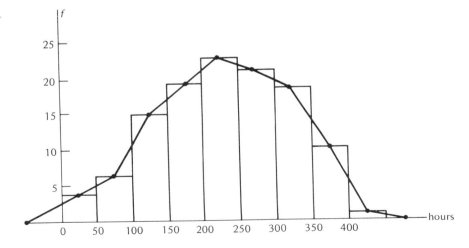

1-56. This would mean that we would have a frequency distribution table with only one class.

1-58.

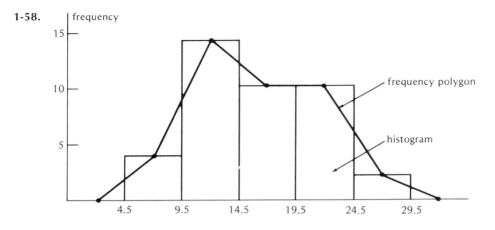

1-60.

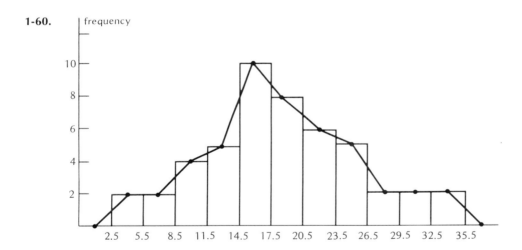

1-62. The largest value is 30. The smallest value is 1. The range is 29. If we would like to have 10 classes, then 3 is a reasonable class size. Certainly, this is not the only possible class size.

Class	Observed Frequency (f)	Class Mark	Upper Class Boundary	Relative Frequency
28–30	4	29	30.5	4/50
25–27	1	26	27.5	1/50
22–24	14	23	24.5	14/50
19–21	5	20	21.5	5/50
16–18	7	17	18.5	7/50
13–15	6	14	15.5	6/50
10–12	1	11	12.5	1/50
7–9	4	8	9.5	4/50
4–6	5	5	6.5	5/50
1–3	3	2	3.5	3/50

Section 1-9

1-65. 2.10 [Hint: The 34th piece of data lies in the 2.10–2.40 class.]

1-67. .79

1-69. 1.48

1-71. .48

1-73. 111.79

1-75. 113.21

1-77. 109.71

Miscellaneous Exercises

1-79. A statistic refers to a value obtained from a sample and a parameter refers to a value obtained from a population.

1-81. a. The 1000 who were surveyed represent our sample, and the population consists of all of the eligible workers in Springfield, Mass.
b. The 1000 interviewed is the sample, and the population consists of all upper-income and lower-income voters in California.

1-83. 8.1

1-85. mean = 8.4, median = 8, mode = 5 and 8, range = 9

1-87.

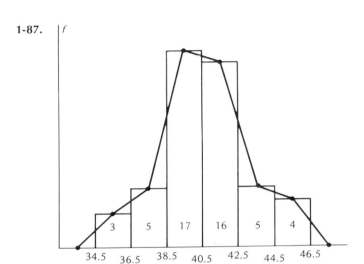

1-89. a and b.

Classes	Frequency, f	Relative Frequency
85–94	4	4/100
75–84	20	20/100
65–74	27	27/100
55–64	26	26/100
45–54	10	10/100
35–44	12	12/100
25–34	1	1/100

ANSWERS AND HINTS TO SELECTED EXERCISES

c.

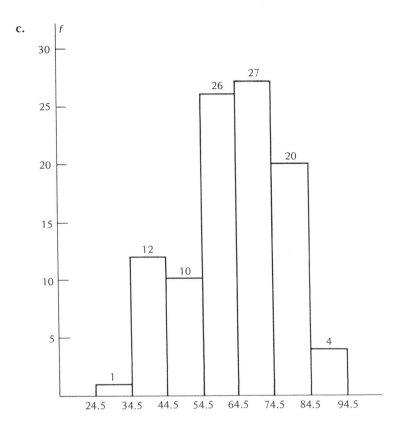

1-91. a. consider: {2, 4, 6}
b. consider: {2, 7, 8}
c. consider: {2, 4, 9}

Chapter 2
Section 2-1

2-1. $s_y^2 = 16$, $s_x^2 = 4$

2-3. $s_w^2 = 21.33$, $s_y^2 = 16$

2-5. Because the two means are equal, the machine with the larger variance (B) is more likely to overfill the cup.

2-7. $\bar{w} = 8$, $s = 0$

2-9. $\bar{y} = 4$, $s = 1.95$

2-11. mean = 100, $s^2 = 11.78$, $s = 3.43$

2-13. $s = 3.45$ [Hint: $\Sigma x_i = 8472$, $\Sigma x_i^2 = 3,588,966$.]

2-15. $s = 7.96$ [Hint: $s^2 = 63.41$.]

2-17. yes

2-19. a. {3, 3, 3, 3}
b. {1, 2, 3, 4, . . .}

Section 2-2

2-20. a. $s_z^2 = \left(\dfrac{1}{2}\right)(1 + 1) = 1$ b. Similar to (a).

2-22. $Z = \left\{ \dfrac{-2\sqrt{1.5}}{1.5}, \dfrac{-\sqrt{1.5}}{1.5}, \dfrac{-\sqrt{1.5}}{1.5}, 0, 0, 0, \dfrac{\sqrt{1.5}}{1.5}, \dfrac{\sqrt{1.5}}{1.5}, \dfrac{2\sqrt{1.5}}{1.5} \right\}$

2-24. There are two solutions: $a = 4$ and $a = -2$. [Hint: Use the definition of the mean and the definition of the variance to make two equations. Solve them simultaneously.]

Section 2-4

2-26. a. At most 1/16 is outside the interval.
b. At least 15/16 is inside the interval.

392 ANSWERS AND HINTS TO SELECTED EXERCISES

2-28. Inside the interval there is at least: for 5—.960; for 5.5—.967; for 6—.972.

2-30. "at most"   **a.** .08   **b.** .07

Section 2-5

2-32. $a_3 = 0$ [Hint: The numerator is 0.]

2-34. $a_3 = -.71$

2-36. a. $a_3 = 1.35$ [Hint: $s = 2.37$.]
b. $Z = \{2.53, 1.27, .42, 0, 0, -.42, -.42, -.42, -.42, -.84, -.84, -.84\}$

2-38. Define $a_3 = 0$.

Miscellaneous Exercises

2-41. This would imply that all of the data values are identical.

2-43. mean = \$2.50 s^2 = \$.07 [Hint: We took each difference squared to four decimal places. mean = \$2.50]

2-45. $Z = \left\{ \dfrac{-4}{\sqrt{10}}, \dfrac{-2}{\sqrt{10}}, 0, \dfrac{2}{\sqrt{10}}, \dfrac{4}{\sqrt{10}} \right\}$
[Hint: Mean = 5, $s = \sqrt{10}$.]

2-47. One possible answer is: 9, 10, 11.

2.49. At least 7500.

Chapter 3
Section 3-1

3-1. $S = \{a, b, c, \ldots, m\}$

3-3. $\{1, 2, 3\}$; $\{6\}$; $\{2, 3, 5\}$

3-5. a. $S = \{BBB, BBG, BGB, BGG, GBB, GBG, GGB, GGG\}$
b. $\{GGG\}$; $\{BBG, BGB, BGG, GBB, GBG, GGB, GGG\}$; $\{BBG, BGB, GBB\}$; $\{BBB, BBG, BGB, GBB\}$

Section 3-2 (Set A)

3-7. 1/5

3-9. 1/4, 1/2, 1/13, 2/13

3-11. a. BBBB
BBBG
BBGB
BBGG
BGBB
BGBG
BGGB
BGGG
GBBB
GBBG
GBGB
GBGG
GGBB
GGBG
GGGB
GGGG
b. In (a) replace B by H and G by T.
c. Exactly the same structure.
d. In (a) replace B by t and G by f.

3-13. 1/8, 5/8, 3/8

3-15. 7/12

3-17. 3/8, 3/8, 1/8, 7/8

3-19. 3/10, 3/5

3-21. 1/2

3-23. 1/3

3-25. a. 1/3   **b.** 2/3   **c.** 2/3

3-27. 1/3 [Hint: We could select two left shoes.]

3-29. a. 7/55   **b.** 27/55   **c.** 19/55   **d.** 0
e. 17/55

3-31. 8/9 [Hint: no y's can occur in 16 ways; exactly one y can occur in 16 ways.]

3-33. 4/9

3-35. 1/6

Section 3-2 (Set B)

3-37. .875

3-39. a. 1/7   **b.** 6/7   **c.** 4/7   **d.** 3/7

3-41. 1/2

3-43. 1/6, 11/18, 0

3-45. 1/6, 5/6 [Hint: 1 is not a prime number.]

Section 3-3

3-47. a. $S = 0$ to 2 inclusive   **b.** 1/4   **c.** 1/2

3-49. .4772, .3414, .0228, .0228, .6828, .0456

3-51. a. .94   **b.** .99   **c.** .95   **d.** .95

Chapter 4
Section 4-1

4-1. 1,860,480 [Hint: (20)(19)(18)(17)(16).]

4-3. 192

4-5. 2^{10}

4-7. a. 9000 **b.** 4536 [Hint: $(9^2)(8)(7)$.]

4-9. 240 [Hint: Take care of the driver first.]

4-11. 60 [Hint: There are three cases.]

4-13. a. 120 **b.** 119 [Hint: Begin by thinking order.]

4-15. 40

4-17. 5^7 [Hint: Assume the three correct ones are marked first.]

Section 4-2

4-18. a. 12 **b.** 1/6 **c.** 180 **d.** 1

4-20. a. 6! **b.** $_6P_2 = 30$ **c.** $_6P_5 = 6!$ **d.** 120

4-22. 40,320 [Hint: The answer is a factorial.]

4-24. 5040

4-26. a. 18 **b.** 8 **c.** 6

4-28. 504 [Hint: This is $_9P_3$.]

4-30. 28,800 [Hint: The answer contains a factor of 2 and a number squared.]

4-32. 5040, 504, 4536 [Hint: The last answer involves the first two answers.]

Section 4-3

4-33. a. 6 **b.** 35 **c.** 28 **d.** 220 **e.** 220 **f.** 252 **g.** 4950 **h.** 1

4-35. 220 [Hint: $_{12}C_3$.]

4-37. 495

4-39. 166,320 [Hint: Multiply two combinations.]

4-41. $n = 9$ [Hint: It may be solved by observation or by solving a simple quadratic equation.]

4-43. 3003

4-45. a. 120 **b.** 10 **c.** 90 [Hint: Multiply two combinations.]

4-47. 28,800 [Hint: There is a 6! in the answer as a factor.]

Section 4-4

4-52. 1/4, 1/2, 5/8, 1/2, 0, 7/8, 1/8

4-54. a. $(1/4)^3$ **b.** $(1/2)(1/4)(1/13) = 1/104$

4-56. 7/12 [Hint: $P(A \cap B) = 0$.]

4-58. 3/5 [Hint: First find the probability that neither A, B, nor C solves it.]

4-60. 13/15

4-62. a. 1/9 **b.** 5/9 [Hint: See (a).]

4-64. 3/4 [Hint: $P(1B \cap 1G) = 0$.]

Section 4-5

4-65. 2/5

4-67. 1/2

4-69. 1/13

4-71. a. 2/7 **b.** 20/21

4-73. $P(A|S) = \dfrac{P(A \cap S)}{P(S)} = P(A)$

Section 4-6

4-75. $\mu = 1/2$, $\sigma^2 = 1/4$, $\sigma = 1/2$

4-77. $\mu = 5/2$, $\sigma^2 = 3/4$, $\sigma = \sqrt{3}/2$

4-79. $\mu = 2.38$, $\sigma^2 = 2.27$

4-81. $\mu = \$750$, $\sigma^2 = \$1,187,500$, $\sigma = \$1089.72$

4-83. $\mu = \$.25$

Miscellaneous Exercises

4-85. The probabilities given by the second analyst total 1.03, which is impossible.

4-87. Because he holds a ticket on every horse, his probability of holding the winning ticket is 1. We did not say he makes money.

4-89. .22, .78

4-91. a. 7/8 **b.** 1/2 **c.** 1/8

4-93. a. 7/25 **b.** 0 **c.** 14/25 **d.** 18/25

4-95. 2/3 [Hint: $P(A \cap B) = 1/6$.]

4-97. 14/25 [Hint: Draw a Venn diagram.]

4-99. a. 10 **b.** 4750 **c.** 70

4-101. 5

4-103. a. 1/8, 3/8, 3/8, 1/8 **b.** 1.5 heads

Chapter 5
Section 5-5

5-1. Elements did not have an equal chance of selection.

5-3. a. There are 5^3.
b. There are 10 in all.

5-5. Assign the first student the number 0001, then 0002, ..., 5000. Using a table of random digits.

5-7. Things to worry about:
1. Only the readers of this one newspaper have been reached.
2. Do the people who take the time to fill out the response form represent the population?

Miscellaneous Exercises

5-9. The fact that the police officer is conducting the survey presents a very serious problem.

5-11. Many families will not be available during regular working hours.

5-13. 800^{50}

5-15. List the 750 customers in alphabetical order and assign the first 001, then 002, ..., 750.

5-17. Division A: 20 people; B: 15 people; C: 10 people; D: 5 people.

5-19. When n constitutes 5% or less of N, the value of the correction factor is less than but nearly equal to 1.

Chapter 6
Section 6-1

6-1. Define the median to be zero.

6-3. a. .9750   b. .2445   c. .6272   d. .8281

6-5. a. $-.50$ [Hint: The desired z score must be to the left of 0.]
b. .60 [Hint: The desired z score must be to the right of 0.]

6-7. $x = 4$ [Hint: See if this answer checks.]

6-9. Our z score is 4. Because the table stops at 3.09, we can only estimate that our final answer is very small.

6-11. $a = 416$ [Hint: Use the standardizing formula.]

6-13. .9974

6-15. 9.34 in. [Hint: We are working under the left tail of our normal distribution.]

6-17. .0764 [Hint: z score $= -1.43$.]

6-19. $\mu = 8856$, $\sigma = 4650$
[Hint: Form two equations using the standardizing formula. Solve these two equations simultaneously.]

6-21. a. .0023   b. .2138   c. .0526

6-23. If $z = \dfrac{a - \mu}{\sigma}$, then
$a - \mu = \sigma z$
$a = \mu + \sigma z$

Miscellaneous Exercises

6-24. a. .1038   b. .8804   c. .0215

6-26. Chebyshev's Theorem is more conservative. It guarantees at least 75% of all of our data within this interval.

6-28. $a_i = 1.59$

6-30. a. .0062 [Hint: $z > 2.50$.]   b. .8882 [Hint: $-1.25 < z < 2.50$.]

6-32. This one rat's performance is more than 5 standard deviations (5.75) from the mean—a highly unlikely result. It may be concluded with a high degree of safety that something is wrong, such as illness, with this animal.

Chapter 7
Section 7-2

7-1. a. Our planet is not the result of an exploding star.
b. Retarded youngsters do not perform better in integrated classes.
c. $x \neq 2$

7-3. a. H_0: $\mu = 25$ lb   b. H_0: $\mu = 25$ lb

7-5. $\mu_{\bar{x}} = 40$, $\sigma_{\bar{x}} = 10$

7-7. $\mu_{\bar{x}} = 100$, $\sigma_{\bar{x}} = 5/2$

7-9. $\sigma = 100$ $\left[\text{Hint: } \sigma_{\bar{x}} = \dfrac{\sigma}{\sqrt{n}}\cdot\right]$

7-11. Conclusion: There is insufficient evidence to reject H_0. [Hint: z-test statistic $= 2.50$; H_a: $\mu \neq \$50,000$.]

7-13. Conclusion: Reject H_0. [Hint: z-test statistic $= 3.60$; H_a: $\mu > \$5.00$.]

7-15. Conclusion: Reject H_0. [Hint: z-test statistic $= 2.67$; H_a: $\mu > 20$.]

ANSWERS AND HINTS TO SELECTED EXERCISES

7-17. Conclusion: There is insufficient evidence to reject H_0. [Hint: z-test statistic = $-.92$; H_a: $\mu < 5$ days (hospital).]

7-19. Conclusion: Reject H_0. [Hint: z-test statistic = -3.00; H_a: $\mu < 48$ days (city).]

Section 7-3

7-21. $.5167 < \mu < .5233$

7-23. $8.44 < \mu < 9.56$

7-25. $26.31 < \mu < 28.95$

Section 7-4

7-27. In a normal distribution the interval $\mu \pm 2\sigma$ contains approximately 95% of our probability

7-29. $n \geq 60$ [Hint: The original answer was 59.91. Let $z = 2.58$.]

7-31. $n \geq 62$

Miscellaneous Exercises

7-33. Begin with $\sigma_{\bar{x}} = \dfrac{\sigma}{\sqrt{n}}$.

7-35. 95%

7-37. a. H_0: $\mu = 50$ b. H_0: $\mu = 50$

7-39. Conclusion: Reject H_0. [Hint: z-test statistic = 2; H_a: $\mu > 12.50$.]

7-41. Conclusion: Reject H_0. [Hint: z-test statistic = 12; H_a: $\mu > \$18{,}000$.]

7-43. Conclusion: Reject H_0. [Hint: z-test statistic = 9; H_a: $\mu \neq 225$.]

7-45. $n \geq 385$ [Hint: The original solution was $n \geq 384.16$.]

Chapter 8
Section 8-2

8-1. $N = 3000$ [Hint: $.89N = 2670$.]

8-3. $N = 5000$ [Hint: $.975N = 4875$.]

8-5. a. .05   b. .925   c. .90   d. .02

8-7. a. .10   b. .90   c. .975   d. .975

8-9. a. 2.093   b. .05   c. 12

8-11. a. 0   b. 0

Section 8-3

8-13. a. when $\bar{x} = \mu$ b. when $\bar{x} < \mu$

8-15. Conclusion: Reject H_0. [Hint: t-test statistic = -3.00; H_a: $\mu \neq 26.4$.]

8-17. Conclusion: There is insufficient evidence to reject H_0. [Hint: t-test statistic = .64; H_a: $\mu > 612$.]

8-19. Conclusion: Reject H_0. [Hint: z-test statistic = 3.19; H_a: $\mu > 93.4$ (note $n = 36$).]

8-21. Conclusion: There is insufficient evidence to reject H_0. [Hint: t-test statistic = -1.77, $s^2 = .103$, mean = 1.50.]

8-23. Conclusion: Reject H_0. [Hint: t-test statistic = 3.50; H_a: $\mu > 20$.]

Section 8-4

8-24. $2.34 < \mu < 3.66$

8-26. $5.58 < \mu < 7.82$

8-28. [Hint: Mean = .30 and $s^2 = .00065$.]
a. The 95% confidence interval for μ: $.30 \pm .03$.
b. The 99% confidence interval for μ: $.30 \pm .05$.
c. The 99% confidence interval is (always) the larger.

8-30. The 95% confidence interval for μ is 270 ± 27.80 or $242.20 < \mu < 297.80$ [Hint: $s = 33.27$]

Miscellaneous Exercises

8-32. student t distribution

8-34. a. .95   b. .925   c. .015

8-36. $1.38 < \mu < 1.82$

8-38. $36{,}469.20 < \mu < 37{,}030.81$

Chapter 9
Section 9-2

9-1. No specific answer

9-3. a. It does appear to lack randomness.
b. The median value is 2. When we convert the data to a's and b's, we obtain: aa bb aa bb aa bb aa bb aa bb aa bb aa bb. There are 14 runs. $\mu = 15$, $\sigma = 2.60$, $n_1 = n_2 = 14$. z-test statistic = $-.38$
Conclusion: There is insufficient evidence to reject H_0.

9-5. Conclusion: There is insufficient evidence to reject H_0. We have 19 runs. [Hint: $\mu = 15.93$, $\sigma = 2.68$, z-test statistic $= 1.15$.]

9-7. Conclusion: Reject H_0. We have 12 runs. [Hint: $\mu = 16.95$, z-test statistic $= -2.00$.]

9-9. Conclusion: There is insufficient evidence to reject H_0. We have 16 runs. [Hint: The median value $= 12.725$, $\mu = 21$, z-test statistic $= -1.60$.]

9-11. Conclusion: There is insufficient evidence to reject H_0. We have 23 runs. [Hint: z-test statistic $= .94$.]

Chapter 10
Section 10-1

10-1. **a.** $\mu = 80$ **b.** $\sigma = 4$ **c.** $n = 200$ **d.** $n = 100$ **e.** 1/5, 4/5 [Hint: Use the quadratic formula.] **f.** $p = .80$

10-3. $p = 1/5$, $\mu = 45$, $\sigma = 6$

10-5. For outcome no. 1, $\mu_1 = np$.
For outcome no. 2, $\mu_2 = n(1 - p)$.
Thus: $\mu_1 + \mu_2 = np + n(1 - p) = n$.

Section 10-2

10-6. (15.5, 16.5)

10-8. .8996 [Hint: $P(84.5 < \text{normal} \times < 94.5)$.]

10-10. .4364

Section 10-3

10-11. Conclusion: Stop. Our sample supports the conjecture in H_0. [Hint: H_a: $p < .10$ (the company).]

10-13. Conclusion: Reject H_0. [Hint; $p > .10$ (the professor), z-test statistic $= 4.72$.]

10-15. Conclusion: There is insufficient evidence to reject H_0. [Hint: To have a chance to reject H_0, we must have more than 80 successes. We do not have this. H_a: $p > .80$.]

10-17. Conclusion: There is insufficient evidence to reject H_0. [Hint: H_a: $p \neq .50$, z-test statistic $= 1.41$.]

10-19. Conclusion: There is insufficient evidence to reject H_0. [Hint: H_a: $p < .30$; z-test statistic $= -1.09$.]

Section 10-4

10-20. Conclusion: Reject H_0. [Hint: H_a: $p \neq .50$; z-test statistic $= 2.14$. There are 11 successes $(+$'s$)$.]

10-22. Conclusion: There is insufficient evidence to reject H_0. [Hint: H_a: $p < .50$; z-test statistic $= -.90$.]

10-24. Conclusion: Reject H_0. [Hint: H_a: $p \neq .50$; z-test statistic $= 1.50$.]

Section 10-5

10-26. $\binom{5}{2} (1/2)^2 (1/2)^3 = .3125$

10-28. .5904 [Hint: It is easier to do a subtraction.]

10-30. 15/64

Miscellaneous Exercises

10-34. $\sigma^2 = 90/7$, $\sigma = 3.59$

10-36. .5120 [Hint: $\mu = 450$ and $\sigma = 15$.]

10-38. **a.** H_0: $p = .4$ **b.** H_0: $p = 2/3$

10-40. Conclusion: There is insufficient evidence to reject H_0. [Hint: H_a: $p < .10$. There is no need to calculate a z-test statistic.]

10-42. Conclusion: There is insufficient evidence to reject H_0. [Hint: H_a: $p \neq 1/2$; z-test statistic $= 1.27$.]

Chapter 11
Section 11-1

11-1. **a.** 5/16 **b.** 5/8

11-3. $\mu = .25$, $\sigma^2 = .003$

11-5. $\mu = .40$, $\sigma^2 = .0067$

11-7. $n = 100$

Section 11-2

11-8. $.52 < p < .59$

11-10. $.42 < p < .46$

11-12. $.42 < p < .64$

Section 11-3

11-14. $n \geq 1305$ [Hint: Original numerical answer $= 1304.65$. Let $z = 2.58$.]

11-16. $n \geq 2377$

11-18. As $n \to \infty$, each end-point approaches $\hat{p}$.

Section 11-4

11-22. $n > z^2/4$

11-24. $n \geq 666$ [Hint: Original numerical answer = 665.64. Let $z = 2.57$]

Miscellaneous Exercises

11-25. $p = .064$

11-27. $\mu_{\text{ratios}} = 2/7$, $\sigma^2_{\text{ratios}} = 10/3087 = .003$

11-29. $.065 < p < .19$ [Hint: The square root in the calculation is .02. Let $z = 2.57$]

11-31. $n \geq 31$ [Hint: Original numerical answer = 30.30.]

Chapter 12
Section 12-2

12-1. a. $\nu = 26$ b. $n_2 = 18$ c. $n = 7$

12-3. Conclusion: There is insufficient evidence to reject H_0. [Hint: $S_p = \sqrt{5/2}$, t-test statistic = $1.55 < 2.776$.]

12-5. Multiply s_1^2 by $n_1 - 1$ and s_2^2 by $n_2 - 1$ and the result follows immediately.

12-7. $S_p = \sqrt{80/13}$. The remainder of the exercise follows as in Exercise 12-4.

12-9. $-4.89 < \mu_2 - \mu_1 < 2.89$

12-11. Conclusion: Reject H_0. [Hint: H_a: $\mu_{\text{special}} > \mu_{\text{untrained}}$ or H_a: $\mu_2 > \mu_1$; $s_2^2 = 3.20$, $s_1^2 = 2.80$; t-test statistic = $3.42 > 1.725$.]

12-13. Conclusion: Reject H_0. [Hint: H_a: $\mu_W > \mu_R$ or H_a: $\mu_2 > \mu_1$; $s_2^2 = \$250$, $s_1^2 = \$200$; t-test statistic = 79.91.]

12-15. Conclusion: Reject H_0. [Hint: H_a: $\mu_1 \neq \mu_2$; t-test statistic = -4.26.]

12-17. Conclusion: Reject H_0. [Hint: H_a: $\mu_A > \mu_B$; t-test statistic = -4.80.]

12-19. Conclusion: There is insufficient evidence to reject H_0. [Hint: H_a: $\mu_A > \mu_B$ or H_a: $\mu_2 > \mu_1$; $S_p = .49$; t-test statistic = 1.80.]

Section 12-3

12-21. Conclusion: Reject H_0. [Hint: H_a: $\mu_2 > \mu_1$; z-test statistic = 2.66.]

12-23. Conclusion: Reject H_0. [Hint: H_a: $\mu_1 \neq \mu_2$; z-test statistic = 4.41, $\sqrt{.03} = .17$.]

12-25. Conclusion: Reject H_0. [Hint: H_a: $\mu_1 \neq \mu_2$; z-test statistic = -4.00, $\sqrt{.12} = .35$.]

12-27. $.42 < \mu_2 - \mu_1 < 1.09$

12-29. $-2.30 < \mu_2 - \mu_1 < -.50$

Section 12-4

12-32. Conclusion: There is insufficient evidence to reject H_0. [Hint: $R_1 = 7.5$, $\sigma_U = 2.29$, z-test statistic = 1.31.]

12-34. Conclusion: There is insufficient evidence to reject H_0. [Hint: $R_1 = 103$, $\sigma_U = 13.23$, z-test statistic = $.15$.]

12-36. Conclusion: There is insufficient evidence to reject H_0. [Hint: $R_1 = 69$, $\sigma_U = 12.96$; z-test statistic = 1.16.]

Miscellaneous Exercises

12-37. $\sigma^2_{(\bar{x}_2 - \bar{x}_1)} = \dfrac{\sigma_1^2 + \sigma_2^2}{n}$

12-39. $S_p = 2.93$

12-41. $-4.64 < \mu_2 - \mu_1 < 4.64$

12-43. $-.21 < \mu_2 - \mu_1 < 1.21$

12-45. $-6.89 < \mu_2 - \mu_1 < 14.89$

12-47. a. $R_1 = 105$, $U = 93$, $\sigma_U = 16.25$
  z-test statistic = 1.66

  b. $R_1 = 159$, $U = 39$, $\sigma_U = 16.25$
  z-test statistic = -1.66

  c. In both (a) and (b) the conclusion is: There is insufficient evidence to reject H_0. Notice the test statistics are negatives of each other.

Chapter 13
Section 13-2 (Set A)

13-1. a. $y = 0$ (the X-axis)
  b. either $x = 0$ (the Y-axis) or $y = x$

13-3. $y = (1/3)(x - 2)$

13-5. $y = (10/13)(x) + 2$ [Hint: $\hat{m} = 10/13$.]

13-7. $y = (-1.71)(x) + 198.38$ [Hint: $\hat{m} = -1.71$.]

13-9. $y = (1.07)(x) - 3.46$

13-11. $y = (9.25)(x) + 1.23$. All work rounded to two decimal places.

Section 13-2 (Set B)

13-13. a. 7   b. 10

13-15. a. 8.22   b. 8.21

13-17. Start with $y = .90x - 2.50$
a. when $x = 1$, $y = -1.60$
b. when $x = 3$, $y = .20$
c. when $x = 7.5$, $y = 4.25$

13-19. a. $y = (10.87)(x) + 54.74$ [Hint: $\hat{m} = 10.87$.]
b. 57.46

13-21. Because the slope of our regression line is negative, we have a decrease in the y variable versus an increase in the x variable (inverse variation). The slope is -1.71. This means that for each unit increase in the x variable, the y variable decreases 1.71 units.

13-23. For each unit increase in the x variable, we have a 1.07 unit increase in the y variable.

Miscellaneous Exercises

13-26. $y = \bar{y}$

13-28. $y = (12.75)(x) + 51.50$

13-30. $y = .76x + 37.84$

Chapter 14
Section 14-2

14-1. no

14-3. $r = \dfrac{(1/4)(14) - (1)(1)}{\sqrt{7/2}\sqrt{15/2}} = .49$

14-5. $r = \dfrac{(1/3)(17.74) - (2)(3.20)}{(1.15)(1.09)} = -.44$

14-7. $\bar{x} = \bar{y} = 0$ and $\sum_{i=1}^{12} x_i y_i = 0$
This means that the numerator of our calculation of r must be zero. Thus, r must be zero.

14-9. $r = .88$

14-11. $r = .79$

14-13. for 3, $r^2 = .24$
for 4, $r^2 = 1.00$
for 5, $r^2 = .19$
for 6, $r^2 = .0018$

14-15. $r = .94$

Section 14-3

14-17. $r_s = 1 - \dfrac{(6)(2)}{(5)(24)} = .90$

14-19. By inspection, $r_s = -1$.

14-21. a. $r_s = .82$ b. direct variation

Section 14-5

14-22. The test demands that $n > 3$.

14-24. Conclusion: Reject H_0. [Hint: z-test statistic = 4.77.]

Miscellaneous Exercises

14-25. a. -1 b. 0

14-27. $\hat{m} = \dfrac{(.85)(.62)}{.71} = .74$

14-29. $r = .85$

14-31. ± 1

14-33. $r = .87$ [Hint: $B_x = 1.12$, $B_y = 2.06$.]

14-35. $r^2 = .55$. This indicates that 55% of the variation in performance can be accounted for by the aptitude test ($r = .74$).

Chapter 15
Section 15-2

15-1. a. 18.307   b. 35.172   c. 30.578

15-3. a. $\nu = 16$ b. $\alpha = .005$

15-5. a. $(O_i - E_i)^2 \geq 0$ for all values of i and $E_i \geq 0$ for all values of i.
b. Our chi-square statistic can equal 0 if, and only if, $O_i = E_i$ for all i.

15-7. Conclusion: There is insufficient evidence to reject H_0. [Hint: Chi-square test statistic = 2.33.]

15-9. a. $\chi^2_{.95,15} > \chi^2_{.99,18}$
b. $\chi^2_{.025,16} < \chi^2_{.01,16}$

15-11. No. Assume that it were possible, then:
$$\sum_{i=1}^{k} E_i = (1/2) \sum_{i=1}^{k} O_i$$
However,
$$\sum_{i=1}^{k} E_i = \sum_{i=1}^{k} O_i = n$$
this is a contradiction.

15-13. The fifth expected cell value is 0. Division by 0 is undefined.

15-15. Conclusion: There is insufficient evidence to reject H_0. [Hint: Chi-square test statistic = 11.40.]

15-17. Conclusion: Reject H_0. [Hint: Chi-square test statistic = 501.10 > 7.815.]

Section 15-4

15-19. a. $\nu = (2)(4) = 8$ b. $\nu = 20$ c. $\nu = 21$

15-21.

					Totals
	3	6	2	1	12
	6	4	3	7	20
	6	5	5	2	18
Totals	15	15	10	10	$n = 50$

15-23.

				Totals
	55	15	30	100
	20	60	40	120
	35	15	30	80
Totals	110	90	100	$n = 300$

15-25. Chi-square statistic = 2.67. [Hint: Two terms in the expansion are 0.]

15-27. Chi-square statistic = 5.16. [Hint: The four terms in the expansion are 1.92 + .66 + 1.92 + .66. Conclusion: Reject H_0.]

15-29. Conclusion: There is insufficient evidence to reject H_0. [Hint: Chi-square test statistic = 2.89.]

15-31. Conclusion: Reject H_0. [Hint: Chi-square test statistic = 27.56 > 12.592.]

15-33. Conclusion: Reject H_0. [Hint: Chi-square test statistic = 7.89 > 7.815.]

15-35. Conclusion: Reject H_0. [Hint: Chi-square test statistic = 20.41 > 18.307.]

Section 15-5

15-36. $4.26 \leq \sigma^2 \leq 11.91$
$2.06 \leq \sigma \leq 3.45$

15-38. $4.85 \leq \sigma^2 \leq 13.56$
$2.20 \leq \sigma \leq 3.68$

15-40. $2.84 \leq \sigma^2 \leq 15.99$

Section 15-7

15-42.

Classes	O_i	Upper Class Boundary	z score of Boundary	Probability Below	Probability Within	E_i
33–35	2	35.5	2.37	.9911	.0161	.77
30–32	2	32.5	1.96	.9750	.0368	1.77
27–29	2	29.5	1.54	.9382	.0696	3.34
24–26	5	26.5	1.12	.8686	.1075	5.16
21–23	6	23.5	.71	.7611	.1470	7.06
18–20	8	20.5	.29	.6141	.1619	7.77
15–17	10	17.5	−.12	.4522	.1576	7.56
12–14	5	14.5	−.54	.2946	.1235	5.93
9–11	4	11.5	−.95	.1711	.0858	4.12
6–8	2	8.5	−1.37	.0853	.0478	2.29
3–5	2	5.5	−1.78	.0375	.0375	1.80
Total	= 48					

mean = 18.38, s^2 = 52.07, s = 7.22, v = 11 − 3 = 8

χ^2 test statistic = 1.965 + .030 + .538 + .005 + .159 + .007 + .788 + .146 + .003 + .037 + .022
= 3.70 < $\chi^2_{.05,8}$ = 15.507

Conclusion: There is insufficient evidence to reject H_0. It does appear that our sample has been drawn from a population that is normally distributed.

Miscellaneous Exercises

15-44. a. v = 72   b. c = 15

15-46. $\chi^2_{.05,15} < \chi^2_{.01,15}$

15-48. a. v = 4
b. Table of Expected Frequencies

110/30	100/30	90/30
55/30	50/30	45/30
165/30	150/30	135/30

15-50. Conclusion: There is insufficient evidence to reject H_0. [Hint: All four numerators are 0.25. Chi-square test statistic = 1.]

15-52. Conclusion: There is insufficient evidence to reject H_0. [Hint: Chi-square test statistic = 4.09.]

15-54. $6.19 \leq \sigma^2 \leq 24.10$

Chapter 16
Section 16-3

16-1. a. 2.62   b. 3.37

16-3. a. v_2 = 20   b. v_1 = 7

16-5. 5/4

16-7. Conclusion: There is insufficient evidence to reject H_0. [Hint: F-test statistic = 1.21 < 10.97.]

16-9. Conclusion: There is insufficient evidence to reject H_0. [Hint: F-test statistic = 2.72 < 4.85.]

16-11. Conclusion: Reject H_0. [Hint: F-test statistic = 19.02 > 3.79.]

ANSWERS AND HINTS TO SELECTED EXERCISES   401

Section 16-5

16-13. $v_2 = (n_1 + n_2 + n_3 + \cdots + n_k) - k$
If $n_1 = n_2 = n_3 = \cdots n_k = n$, then
$v_2 = k(n - 1)$

16-15. Conclusion: Reject H_0. [Hint: F-test statistic $= 5.95$, BVE $= 1713$ and WVE $= 287.67$.]

16-17. Conclusion: Reject H_0. [Hint: F-test statistic $= 42.45$, BVE $= 26{,}000$, WVE $= 612.50$.]

16-19. Conclusion: Reject H_0. [Hint: F-test statistic $= 13.31$, BVE $= 180$, WVE $= 13.52$.]

16-21. Conclusion: Reject H_0. [Hint: F-test statistic $= 62.13$.] All work to two decimal places.

Section 16-6

16-23. Conclusion: There is insufficient evidence to reject H_0. [Hint: The correction factor $C = .98$. Because the value of C is so close to 1, it could be dropped. However, we did consider it here in the calculation of the H-test statistic. $H = 5.78$.]

16-25. Conclusion: Reject H_0. [Hint: H-test statistic $= 10.82$. We dropped the correction factor $C = .999$.]

16-27. Compute only

$$\sum_{i=1}^{4} \frac{R_i^2}{n_i} = 1702.67$$

This is the same result found in Exercise 16-14. Thus the value of H, our test statistic, is unchanged.

Miscellaneous Exercises

16-29. Because the grand mean is an average of the k sample means it must be greater than or equal to the smallest sample mean and less than or equal to the largest of the sample means.

16-31. Yes, it is possible for WVE (although highly unlikely) to be zero. In this unlikely case, we may derive a conclusion from inspecting the k sample means.

16-33. Conclusion: Reject H_0. [Hint: F-test statistic $= 2.40$.]

16-35. Conclusion: There is insufficient evidence to reject H_0. [Hint: H-test statistic $= 2.30$. B.V.E $= 28.13$ and W.V.E $= 12.22$]

16-37. Conclusion: There is insufficient evidence to reject H_0. [Hint: F-test statistic $= 7.69$. All work to 4 decimal places.]

Index

Alpha (α), 142
Alternative hypothesis, 140
Analysis of variance, 341
Average (see mean)

Bell curve, 117
Bernoulli, J., 193
Beta (β), 146
Between variance estimate, 343
Bias
  estimator, 114
  sample, 111
Binomial experiment, 192
  failure, 192
  mean, 195
  success, 192
  trial, 192
  variance, 195
Binomial probability, 216
Birthday problem, 99

Cell, 298, 309
Central Limit Theorem
  for u, 138
  for p, 224
Chebyshev's Theorem, 50
Chi-square (χ^2)
  distribution, 298, 301
  test statistic, 300
Chi-square test for population normality, 327
Class
  boundary, 23
  frequency, 23
  interval (or range), 23
  limit, 23
  mark, 24
Coefficient of determination, 278
Combinations, 85
Comparing several means, 341
Comparing two means, 237
Comparing two variances, 326
Complementary events, 69
Computational formula for s^2, 43
Conditional probability, 92
Correlation
  coefficient, 279, 280, 285
  perfect negative correlation, 281
  perfect positive correlation, 281
  rank (Spearman r), 289
Contingency table, 309, 310
Continuous data, 19
Continuous sample space, 60
Critical point(s), 143

Critical region
  definition, 143, 146
  how to select, 147

Deciles, 33
Degrees of freedom, 168
  for χ^2, 312
  for F, 335, 344, 347
  for t, 171
DeMoivre, A., 117
Dependent variable, 234, 257
Descriptive statistics, 5
Design of experiment, 253
Difference between two sample means, 234
  confidence interval, 241, 246
  mean of, 235
  point estimate, 240
  standard deviation, 235
  test for, 237, 245
Discrete data, 19
Discrete sample space, 60
Distribution free test (see nonparametric test)

Empirical rule, 123
Empty event, 59
Error
  type I, 145
  type II, 145
Estimation (see point estimate and interval estimate)
Events
  complementary, 69
  independent, 89
  mutually exclusive, 88
Exact binomial probabilities, 216

F Distribution
  definition, 335
  degrees of freedom, 335, 344, 347
  test statistic, 337
    equal sample sizes, 344
    unequal sample sizes, 347
Factorial notation, 81
Fermat, P., 61
Finite collection, 4
Fisher, Sir Ronald, 340
Fisher z-transformation, 292, 293
Frequency
  expected, 298
  observed, 298
  relative, 25
Frequency distribution table, 23
Frequency polygon, 27

INDEX

Galton, Sir Francis, 258
Gauss, C., 117
Gaussian curve (see bell curve)
Goodness-of-fit, 302
Gosset, W. S., 171
Grand mean, 342
Grouped data, 16

Histogram, 27
Hypothesis
 alternative, 140
 null, 140, 175, 204
Hypothesis testing
 for μ, 139, 175
 for p, 204
 for $\mu_2 - \mu_1$, 237, 244–246
 for $\sigma_1^2 = \sigma_2^2$, 338, 339
 nonparametric testing, 184, 211, 249, 289, 357
 one tail test, 146
 two tail test, 143

Independence of classification, 309
Independent events, 89
Independent variable, 234, 257
Index numbers, 6
Inferential statistics, 6
Infinite collection, 5
Intersection of events, 67
Interval estimate
 for μ, 157, 180
 for p, 225
 for $\mu_2 - \mu_1$,
  at least one sample size is small, 241
  both sample sizes are large, 246
 for σ^2, 323
 for σ, 323

Kruskal-Wallis Test, 357

Least squares method, 263
Legendre, A., 263
Level of significance, 142
Line of regression, 258
 equation of line, 263
 formula for slope ($\hat{m}$), 263, 285
Linear correlation, 279, 280, 285, 299

Mann-Whitney U test, 249
 mean, 250
 variance, 250
Marginal frequencies, 326
Mathematical expectation, 95
Mean
 for binomial distribution (success numbers), 195
 for proportions (success ratios), 224
 interval estimate for μ, 157, 175
 point estimate of μ, 156
 population, 17
 sample, 12

Measures of central tendency, 15
Multiple regression, 273
Multiplication principle, 78
Mutually exclusive events, 88

Nonparametric methods, 184
Nonparametric tests
 Kruskal-Wallis test, 357
 Mann-Whitney U Test, 249
 Runs test, 185
 Sign test, 208
 Spearman r, 289
Nonlinear regression, 273
Normal curves
 area below, 120
 definition, 117
 standard normal, 119
Normal curve approximation to the binomial distribution, 198
Null event (see empty event)
Null hypothesis, 140

Observed frequency, 298
One-tail test, 146

Parameter, 2, 5
Pascal, B., 61
Pearson, K., 277
Pearson r (see linear correlation coefficient)
Percentile, 32
Permutations, 80
Point estimate
 for μ, 156
 for p, 223
 for $\mu_2 - \mu_1$, 240
 for σ^2, 156
 for σ, 156
Pooled estimate of the variance (Sp), 236
Population
 definition, 2, 3
 mean, 17, 96
 standard deviation, 45
 variance, 45, 97
Post hoc comparisons, 351
Probability
 area, 71
 curves, 70
 definition, 61
Proportions (see success ratios)

Quartiles, 33

Random digits
 explanation, 108, 109
 table, 384, 385
Random sample, 108
Range, 3, 14
Ranking data
 Mann-Whitney U Test, 249
 Spearman r, 289
 Kruskal-Wallis, 357

Raw data, 21, 181
Reduced sample space, 93
Regression line, 260
Regression plane, 273
Rejection region (see critical region)
Rho (ρ), 293
Run, 185
Runs test, 185

Sample
  definition, 2, 4
  large, 138
  mean, 12
  median, 13
  mode, 13
  size, 162, 163, 227
  small, 168
  standard deviation, 40
  range, 3, 14
  variance, 40
Sample point, 58
Sample space, 58
Sampling distribution, 107
  for the difference between sample means, 235
  for F ratios, 335
  for H's, 357
  for means (large), 134
  for means (small), 171
  for r's, 292, 293
  for runs, 185
  for success numbers, 192
  for success ratios (proportions), 222, 223
  for U's, 249
Sampling method
  cluster, 109
  random, 108, 109
  stratified, 109
  systematic, 109
  with replacement, 104, 105
  without replacement, 104, 106
Scatter diagram, 259
Sign test, 208
Significant result, 142
Skewness, 52

Slope (see line of regression)
Snedecor, G., 340
Spearman C., 289
Spearman r, 289
Standard deviation
  pooled, 236
  population, 45
  sample, 41
  unit of measure, 44
Standard error, 139
Standard normal probability curve, 119
Standardizing formula, 47, 48
Statistic, 3, 5
Student t distribution, 170
Success numbers, 192, 193
Success ratios, 222, 223
Summation notation, 6

t distribution (see student t distribution)
Table of random numbers, 384, 385
Test of hypothesis (see hypothesis testing)
Test statistic, 141
Tree diagram, 105
Trial, 192
Two-tail test, 143
Type I error, 145
Type II error, 145

Unbiased estimator, 114
Union of events, 67
Upper quartile, 33

Variance, 40, 45
Variation
  direct, 286
  inverse, 286

Wallis, W., 356
Within variance estimate, 343
World series problem, 99

Yates' Correction, 317
Yates, F., 318

z-score, 47, 120